AF411324

Food Bioactives: Chemical Challenges and Bio-Opportunities

Food Bioactives: Chemical Challenges and Bio-Opportunities

Editors

Severina Pacifico
Simona Piccolella

MDPI • Basel • Beijing • Wuhan • Barcelona • Belgrade • Manchester • Tokyo • Cluj • Tianjin

Editors

Severina Pacifico
Department of
Environmental, Biological
and Pharmaceutical Sciences
and Technologies
University of Campania
"Luigi Vanvitelli"
Caserta
Italy

Simona Piccolella
Department of
Environmental, Biological
and Pharmaceutical Sciences
and Technologies
University of Campania
"Luigi Vanvitelli"
Caserta
Italy

Editorial Office
MDPI
St. Alban-Anlage 66
4052 Basel, Switzerland

This is a reprint of articles from the Special Issue published online in the open access journal *Molecules* (ISSN 1420-3049) (available at: www.mdpi.com/journal/molecules/special_issues/Food_Bioactives_Chemical_Challenges_Bio-Opportunities).

For citation purposes, cite each article independently as indicated on the article page online and as indicated below:

LastName, A.A.; LastName, B.B.; LastName, C.C. Article Title. *Journal Name* **Year**, *Volume Number*, Page Range.

ISBN 978-3-0365-5160-9 (Hbk)
ISBN 978-3-0365-5159-3 (PDF)

Contents

Editorial

Editorial to the Special Issue "Food Bioactives: Chemical Challenges and Bio-Opportunities"

Severina Pacifico * and Simona Piccolella

Department of Environmental, Biological and Pharmaceutical Sciences and Technologies, University of Campania "Luigi Vanvitelli", Via Vivaldi 43, 81100 Caserta, Italy; simona.piccolella@unicampania.it
* Correspondence: severina.pacifico@unicampania.it; Tel.: +39-0823-274578

Citation: Pacifico, S.; Piccolella, S. Editorial to the Special Issue "Food Bioactives: Chemical Challenges and Bio-Opportunities". *Molecules* **2021**, *26*, 2517. https://doi.org/10.3390/molecules26092517

Received: 19 April 2021
Accepted: 20 April 2021
Published: 26 April 2021

Publisher's Note: MDPI stays neutral with regard to jurisdictional claims in published maps and institutional affiliations.

This Special Issue, entitled "Food Bioactives: Chemical Challenges and Bio-Opportunities", was born with the aim of attracting contributions on analytical challenges in food bioactives' chemistry and bioactivity, which form the basis of proper bio-opportunities. To date, it represents a collection of nine original research papers and one review article that extended scientific knowledge regarding the chemical constituents of edible plant extracts and their role as health-promoting agents. Innovation and originality in the research scope, design of experiments and results were highlighted in all the subtopics covered by the enclosed articles, which included medicinal and aromatic plant (MAP) ingredients, ancient foods nowadays re-valued for their functionality, compounds derived from food processing procedures, analytical methods related to food quality and agro-wastes as innovative and renewable sources for nutraceuticals.

MAPs are well-known species widely used in several sectors, e.g., pharmaceutical, cosmetic, liquor and food industries, due to their richness in biologically active ingredients, which can be classified based on their chemical features and/or therapeutic effects [1]. *Aloe vera* is one of the most used plants with a great commercial spread due to the multiple health-beneficial effects exerted by its pulp or gel constituents. The awareness that the soluble and fermentable fiber components of *Aloe* possess prebiotic effects able to modulate the composition of the microbiota, making them useful in the food industry, led Tornero-Martínez et al. to obtain new insights into the changes in the fiber and free phenols during in vitro digestion and colon fermentation of aloe gel and polysaccharide extract [2]. Their results showed that the behavior of the latter was similar to that of lactulose, suggesting the possible use as a prebiotic.

A novel perspective in the exploitation of MAPs was represented by the paper of Ielciu and co-workers, in which fresh young shoots of *Rosmarinus officinalis* L. were macerated in 90% ethanol to obtain a tincture, which was tested against liver tissue damages [3]. The phytochemical analysis led to the identification and quantification of polyphenols and terpenes by HPLC–UV–MS and GC–MS, respectively. The authors highlighted that this rosemary organ, besides the most studied leaves, could also represent a valuable plant material for the formulation of products with hepatoprotective properties through antioxidant mechanisms, which they investigated in rats with experimentally induced hepatotoxicity.

In addition to MAP-derived extracts, pure thymol and carvacrol, two phenolic terpenoids mainly derived from essential oils of thyme and oregano, were investigated for their antibacterial activity against *Staphylococcus aureus* [4]. Both proved to be promising antibacterial agents, which could likely be involved in counteracting bacterial resistance to antibiotics.

A full awareness of the role played by a healthy diet, as part of a healthy lifestyle, in countering or slowing down chronic and degenerative diseases has strongly increased the interest in food bioactives and the return of ancient but nowadays considered functional foods. This is the case of hemp seeds, which represented a significant source of nutrition for thousands of years and have recently been re-evaluated from a nutraceutical perspective [5]. In this context, Nigro et al. deeply investigated the chemical diversity of hemp seed phenylpropanoid amides and their derivative lignanamides by means of high-resolution tandem mass spectrometry (HR-MS/MS) techniques. They highlighted that these molecules were able to inhibit U-87 glioblastoma cell line survival and migration, inducing apoptosis while suppressing autophagy, whereas they interestingly did not exert cytotoxicity in nontumorigenic human fibroblasts [6].

The investigation of food metabolic composition is very challenging, as it is a very complex biological matrix. Thus, proper approaches that employ chromatographic separation coupled with UV and mass spectrometry detection, as described in the above-mentioned paper, are required. Furthermore, there is an urgent need to make the methods as sustainable as possible, e.g., by replacing chlorinated solvents in the extraction protocols of less polar compounds. In the light of the above, Cortés-Herrera et al. developed a greener alternative to the recovery and simultaneous analysis of carotenoids and fat-soluble vitamins from Costa Rican avocados, fruits of high economic value and of dietary interest, especially in Latin American cuisine [7]. Analytical methods, which are becoming increasingly powerful, can also enhance our knowledge about food quality indices; when jointly used with chemometric tools, they allowed discriminating between different *Pistacia vera* oils according to their quality profiles [8]. The proposed methodology had a number of advantages, including being accurate and not time- or solvent-consuming, and proved to be promising in fraud detection related to plant-derived oils.

The role that dietary substances, to which nutraceutical attributes are increasingly entrusted, play in halting or reversing oxidative stress-related diseases has been ever more deeply investigated, especially in relation to cancer [9]. The review article enclosed in the present paper collection focused its attention on cytotoxic, antitumor and antimetastatic properties of *Urtica dioica* L. extracts; it was based on previous investigations employing both human cancer cell lines and in vivo models, with particular attention devoted to breast cancer [10]. The authors concluded that thanks to its richness in bioactive compounds, *U. dioica* could be used as a promising source of nutraceuticals useful as chemopreventive and/or chemoprotective agents in cancer disease onset and development. Breast cancer is considered the most prevalent cancer among women and one of the main causes of death worldwide. It was also the focus of another research article in this Special Issue, which dealt with a promising cytotoxic pistachio hull ethyl acetate extract that could be considered in the future as part of anticancer drug treatment [11].

It is worth noting that plant sources for nutraceutical compounds are not limited to the edible organs of the plant itself; interestingly, such compounds can also be recovered from by-products and wastes, giving them new life in the prevention of human diseases. Piccolella et al. took into consideration wine production wastes, focusing in particular on the leaves of a *Vitis vinifera* cultivar named "Greco di Tufo", which represents a great resource of the Campania Region (Italy) territory, as it received the quality classification "Controlled and Guaranteed Designation of Origin" (DOCG acronym in Italy) [12]. The authors applied an extraction/fractionation protocol that favored the obtainment of a mixture of flavonol glycosides and glycuronides variously oxygenated at the B-ring. The isolation of the most abundant compound led to its full chemical characterization by spectroscopic and HR-MS/MS tools. Then, the absence of cytotoxicity in preliminary tests using central nervous system cell lines paved the way for further exploitation of their neuroprotective potential.

Bioactives' presence and abundance are strictly related to their food source. However, food could contain other constituents, beyond naturally occurring compounds, whose presence is processing-induced. For example, during deep-frying processes, vegetal oils become a source of lipid peroxidation products due to the exposure to high temperatures. The study of Zaunschirm et al. aimed at investigating the impact of linoleic acid and its primary and secondary peroxidation products on genomic and metabolomic pathways in human gastric cells (HGT-1). They demonstrated that these substances were able to influence the pathways related to amino acid metabolism, thus affecting metabolic health [13].

In conclusion, the contributions in this Special Issue underscore the potential of exploiting plants, beyond nutrition, as "biofactories" of compounds useful for human health and well-being.

Funding: This research received no external funding.

Conflicts of Interest: The authors declare no conflict of interest.

References

1. Piccolella, S.; Crescente, G.; Pacifico, F.; Pacifico, S. Wild aromatic plants bioactivity: A function of their (poly) phenol seasonality? A case study from Mediterranean area. *Phytochem. Rev.* **2018**, *17*, 785–799. [CrossRef]
2. Tornero-Martínez, A.; Cruz-Ortiz, R.; Jaramillo-Flores, M.E.; Osorio-Díaz, P.; Ávila-Reyes, S.V.; Alvarado-Jasso, G.M.; Mora-Escobedo, R. In vitro Fermentation of Polysaccharides from *Aloe vera* and the Evaluation of Antioxidant Activity and Production of Short Chain Fatty Acids. *Molecules* **2019**, *24*, 3605. [CrossRef] [PubMed]
3. Ielciu, I.; Sevastre, B.; Olah, N.-K.; Turdean, A.; Chișe, E.; Marica, R.; Oniga, I.; Uifălean, A.; Sevastre-Berghian, A.C.; Niculae, M.; et al. Evaluation of Hepatoprotective Activity and Oxidative Stress Reduction of *Rosmarinus officinalis* L. Shoots Tincture in Rats with Experimentally Induced Hepatotoxicity. *Molecule* **2021**, *26*, 1737. [CrossRef] [PubMed]
4. Sousa Silveira, Z.d.; Macêdo, N.S.; Sampaio dos Santos, J.F.; Sampaio de Freitas, T.; Rodrigues dos Santos Barbosa, C.; Júnior, D.L.d.S.; Muniz, D.F.; Castro de Oliveira, L.C.; Júnior, J.P.S.; Cunha, F.A.B.d.; et al. Evaluation of the Antibacterial Activity and Efflux Pump Reversal of Thymol and Carvacrol against *Staphylococcus aureus* and Their Toxicity in *Drosophila melanogaster*. *Molecules* **2020**, *25*, 2103. [CrossRef] [PubMed]
5. Crescente, G.; Piccolella, S.; Esposito, A.; Scognamiglio, M.; Fiorentino, A.; Pacifico, S. Chemical composition and nutraceutical properties of hempseed: An ancient food with actual functional value. *Phytochem. Rev.* **2018**, *17*, 733–749. [CrossRef]
6. Nigro, E.; Crescente, G.; Formato, M.; Pecoraro, M.T.; Mallardo, M.; Piccolella, S.; Daniele, A.; Pacifico, S. Hempseed Lignanamides Rich-Fraction: Chemical Investigation and Cytotoxicity towards U-87 Glioblastoma Cells. *Molecules* **2020**, *25*, 1049. [CrossRef] [PubMed]
7. Cortés-Herrera, C.; Chacón, A.; Artavia, G.; Granados-Chinchilla, F. Simultaneous LC/MS Analysis of Carotenoids and Fat-Soluble Vitamins in Costa Rican Avocados (*Persea Americana* Mill.). *Molecules* **2019**, *24*, 4517. [CrossRef] [PubMed]
8. Valasi, L.; Arvanitaki, D.; Mitropoulou, A.; Georgiadou, M.; Pappas, C.S. Study of the Quality Parameters and the Antioxidant Capacity for the FTIR-Chemometric Differentiation of *Pistacia Vera* Oils. *Molecules* **2020**, *25*, 1614. [CrossRef] [PubMed]
9. Piccolella, S.; Pacifico, S. Plant-Derived Polyphenols: A Chemopreventive and Chemoprotectant Worth-Exploring Resource in Toxicology. *Adv. Mol. Toxicol.* **2015**, *9*, 161–214. [CrossRef]
10. Esposito, S.; Bianco, A.; Russo, R.; Di Maro, A.; Isernia, C.; Pedone, P.V. Therapeutic Perspectives of Molecules from *Urtica dioica* Extracts for Cancer Treatment. *Molecules* **2019**, *24*, 2753. [CrossRef] [PubMed]
11. Seifaddinipour, M.; Farghadani, R.; Namvar, F.; Bin Mohamad, J.; Muhamad, N.A. In Vitro and In Vivo Anticancer Activity of the Most Cytotoxic Fraction of Pistachio Hull Extract in Breast Cancer. *Molecules* **2020**, *25*, 1776. [CrossRef] [PubMed]
12. Piccolella, S.; Crescente, G.; Volpe, M.G.; Paolucci, M.; Pacifico, S. UHPLC-HR-MS/MS-Guided Recovery of Bioactive Flavonol Compounds from Greco di Tufo Vine Leaves. *Molecules* **2019**, *24*, 3630. [CrossRef] [PubMed]
13. Zaunschirm, M.; Pignitter, M.; Kopic, A.; Keßler, C.; Hochkogler, C.; Kretschy, N.; Somoza, M.M.; Somoza, V. Exposure of Human Gastric Cells to Oxidized Lipids Stimulates Pathways of Amino Acid Biosynthesis on a Genomic and Metabolomic Level. *Molecules* **2019**, *24*, 4111. [CrossRef] [PubMed]

Article

Evaluation of Hepatoprotective Activity and Oxidative Stress Reduction of *Rosmarinus officinalis* L. Shoots Tincture in Rats with Experimentally Induced Hepatotoxicity

Irina Ielciu [1], Bogdan Sevastre [2,*], Neli-Kinga Olah [3,4], Andreea Turdean [3], Elisabeta Chişe [5], Raluca Marica [2], Ilioara Oniga [6], Alina Uifălean [7], Alexandra C. Sevastre-Berghian [8], Mihaela Niculae [9], Daniela Benedec [6,*] and Daniela Hanganu [6]

[1] Department of Pharmaceutical Botany, Iuliu Haţieganu University of Medicine and Pharmacy, 400010 Cluj-Napoca, Romania; irina.ielciu@umfcluj.ro

[2] Department of Clinic and Paraclinic Sciences, University of Agricultural Sciences and Veterinary Medicine, 400372 Cluj-Napoca, Romania; raluca.marica@usamvcluj.ro

[3] PlantExtrakt, 407059 Cluj-Napoca, Romania; neliolah@yahoo.com (N.-K.O.); andreea.turdean@plantextrakt.ro (A.T.)

[4] Department of Pharmaceutical Industry, Faculty of Pharmacy, Vasile Goldiş Western University of Arad, 310414 Arad, Romania

[5] Department of Pharmaceutical Chemistry, Faculty of Pharmacy, Vasile Goldiş Western University of Arad, 310414 Arad, Romania; c_elisabeta@yahoo.com

[6] Department of Pharmacognosy, Iuliu Haţieganu University of Medicine and Pharmacy, 400010 Cluj-Napoca, Romania; ioniga@umfcluj.ro (I.O.); dhanganu@umfcluj.ro (D.H.)

[7] Department of Pharmaceutical Analysis, Faculty of Pharmacy, Iuliu Haţieganu University of Medicine and Pharmacy, 400349 Cluj-Napoca, Romania; alina.uifalean@umfcluj.ro

[8] Department of Physiology, Faculty of Medicine, Iuliu Haţieganu University of Medicine and Pharmacy, 400006 Cluj-Napoca, Romania; berghian.alexandra@umfcluj.ro

[9] Department of Clinical Sciences, Division and Infectious Diseases, University of Agricultural Sciences and Veterinary Medicine, 400374 Cluj-Napoca, Romania; mihaela.niculae@usamvcluj.ro

* Correspondence: bogdan.sevastre@usamvcluj.ro (B.S.); dbenedec@umfcluj.ro (D.B.)

Citation: Ielciu, I.; Sevastre, B.; Olah, N.-K.; Turdean, A.; Chişe, E.; Marica, R.; Oniga, I.; Uifălean, A.; Sevastre-Berghian, A.C.; Niculae, M.; et al. Evaluation of Hepatoprotective Activity and Oxidative Stress Reduction of *Rosmarinus officinalis* L. Shoots Tincture in Rats with Experimentally Induced Hepatotoxicity. *Molecules* **2021**, 26, 1737. https://doi.org/10.3390/molecules26061737

Academic Editor: Akihito Yokosuka

Received: 18 February 2021
Accepted: 16 March 2021
Published: 20 March 2021

Publisher's Note: MDPI stays neutral with regard to jurisdictional claims in published maps and institutional affiliations.

Abstract: *Rosmarinus officinalis* L. is a widely known species for its medicinal uses, that is also used as raw material for the food and cosmetic industry. The aim of the present study was to offer a novel perspective on the medicinal product originating from this species and to test its hepatoprotective activity. The tested sample consisted in a tincture obtained from the fresh young shoots. Compounds that are evaluated for this activity are polyphenols and terpenoids, that are identified and quantified by HPLC–UV–MS and GC–MS. Antioxidant activity was assessed in vitro, using the DPPH, FRAP and SO assays. Hepatoprotective activity was tested in rats with experimentally-induced hepatotoxicity. In the chemical composition of the tincture, phenolic diterpenes (carnosic acid, carnosol, rosmanol, rosmadial) and rosmarinic acid were found to be the majority compounds, alongside with 1,8-cineole, camphene, linalool, borneol and terpineol among monoterpenes. In vitro, the tested tincture proved significant antioxidant capacity. Results of the in vivo experiment showed that hepatoprotective activity is based on an antioxidant mechanism. In this way, the present study offers a novel perspective on the medicinal uses of the species, proving significant amounts of polyphenols and terpenes in the composition of the fresh young shoots tincture, that has proved hepatoprotective activity through an antioxidant mechanism.

Keywords: *Rosmarinus officinalis* L.; fresh young shoots tincture; polyphenols; terpenes; antioxidant; hepatoprotective

1. Introduction

Among the most important flowering plant families, the Lamiaceae family is one of the largest, comprising numerous species that are known for their biological activities or for their use in different branches related to economy [1–3]. Species belonging to the genera *Salvia* [4], *Melissa* [5] or *Thymus* [6] are the most well known, being widely used in different forms for the treatment of numerous pathologies, but also for their nutritional or economical values [4–6].

Rosmarinus is an important genus of the Lamiaceae family, comprising 2 species, *Rosmarinus eriocalyx* Jordan and Fourr. and *Rosmarinus officinalis* L., that are largely distributed in habitats from Southern and Northern Africa, Western Asia, Anatolia and the Mediterranean basin [7,8]. The most studied species is *Rosmarinus officinalis* L., rosemary, a worldwide cultivated plant, known for its culinary use and pharmacological properties that made it famous in traditional medicine. It has been reported that its leaves present significant therapeutic applications in managing a wide range of diseases such as diabetes mellitus, respiratory disorders, stomach problems and inflammatory diseases [9]. It is also used in the food industry, as a food flavoring and preservative due to its antioxidant and antimicrobial properties. *R. officinalis* is also used as raw material in cosmetic products [10].

Hepatic diseases are a main threat to public health, indicating problems to the hepatic tissue or to the liver functions, which can be caused by different factors, such as viruses or bacteria, autoimmune diseases, or by the external action of different chemicals (drugs or toxic compounds) [11,12]. Nowadays, modern medicine offers alternatives for the treatment of these pathologies, but despite the advances, few effective drugs that offer protection and regeneration of hepatic cells exist [13]. Moreover, existing treatments can cause adverse effects which make the therapy of these pathologies even harder [12]. Thereby, the need for identifying novel alternatives for the treatment of hepatic diseases and for the protection of the liver appears to be important, in order to develop novel agents with high efficiency and superior safety profile [11–13]. Mechanisms that underlie the hepatoprotective activity are strongly related to the capacity of antioxidants to scavenge reactive oxygen species (ROS) that are produced by the metabolic conversion of xenobiotics and induce oxidative stress and damage to the liver tissue [14].

Hepatoprotective activity is amongst the biological activities that is reported for *R. officinalis* extracts in a model of azathioprine-induced toxicity in rats [15], acetaminophen-induced liver damage [16,17], gentamicin-treated rats [18], hepatic damage induced by hypotermic-ischemia in rats [19] and alcoholic liver disease [20], being assigned to the essential oil composition [21] and to its composition in polyphenols [17–19], among which rosmarinic acid is the most representative [16]. Nevertheless, this biological activity is less documented and, to the best of our knowledge, there is no clear evidence on the compounds that are responsible for this activity. Moreover, there is also little evidence on the underlying mechanism of action, which is supposed to be due to the reduction of ROS [14,15]. On the other side, in vitro antioxidant activity of *R. officinalis* is largely studied, being reported to numerous vegetal products of the species: fresh aerial parts [9], flower extracts [7] or essential oil [10,22]. Compounds that are responsible for this activity are polyphenols [7], but also terpenes from the essential oil [10,22].

Rosmarinic acid (RA) is a phenolic compound firstly isolated from *Rosmarinus officinalis* L., having remarkable pharmacological activities. It is commonly found in Lamiaceae species, such as *Melissa* sp., *Salvia* sp., *Origanum* sp. or *Thymus* sp. Its pharmacological importance is mainly due to its antioxidant, anti-inflammatory, antimicrobial and anti-diabetic properties [23]. Its hepatoprotective activity is also reported and is assigned to an antioxidant mechanism [24]. Other polyphenols that are found in the composition of the species are caffeic acid, chlorogenic acid, *p*-coumaric acid, quinic acid, kaempferol, quercetin, rutin and apigenin [7,25]. Among the phenolic terpenes, those such as carnosol and carnosic acid are also reported to be found in the composition of the species [26–28]. Essential oil of *R. officinalis* is rich in a large variety of monoterpenes such as borneol, camphor, linalool, α-and β-pinene, camphene, 1,8-cineole [22,29–31].

In this context, the main objective of the present study was to evaluate the hepatoprotective activity of a fresh young shoots of *R. officinalis* tincture against carbon tetrachloride (CCl_4)-induced hepatotoxicity in rats. Compounds that are analyzed for this activity are the phenolic compounds and terpenoids. Moreover, the present study aims to offer important details on the mechanism underlying this activity, which appears to be due to the antioxidant properties of the tested extract. Finally, the present study also aimed to prove the significant potential of fresh young shoots as an important source of compounds exhibiting antihepatotoxic activity by an antioxidant mechanism.

2. Results and Discussion

The tincture, obtained from fresh young shoots macerated with 90% *v/v* ethanol, was analyzed from the physico-chemical point of view, according to European Pharmacopoeia (EPh) and German Homeopathic Pharmacopoeia (GHPh). Thus, the tincture presented a dark brown color, being an aromatic liquid with a relative density of 0.920. The value of the dry residue was 2.38% and the ethanol content was 55% vol. The identity according to the GHPh was performed by Thin Layer Chromatography (TLC) highlighting the main terpenoidic compounds.

2.1. Phytochemical Analysis of Tincture

The tincture was analyzed by HPLC–UV–MS for the identification and quantification of phenolic compounds. Results were expressed as µg polyphenol/g dry vegetal material (d.w.) (Table 1).

Table 1. Phenolic compounds identified in *R. officinalis* extract by HPLC–UV–MS.

Peak. No.	Compound	Structural Class	Retention Time R_t (min)	UV λ_{max} (nm)	$[M + H]^+$ (m/z)	Concentration (µg/g d.w.)
1.	Syringic acid (3,5-Dimethoxy-4-hydroxybenzoic acid)	Hydroxybenzoic acid	3.30	265	198	355.84 ± 0.25
2.	Hesperidin (Hesperetin-rutinoside)	Flavanone	15.97	280	611	72.36 ± 0.15
3.	Nepetrin (Nepetin-glucoside)	Flavone	16.85	350, 265	479	259.65 ± 0.85
4.	Luteolin-glucuronide	Flavone	17.82	350, 260	463	698.70 ± 0.05
5.	Homoplantaginin (Hispidulin-glucoside)	Flavone	18.14	340, 260	463	364.31 ± 0.66
6.	Rosmarinic acid	Hydroxycinnamic acid	18.91	330	360	406.29 ± 0.95
7.	Luteolin-acetyl-glucuronide	Flavone	19.40	350, 260	505	146.82 ± 1.15
8.	Carnosol	Phenolic terpene	20.13	330	331	368.25 ± 0.33
9.	Luteolin	Flavone	21.74	350, 260	287	95.71 ± 1.02
10.	Nepetin	Flavone	21.89	350, 265	317	155.04 ± 0.98
11.	Rosmanol	Phenolic terpene	22.91	330	347	308.4 ± 1.02
12.	Rosmadial	Phenolic terpene	23.41	330	345	777.95 ± 0.85
13.	Cirsimaritin	Flavone	24.00	330, 260	315	713.7 ± 0.96
14.	Carnosic acid	Phenolic terpene	25.39	270	332	804.27 ± 0.89

Note: Values represent the mean ± standard deviations of three independent measurements.

High amounts of polyphenols were identified in the composition of the fresh young shoots tincture. Phenolic terpenes were found to be the majority class of compounds. Among these, the one that was found in the highest concentration was carnosic acid, followed by rosmadial, carnosol and rosmanol. Among flavones, luteolin-glucuronide, cirsimaritin and homoplantaginin were also found in significant amounts. Rosmarinic acid is the hydroxycinnamic acid that was found in the highest amounts. Another compound that was found in significant amounts is syringic acid, a hydroxybenzoic acid.

Polyphenolic compounds from *R. officinalis* were reported in scientific literature, but the raw material was in general represented by the leaves, not the young shoots, which emphasize even more the originality of the present study. Carnosic acid and carnosol, together with rosmarinic acid are found in the composition of the species, but their presence is cited in the composition of the leaves [26,28,32], that are mostly dried [26,28]. Luteolin-glucuronide is also the flavonoidic compound that is mostly cited in the composition of the same parts of the species [28,32,33]. Presence of cirsimaritin, rosmanol, rosmadial and homoplantaginin is also reported for the leaves [32]. Differences can be found in the amounts of these compounds, significantly higher in the composition of fresh young shoots. Moreover, the novelty of the present study consists nevertheless in the vegetal material that is studied, which is represented by fresh young shoots, which, to the best of our knowledge, was not previously studied. Taking into consideration the fact that high amounts of these polyphenols were found in the composition of the tested tincture, important correlation could be found with the tested biological activities.

In addition, terpenoids that are found in the essential oil were also found in significant amounts in the composition of the tested tincture, being solubilized by the extraction solvent, concentrated ethanol. The terpenoids were evaluated by a GC–MS method and the main components identified and quantified were 1,8-cineole, camphene, linalool, borneol, and terpineol (Figure 1-left). These compounds were also previously identified in the composition of the essential oil belonging to this species [7,14,31,34,35]. The antioxidant and hepatoprotective activities of the tested tincture can also be attributed to the presence of high percentage of 1,8-cineole (27.5 mg/100 g tincture), which is found to be the majoritary compound in the composition of the tested tincture (Figure 1-right). These compounds previously reported hepatoprotective activity [10,12,14] and the high amount that is found in the tested tincture, together with the phenolic compounds, may synergistically act in order to exert these biological activities. The obtained results of the GC–MS analysis are presented in Table 2.

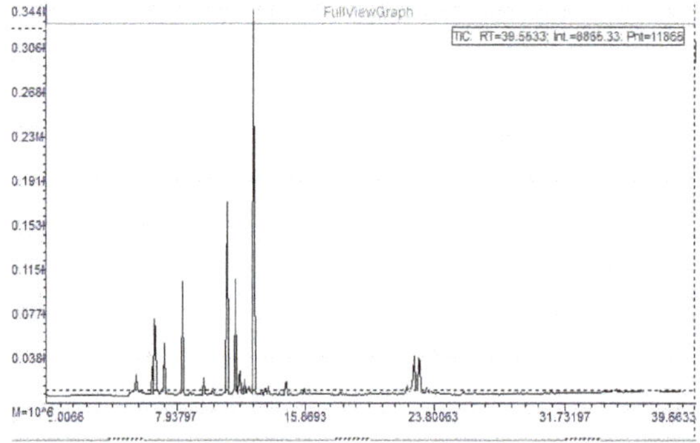
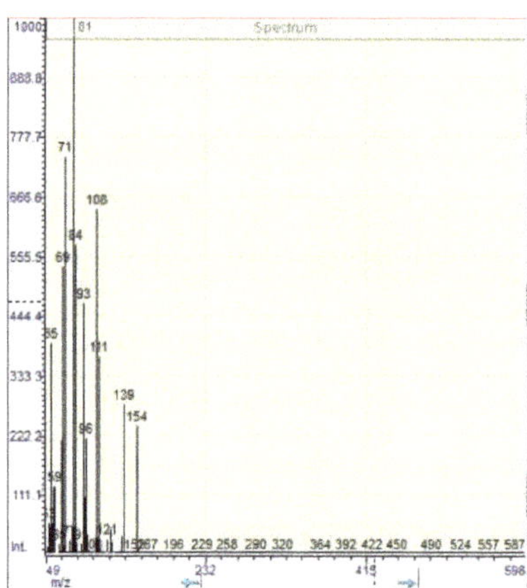

Figure 1. Left—GC chromatogram of *Rosmarinus officinalis* tincture. **Right**—MS spectra of separated 1,8-cineole.

Table 2. GC–MS parameters for the main identified and quantified terpenoid compounds of *R. officinalis* tincture.

Terpenic Compound	Retention Time, min	Match Factor	Calibration Curve/R^2	Detection (DL)/Quantification (QL) Limits, µg/mL	Content %(mg/g Tincture)
Camphene	6.73	832	A = 16.353·c + 193,775 R^2 = 0.9971	DL = 23.7 QL = 47.4	0.34 ± 0.005
1,8-cineole	8.35	836	A = 13.712·c + 304.523 R^2 = 0,9887	DL = 44.4 QL = 88.8	25.7 ± 0.308
Linalool	9.72	868	A = 13,796·c − 136,712 R^2 − 0.9987	DL = 39.6 QL = 59.5	11 4 ± 0.148
Borneol	11.64	783	A = 14.380·c − 86.930 R^2 = 0,98920	DL = 24.2 QL = 36.3	19.3 ± 0.212
Terpineol	12.19	865	A= 8324.8·c + 4,000,000 R^2 = 0.9940	DL = 961 QL = 1922	1.9 ± 0.032

Values represent the mean ± standard deviations of three independent measurements.

2.2. Antioxidant Activity: In Vitro Assays

The antioxidant capacity was evaluated by three in vitro methods: 2,2-diphenyl-picrylhydrazil (DPPH), ferric-reducing antioxidant power (FRAP) and superoxid (SO) anion radical scavenging assay, assessing therefore by different mechanisms of the antioxidant capacity of the tested sample [36–38] (Table 3).

Table 3. Polyphenols content and antioxidant activity of the fresh young shoots *R. officinalis* tincture.

Sample	TPC (mg GAE/g)	Flavonoid Total (mg RE/g)	Rosmarinic Acids (mg RAE/g)	DPPH (IC50, µg/mL)	FRAP (µm TE/g)	SO Scavenging (µm TE/g)
R. officinalis tincture	60.18 ± 0.42	33.01 ± 0. 24	25.40 ± 0. 84	31.85 ± 1.81	257.88 ± 1.74	99.70 ± 0. 65

Values represent the mean ± SD of three independent measurements. TPC = total polyphenols content; SO = superoxide anion radical; GAE = gallic acid equivalents; RE = rutin equivalents; RAE = rosmarinic acid equivalents; TE = Trolox equivalents.

The total polyphenols, flavonoids and rosmarinic acids content of the *R. officinalis* tested tincture could be correlated with the antioxidant effects, as they showed significantly important amounts, compared to the dried leaves or extracts [39,40].

The significant content of polyphenolic compounds in the tested extract is highly related to its antioxidant potential. Synergistic, additive or antagonistic interactions between them can be revealed in this activity of neutralization of free radicals [41]. Most of the existing studies assign the antioxidant effect of the species to the essential oil content [22,31], but as high amounts of polyphenols are evaluated in different studies, it seems that they are very important for the antioxidant activity of the extracts [36].

The DPPH radical scavenging activity of the extract was high (IC_{50} = 31.85 ± 1.81 µg/mL). Previous studies have reported the antioxidant activity by the same method [7,40], testing the flowers [7] or the dried leaves [40] extracts. Other studies that were performed on fresh plants showed significantly higher value of IC_{50}, indicating therefore a poorer effect of the whole mature plants [25].

The tested tincture has demonstrated a very good ferric ion-reducing antioxidant capacity with a value of 257.88 ± 1.74 µm TE/g, higher than other data revealed by other studies performed on leaves [26]. Antioxidant activity by FRAP method of *R. officinalis* extract was evaluated by Gîrd et al. as EC_{50} = 285.25 ± 0.88 µg/mL, but the study was performed on dried leaves [40].

In the case of superoxide (SO) scavenging assay, the decrease of absorbance at 560 nm with antioxidant compounds indicated the consumption of superoxide anion in the reaction mixture. The phenolic and terpenoid compounds of the tested tincture have the ability to neutralize superoxide radicals, which are extremely aggressive in liver tissue, causing significant cell damages. At the same time, superoxide radicals can be inhibited by the endogenous antioxidant enzyme, superoxide dismutase (SOD), whose level can be increased by the same compounds. In this way, the tested compounds synergistically act and help to link the antioxidant activity proven by in vitro experiments with those tested in vivo. The results of in vitro experiments show a good effect of removing superoxide radicals of *R. officinalis* fresh young shoots tincture, which was consistent with the in vivo results indicating an increase in SOD activity [42].

As it can be observed, the comparison with scientific literature is difficult, since different extraction methods and experimental protocols have been used and different plant material represented the starting point. However, taking into consideration all of the above, it becomes clearer that the fresh young shoots represent an important source of polyphenols and terpenes with promising antioxidant potential. Moreover, this antioxidant potential that is hereby proved is directly correlated with the hepatoprotective activity in vivo.

2.3. Hepatoprotective Activity: Animal Studies

The protective effect of *R. officinalis* extract was investigated using liver toxicity model induced by carbon tetrachloride (CCl_4) (1 mL/kg).

The CCl_4-hepatotoxicity model is extensively used to evaluate the hepatoprotective effects of drugs and plant extracts. It has been reported that one of the principal causes of CCl_4-induced liver injury is the lipid peroxidation which is induced and accelerated by free radical derivatives of CCl_4 [43]. CCl_4 exerts its hepatotoxic effect through covalent binding of CCl_4 metabolites and reaction with oxygen to initiate lipid peroxidation [44]. The extent of liver oxidative damage was investigated by measuring the level of malondialdehyde (MDA) and the activity of antioxidant enzymes as glutathione peroxidase (GPx), catalase (CAT) and superoxide dismutase (SOD). Liver injury was assessed by measuring of plasma activity of transaminases, as alanine aminotransaminase (ALT), aspartate aminotransaminase (AST), gamma-glutamyl transferase (GGT), while albumins reflected the liver function.

All animals survived up to the end of the study; they showed no clinical signs and maintained a body weight gain, food and water consumption similar to control group. In mice receiving CCl_4 and no therapy, the liver injury was reflected in elevated activity of plasma transaminases, ALT, AST, and GGT (Figure 2), but albumin levels maintained close to the values of the control group, suggesting the absence of liver failure. However, the total proteins were elevated, based on globulin fraction, probably as a result of liver inflammatory reaction. *R. officinalis* tincture administration revealed a protective effect, visible at the doses of 50 and 500 mg/kg b.w. However, the higher dose of 500 mg/kg b.w. did not provide a better protective effect, thus no dose-dependent effect has been found. When comparing the two time intervals, ASAT and GGT had a slightly increasing trend, but only the variation of ASAT was statistically relevant ($p < 0.05$), while ALAT remained at the same level. Similarly, the plasma proteins remained at the same levels. In the groups receiving therapy, the values remained also very similar.

The protective effect of *R. officinalis* extract seems to be exerted by antioxidant effect. Expectedly, CCl_4 induced inhibition of antioxidant GPx, CAT, SOD and a threefold increase of MDA levels. The doses of 50 and 500 mg/kg b.w. of *R. officinalis* partially restored the activity of antioxidant enzymes and alleviated the lipid peroxidation. The values remained significantly altered compared to those of the control group. Once again, the higher dose of 500 mg/kg did not provide a better protection than the dose of 50 mg/kg did (Figure 3). In accordance to transaminase levels, the oxidative stress markers showed no significant variations between the two time intervals.

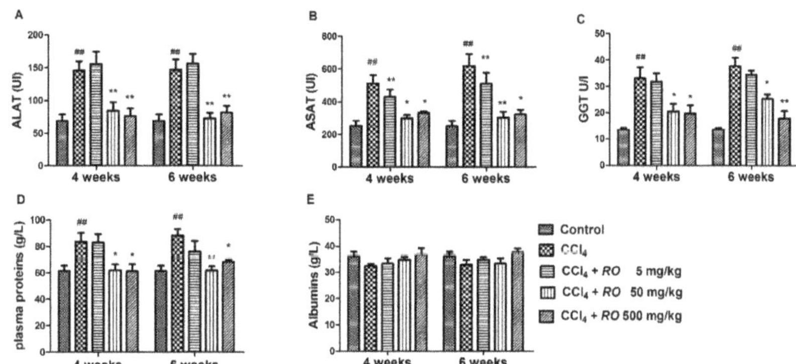

Figure 2. Effects of *R. officinalis* tincture on: alanine aminotransferase (ALAT) (**A**); aspartate amino-transferase (ASAT) (**B**); gamma glutamyl transferase (GGT) (**C**) activity and plasma total proteins (**D**); and albumins (**E**) concentration (mean $\pm$ SD, 5 animals/group). ## $p < 0.01$, compared to the Control group; * $p < 0.05$, ** and $p < 0.01$ compared to the CCl_4 alone treated group.

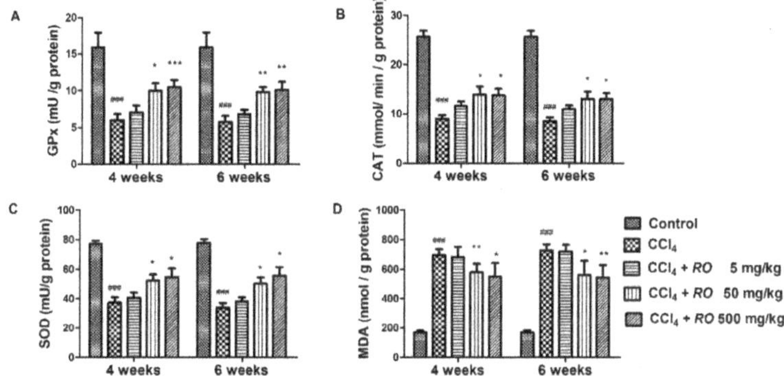

Figure 3. Effects of *R. officinalis* tincture on: glutathione peroxidase (GPx) (**A**); catalase (CAT) (**B**); superoxide dismutase (SOD) (**C**); and malondialdehyde (MDA) (**D**) (mean $\pm$ SD, 5 animals/group). ### $p < 0.001$, compared to the Control group; * $p < 0.05$, ** $p < 0.01$ and *** $p < 0.001$ compared to the CCl_4 alone treated group.

Histological examination revealed necrosis and inflammatory cells infiltrate as the predominant features in the groups receiving CCl_4, in absence of *R. officinalis* therapy. Hence, around portal spaces, most of the hepatic lobules exhibited multifocal hepatocyte necrosis surrounded by inflammatory infiltrate mainly composed of neutrophils and macrophages. Another important feature of hepatic toxicity was the proliferation of fibrous tissue between portal spaces. In the groups receiving *R. officinalis* therapy, we observed an improvement in the hepatic regeneration, dependent to the dose. The main effect of the extract was the downregulation of the inflammatory reaction; better manifested in the group receiving the dose of 50 mg/b.w. for 6 weeks long (Figure 4G). Although the inflammation was not quantitatively assessed, we noticed a marked decrease in the number of neutrophils and macrophages. Notably, this improvement was less obvious in the first (5 mg/b.w.) and the third groups (500 mg/b.w.), suggesting that, in high doses, the extract may exert a toxic effect on the liver or may interfere with the regeneration processes. The *R. officinalis* therapy improved also the regeneration process, manifested by the presence of mitotic figures, indicating the restitution of the hepatic architecture and function (Figure 4C). Reduction of the hepatocyte necrosis and fibrosis were also noticed, mainly for the dose of 50 mg/b.w. Once again, the benefits were minimal in the case of animals receiving the doses of 5, and 500 mg/b.w. (Figure 4D,H).

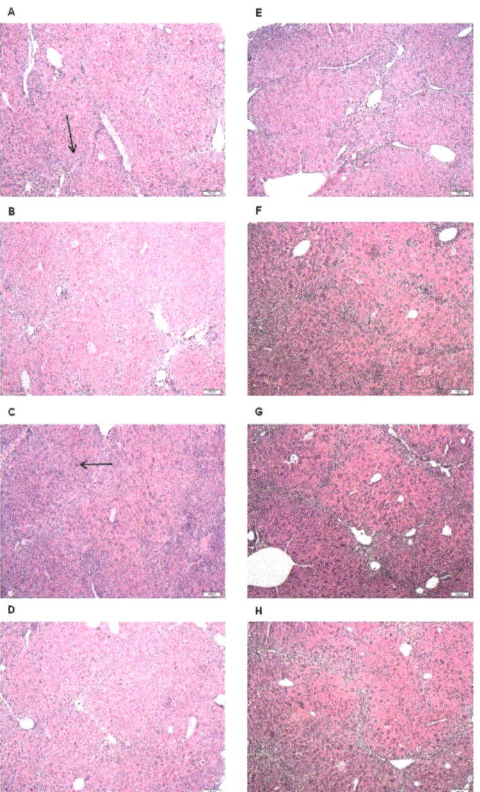

Figure 4. Effects of *R. officinalis* tincture on the histologic aspect of the liver: (**A,E**) the negative control groups receiving only carbon tetrachloride (1 mL/b.w.) showed inflammatory infiltrate (black arrow) and hepatocyte necrosis; the groups receiving therapy with the extract in a dose of: 5 mg/b.w., (**B,F**); 50 mg/b.w.; (**C,G**); and 500 mg/b.w. (**D,H**). Duration: (**A–D**) four weeks; (**E–H**) six weeks. Hematoxylin & Eosin stain; Bar, 100 μm.

3. Materials and Methods

3.1. Chemicals and Reagents

Acetonitrile for the HPLC-gradient was provided by Merck (Darmstadt, Germany) and water was purified with a Direct-Q UV system by Millipore (Darmstadt, Germany). Luteolin was purchased from Sigma (Darmstadt, Germany) and all other chemicals used were obtained from Alfa-Aesar, Karlsruhe, Germany.

3.2. Plant Material, Preparation and Characterization of Fresh Young Shoots Tincture

The vegetal material was harvested from the ecological culture of PlantExtrakt (Rădaia, Cluj county, Romania), in September 2020. Voucher specimens are deposited in the herbarium of the Pharmacognosy Department of the Faculty of Pharmacy Cluj-Napoca (Voucher no. 159). The *R. officinalis* tincture was prepared according to the German Homeopathic Pharmacopoeia (GHPh) and European Pharmacopoeia (EPh) specifications for tincture preparation (method 1.1.5). Fresh, young shoots were extracted by cold maceration with 90% *v/v* ethanol. One part of the crushed vegetal material was mixed with 1.4 parts of ethanol for 10 days. Afterwards, the extract was filtered. For the physicochemical characterization of the obtained tincture, organoleptic properties, relative density, residue at evaporation and ethanol content were assessed according to the methods of the EPh [45]. Before use in the animal experiment, the tincture was evaporated until the complete removal of the alcohol and immediately given to animals.

The relative density was performed according to the European Pharmacopoeia (EPh) using a Mettler Toledo (Greifensee, Switzerland) digital densimeter. The dry residue was performed according to the European Pharmacopoeia (EPh) using Kern analytical scale (Berlin, Germany) and a Memmert drying cabinet (Schwabach, Germany). The ethanol content was assessed according to European Pharmacopoeia (EPh) using a Neo-Clevenger apparatus and a Mettler Toledo digital densimeter.

3.3. Quantification of Total Polyphenols, Flavonoids and Phenolic Acids Content

The total phenolic content (TPC) was evaluated by a spectrophotometric method using the Folin–Ciocâlteu reagent, according to the European Pharmacopoeia. TPC values were calculated using the calibration curve of gallic acid (R^2 = 0.9931) and expressed as mg gallic acid equivalents (GAE)/g fresh vegetal material. Quantitative determination of flavonoids (TFC) was performed by a spectrophotometric method using aluminum chloride. TFC values were determined using an equation obtained from a calibration curve of rutoside (R^2 = 0.9992) and expressed as mg of rutoside equivalents (RE)/g fresh vegetal material. Quantitative determination of phenolic acids (TPA) was analyzed by a spectrophotometrical method according to the 10th Edition of the Romanian Pharmacopoeia (*Cynarae folium* monograph). TPA results were expressed as mg rosmarinic acid equivalents (RAE)/g fresh vegetal material, calculated using a rosmarinic acid calibration curve graph (R^2 = 0.9956). All experiments were performed in triplicate [46,47].

3.4. HPLC/DAD/ESI+ Analysis

This analysis was performed using a HP-1200 liquid chromatograph equipped with a quaternary pump, autosampler, DAD detector and MS-6110 single quadrupole API-electrospray detector (Agilent-Technologies Inc., Santa Clara, CA, USA). Positive ionization mode was used for the detection of phenolic compounds. Different fragmentors, in the range 50–100 V, were applied. The separation of compounds was carried out on an Eclipse XDB-C18 (5 μm; 4.5 × 150 mm i.d.) column (Agilent). Mobile phase consisted in water acidified with acetic acid 0.1% (A) and acetonitrile acidified with acetic acid 0.1% (B). Elution was performed in a multistep linear gradient, with the following composition: 5% B for 2 min; from 5% to 90% of B in 20 min, hold for 4 min at 90% B, then 6 min to arrive at 5% B. Flow rate was 0.5 mL/min and oven temperature 25 ± 0.5 °C. Detection of positively charged ions was performed by mass spectrometry, using the Scan mode and the following conditions: gas temperature 3500 °C, nitrogen flow 7 L/min, nebulizer pressure 35 psi, capillary voltage 3000 V, fragmentor 100 V and *m/z* 120–1200. Recording of chromatograms was performed at λ = 280 and 340 nm. The data acquisition was carried out with the Agilent ChemStation software [48,49].

3.5. GC–MS Analysis

For this analysis, a Dani Master GC–MS System, equipped with a SH-Rxi-5 ms column with 30 m × 0.25 mm × 0.25 μm was used. Nitrogen was used as carrier gas, with 10 mL/min flow rate and the temperature gradient in Table 4.

Table 4. Composition of the GC-MS gradient of temperature.

Time	Temperature	Rate
0 min	80 °C	0 °C/min
7 min	220 °C	20 °C/min
11 min	240 °C	5 °C/min
24 min	240 °C	0 °C/min

The EIS-MS detector was used to identify compounds with molecular weight from 50 to 600 daltons and the ion source was operated at 200 °C. Five µL tincture diluted 1 to 10 with absolute ethanol and 0.1 µL references having concentrations from 10 to 1280 µg/mL were injected. The terpenoids were identified based on the match factor (higher than 750), using the NIST MS 2.2 spectra database. The content of main terpenes was determined using the calibration curve method [45].

3.6. Antioxidant Activity Assays
3.6.1. DPPH Radical Scavenging Activity

The first assay that was used is the DPPH bleaching assay, a spectrophotometric method based on the reaction of the DPPH reagent and antioxidants. Two mL of tincture at different concentrations were added to 2 mL of a DPPH methanolic solution at a concentration of 0.1 g/L and maintained at 40 °C in a thermostatic bath for 30 min. Changes in absorbance were measured at 517 nm and inhibition of the DPPH radical was calculated using the following formula: DPPH scavenging ability % = (A control − A sample/A control) × 100, A control is the absorbance of the control, which is composed of the DPPH radical solution + methanol (a mixture containing all reagents except the tincture) and A sample is the absorbance of DPPH radical + sample tincture. The percentage of DPPH decrease was expressed in Trolox equivalents (TE, R^2 = 0.998). The DPPH radical scavenging activity of the tincture was expressed as IC50 (µg/mL). The assays were performed in triplicate [46,47,50].

3.6.2. Ferric-Reducing Antioxidant Power Assay (FRAP)

The FRAP method is a spectrometric method that is based on the change of color of a complex of the 2,4,6-tri(2-pyridyl)-1,3,5-triazine (TPTZ) radical with the Fe^{3+} ion, which is assessed by the reduction of the ferric ion to the ferrous ion (Fe^{2+}) in this complex [51]. The FRAP reagent is a mixture of 2.5 mL of a 10 mm TPTZ solution in 40 mM HCl, which are mixed with 2.5 mL 20 mm ferric chloride solution and 25 mL of acetate buffer at a pH of 3.6. Four mL of the tincture were diluted to 1.8 mL with water and mixed with 6 mL of the FRAP reagent. The blank solution was prepared similarly, using water instead of the tincture. The antioxidant capacity was assessed in correlation with the color change, by measuring absorbance at 450 nm. Trolox was used as a reference, using a calibration curve (R^2 = 0.992). Results were expressed as µm Trolox equivalents/100 mL extract. The assays were performed in triplicate [52].

3.6.3. Superoxide Radical (SO) Scavenging Activity Assay

The SO assay evaluated the ability of the tincture to inhibit the formation of the formazan by the reduction of the nitro blue tetrazolium (NBT) radical, by scavenging the superoxide radicals generated in the riboflavin-light-NBT system. The percentage of SO radical scavenging activity was calculated using the following formula: % Superoxide radical scavenging activity = (A_0 − A_1/A_0) × 100, where A_0 = absorbance of control (blank) and A_1 = absorbance of tincture. The superoxide anion radicals (SO) were generated in a mixture of 2.0 mL of Tris-HCl buffer (16 mm, pH 8.0) with 2.0 mL of nitroblue tetrazolium (NBT, 0.3 mm) and 2.0 mL nicotinamide adenine dinucleotide solution (NADH, 0.936 mm). Then, 0.1 mL tincture was diluted to 100.0 mL with water and 1 mL of this solution was added to this mixture. Next, 2.0 mL phenazine methosulfate solution (PMS, 0.12 mm) were then added to all this to initiate the reaction and the mixture was incubated at 250 °C for 5 min. Absorbance was measured at 560 nm using a blank prepared from 2.0 mL Tris-HCl buffer, mixed with 2.0 mL NBT and 2.0 mL NADH solution, 4.0 mL water and 2.0 mL PMS solution. As a reference, 4.0 mL of 1.152 mg/mL Trolox solution was used and the results were expressed as µm Trolox equivalents/g. This solution was added to 2.0 mL Tris-HCl buffer, mixed with 2.0 mL NBT solution, 2.0 mL NADH solution and 2 mL PMS solution (0.12 mm). All tests were performed in triplicate [53].

3.6.4. Animal Studies

The in vivo studies were conducted on outbreed Swiss mice, six months old, of 30.34 ± 2.87 g body weight. The animals originated from and were maintained throughout the study in the Establishment for Laboratory Animals of University of Agricultural Science and Veterinary Medicine Cluj Napoca. They were housed in conventional standard laboratory conditions (temperature 25 ± 1 °C, relative humidity $55 \pm 5\%$, and 12 h light/dark cycle), five animals per cage. The animals benefited from environmental enrichment and had free access to granular standard food and water.

Housing conditions and the procedures complied with the Directive 2010/63/EU and national legislation, Law 43/2014. The project was approved by the Committee for Bioethics and Research Ethics of UASVM (accord No. 68/30.05.2017), and the Veterinary State Authorities (project authorization No. 73/14.06.2017).

Toxicity studies followed the Organization for Economic Cooperation and Development (OECD) Guidelines for the Testing of Chemicals, Test No. 425: Acute Oral Toxicity: Up-and-Down Procedure [54]. Five male and five female Swiss mice received a dose of 2000 mg/b.w. orally; all animals survived and showed no signs of toxicity. Fourteen days later, the animals were euthanized, then subjected to gross examination and displayed no alteration in the internal organs. Furthermore, liver and kidney showed a normal histological architecture.

The hepatoprotective effect was investigated using forty-five Swiss female mice. First, the animals were divided into five groups. Five animals were allocated to the control group, a group receiving placebo oral therapy (vegetable oil); that group was euthanized in the end of the study; they provided the reference values for plasma biochemistry and histopathology. The remaining forty animals were all subjected to liver insufficiency induction protocol; they received carbon tetrachloride (CCl_4) in a dose of 1 mL/kg diluted in vegetable oil. CCl_4 was administrated orally, three times a week, on days 1, 3 and 5, up to the end of the protocol, using a lubricated plastic probe. Those animals were divided into four equal groups; one group, the CCl_4 group, received placebo therapy, and the remaining three groups were treated with *Rosmarinus officinalis* extract in doses of 5, 50 and 500 mg/kg b.w. The extract was administered on the same days as CCl_4, four hours later. To prevent any influence of ethylic alcohol, before use, the alcoholic extract was maintained into a rotary evaporator until the entire amount of alcohol was removed, then it was reconstituted with distillate water and immediately administered to the animals. Each group of ten mice was further subdivided into two subgroups, five animals each, one was euthanized at four weeks, and the other was maintained for another two weeks.

In the end, the blood was collected from retroocular sinus, using deep isoflurane (3%) narcosis. The animals were considered dead when heart beats and respiratory movements stopped, the irreversibility of the phenomenon was assured by cervical dislocation, immediately followed by removal of internal organs. The blood was collected in clot activator vacutainers; after clotting, it was centrifuged (1465 g) for 15 min. The serum was immediately removed, stored at -20 °C and thawed just before use.

3.7. Oxidative Stress Markers and Plasma Biochemistry

In order to assess oxidative stress markers, extraction of total protein from liver tissue was done by homogenization with a phosphate buffer solution (at 10 mm, pH 7.4). The obtained protein extracts were analyzed for total protein content, catalase (CAT), superoxide dismutase (SOD) activity, glutathione peroxidase (GPx) activity, and lipid peroxidation. The activity of CAT, XOD and GPx was determined using the corresponding assay kits (BioVision, Milpitas, CA, USA), according to the manufacturer's specifications. The concentration of malonyl dialdehyde (MDA) was determined using reaction with thiobarbituric acid (TBA). The results were measured by a METERTECH Spectrophotometer SP-830 Plus.

Serum chemistry was measured using screen point semi-automatic analyzer STAT-FAX 1904 Plus Global Medical Instrumentation Inc. (Ramsey, MN, USA) using special kits for plasma biochemistry (Diagnosticum Zrt. Hungary, Budapest) according to the producer specifications.

3.8. Histology

The tissue fragments were fixed in phosphate-buffered formalin 10%, pH 7, for 24 h, embedded in paraffin wax, cut to a thickness of 2–3 μm and stained with hematoxylin and eosin (H&E). The examination of the histopathological slides was performed using an Olympus image processing and retrieval system, the Olympus Cell B image acquisition and processing program. The histological examination was performed by an experienced pathologist RM, unaware of the therapy received by each group. Histopathological examination was focused on classic effects of CCl_4 toxicity, such as circulatory, inflammatory, necrotic and fibrotic lesions.

3.9. Statistical Analysis

All data are reported as the mean $\pm$ SD ($n = 5$). To assume Gaussian distribution, normality distribution was checked by Shapiro–Wilk normality test two-way ANOVA, followed by Bonferroni post-test which was done for pairwise comparisons. Statistical significance was set at $p < 0.05$ (95% confidence interval). Statistical values and figures were obtained in GraphPad Prism version 5.0 for Windows, GraphPad Software, San Diego California USA.

4. Conclusions

The present study was conducted in order to offer a novel perspective on the species *R. officinalis*. The analyses that were performed on a fresh young shoots tincture highlighted the presence of polyphenols and terpenoids in *R. officinalis* fresh young shoots in significant amounts, showing, therefore, that the tested vegetal material may represent an important medicinal product. Moreover, this vegetal medicinal product was tested in order to evaluate its capacity to exhibit hepatoprotective activity by an antioxidant mechanism. Therefore, due to its chemical composition, the studied vegetal product belonging to the rosemary could be an important raw material for pharmaceutical formulations, contributing to the improvement of human health through its antioxidant and hepatoprotective properties, with significant effects against free radical damages, that may be useful for the protection against liver tissue damage.

Author Contributions: Conceptualization, I.I., B.S., N.-K.O., I.O., D.B. and D.H.; methodology, I.I., B.S., N.-K.O., A.T., E.C., A.U., A.C.S.-B., M.N., R.M. and D.H.; software, B.S., N.-K.O., A.T., E.C., M.N. and R.M.; validation, I.I., B.S., N.-K.O., A.T., E.C., M.N., R.M, A.U. and D.H.; formal analysis, I.I., B.S., N.-K.O., A.T., E.C., A.U., A.C.S.-B., M.N., R.M. and D.H.; investigation, I.I., B.S., N.-K.O., A.T., E.C., A.U., A.C.S.-B., M.N., R.M. and D.H.; resources, B.S., N.-K.O., A.T., E.C., I.O. and D.H.; writing: original draft preparation, I.I., B.S., R.M., D.B. and D.H.; writing: review and editing, N.-K.O., I.O., D.B. and D.H.; visualization, II, B.S., N.-K.O., I.O., D.B. and D.H.; supervision, N.-K.O., I.O., D.B. and D.H. All authors have read and agreed to the published version of the manuscript.

Funding: This research received no external funding.

Institutional Review Board Statement: The study was conducted according to the guidelines of the Declaration of Helsinki, and approved by the Committee for Bioethics and Research Ethics of the University of Agricultural Science and Veterinary Medicine Cluj-Napoca, Romania (protocol No. 68/30.05.2017).

Informed Consent Statement: Not applicable.

Data Availability Statement: The data presented in this study are available on request from the corresponding author.

Acknowledgments: This paper was published under the frame of European Social Found, Human Capital Operational Programme 2014–2020, project no. POCU/380/6/13/125171.

Conflicts of Interest: The authors declare no conflict of interest.

Sample Availability: Samples of the compounds are not available from the authors.

References

1. Niculae, M.; Hanganu, D.; Oniga, I.; Benedec, D.; Ielciu, I.; Giupana, R.; Sandru, C.D.; Ciocarlan, N.; Spinu, M. Phytochemical profile and antimicrobial potential of extracts obtained from *Thymus Marschallianus* Willd. *Molecules* **2019**, *24*, 3101. [CrossRef]
2. Benedec, D.; Hanganu, D.; Oniga, I.; Tiperciuc, B.; Olah, N.-K.; Raita, O.; Bischin, C.; Silaghi-Dumitrescu, R.; Vlase, L. Assessment of rosmarinic acid content in six Lamiaceae species extracts and their antioxidant and antimicrobial potential. *Pak. J. Pharm. Sci.* **2015**, *28*, 2297–2303.
3. Mulas, M. Traditional uses of Labiatae in the Mediterranean area. *Acta Hortic.* **2006**, *723*, 25–32. [CrossRef]
4. Ghorbani, A.; Esmaeilizadeh, M. Pharmacological properties of *Salvia officinalis* and its components. *J. Tradit. Complement. Med.* **2017**, *7*, 433–440. [CrossRef]
5. Shakeri, A.; Sahebkar, A.; Javadi, B. *Melissa officinalis* L.—A review of its traditional uses, phytochemistry and pharmacology. *J. Ethnopharmacol.* **2016**, *188*, 204–228. [CrossRef] [PubMed]
6. Salehi, B.; Abu-Darwish, M.S.; Tarawneh, A.H.; Cabral, C.; Gadetskaya, A.V.; Salgueiro, L.; Hosseinabadi, T.; Rajabi, S.; Chanda, W.; Sharifi-Rad; et al. *Thymus* spp. plants—Food applications and phytopharmacy properties. *Trends Food Sci. Technol.* **2019**, *85*, 287–306. [CrossRef]
7. Karadağ, A.E.; Demirci, B.; Çaşkurlu, A.; Demirci, F.; Okur, M.E.; Orak, D.; Sipahi, H.; Başer, K.H.C. *In vitro* antibacterial, antioxidant, anti-inflammatory and analgesic evaluation of *Rosmarinus officinalis* L. flower extract fractions. *S. Afr. J. Bot.* **2019**, *125*, 214–220. [CrossRef]
8. Bendif, H.; Boudjeniba, M.; Djamel Miara, M.; Biqiku, L.; Bramucci, M.; Caprioli, G.; Lupidi, G.; Quassinti, L.; Sagratini, G.; Vitali, L.A.; et al. *Rosmarinus eriocalyx*: An alternative to *Rosmarinus officinalis* as a source of antioxidant compounds. *Food Chem.* **2017**, *218*, 78–88. [CrossRef] [PubMed]
9. Bakirel, T.; Bakirel, U.; Keleş, O.Ü.; Ülgen, S.G.; Yardibi, H. In Vivo assessment of antidiabetic and antioxidant activities of rosemary (*Rosmarinus officinalis*) in alloxan-diabetic rabbits. *J. Ethnopharmacol.* **2008**, *116*, 64–73. [CrossRef]
10. Borges, R.S.; Sánchez Ortiz, B.L.; Matias Pereira, A.C.; Keita, H.; Tavares Carvalho, J.C. *Rosmarinus officinalis* essential oil: A review of its phytochemistry, anti-inflammatory activity, and mechanisms of action involved. *J. Ethnopharmacol.* **2019**, *229*, 29–45. [CrossRef] [PubMed]
11. Farghali, H.; Canová, N.K.; Zakhari, S. Hepatoprotective properties of extensively studied medicinal plant active constituents: Possible common mechanisms. *Pharm. Biol.* **2015**, *53*, 781–791. [CrossRef] [PubMed]
12. Madrigal-Santillán, E.; Madrigal-Bujaidar, E.; Álvarez-González, I.; Sumaya-Martínez, M.T.; Gutiérrez-Salinas, J.; Bautista, M.; Morales-González, Á.; García-Luna, Y.; González-Rubio, M.; Aguilar-Faisal, J.L.; et al. Review of natural products with hepatoprotective effects. *World J. Gastroenterol.* **2014**, *20*, 14787–14804. [CrossRef] [PubMed]
13. Zhang, A.; Sun, H.; Wang, X. Recent advances in natural products from plants for treatment of liver diseases. *Eur. J. Med. Chem.* **2013**, *63*, 570–577. [CrossRef]
14. Rašković, A.; Milanović, I.; Pavlović, N.; Ćebović, T.; Vukmirović, S.; Mikov, M. Antioxidant activity of rosemary (*Rosmarinus officinalis* L.) essential oil and its hepatoprotective potential. *BMC Complement. Altern. Med.* **2014**, *14*, 1–9. [CrossRef]
15. Amin, A.T.; Hamza, A. Hepatoprotective effects of *Hibiscus*, *Rosmarinus* and *Salvia* on azathioprine-induced toxicity in rats. *Life Sci.* **2005**, *77*, 266–278. [CrossRef]
16. Lucarini, R.; Bernardes, W.A.; Tozatti, M.G.; da Silva Filho, A.A.; Silva, M.L.A.; Momo, C.; Crotti, A.E.M.; Martins, C.H.G.; Cunha, W. Hepatoprotective effect of *Rosmarinus officinalis* and rosmarinic acid on acetaminophen-induced liver damage. *Emir. J. Food Agric.* **2014**, *26*, 878–884. [CrossRef]
17. Fadlalla, E.A.S.; Galal, S.M. Hepatoprotective and reno-protective effects of artichoke leaf extract and rosemary extract against Paracetamol induced toxicity in Albino Rats. *J. Pharm Res. Int.* **2020**, *32*, 67–81. [CrossRef]
18. Hegazy, A.M.; Abdel-Azeem, A.S.; Zeidan, H.M.; Ibrahim, K.S.; El-Sayed, E.M. Hypolipidemic and hepatoprotective activities of rosemary and thyme in gentamicin-treated rats. *Hum. Exp. Toxicol.* **2018**, *37*, 420–430. [CrossRef]
19. Bahri, S.; Ben Ali, R.; Abdennabi, R.; Ben Said, D.; Mlika, M.; Ben Fradj, M.K.; El May, M.V.; Jameleddine, S.B.K. Comparison of the protective effect of *Salvia officinalis* and *Rosmarinus officinalis* infusions against hepatic damage induced by hypotermic-ischemia in Wistar rats. *Nutr. Cancer* **2020**, *72*, 283–292. [CrossRef]
20. Martínez-Rodríguez, J.L.; Gutiérrez-Hernández, R.; Reyes-Estrada, C.A.; Granados-López, A.J.; Pérez-Veyna, O.; Arcos-Ortega, T.; López, J.A. Hepatoprotective, antihyperlipidemic and radical scavenging activity of hawthorn (*Crataegus oxyacantha*) and rosemary (*Rosmarinus officinalis*) on alcoholic liver disease. *Altern. Ther. Health Med.* **2019**, *25*, 54–63.
21. El-Hadary, A.E.; Elsanhoty, R.M.; Ramadan, M.F. In vivo protective effect of *Rosmarinus officinalis* oil against carbon tetrachloride (CCl4) -induced hepatotoxicity in rats. *PharmaNutrition* **2019**, *9*, 1–7. [CrossRef]
22. Selmi, S.; Rtibi, K.; Grami, D.; Sebai, H.; Marzouki, L. Rosemary (*Rosmarinus officinalis*) essential oil components exhibit anti-hyperglycemic, anti-hyperlipidemic and antioxidant effects in experimental diabetes. *Pathophysiology* **2017**, *24*, 297–303. [CrossRef] [PubMed]
23. Ngo, Y.L.; Lau, C.H.; Chua, L.S. Review on rosmarinic acid extraction, fractionation and its anti-diabetic potential. *Food Chem. Toxicol.* **2018**, *121*, 687–700. [CrossRef]

24. Elufioye, T.O.; Habtemariam, S. Hepatoprotective effects of rosmarinic acid: Insight into its mechanisms of action. *Biomed. Pharm.* **2019**, *112*, 1–11. [CrossRef]
25. Maldini, M.; Montoro, P.; Addis, R.; Toniolo, C.; Petretto, G.L.; Foddai, M.; Nicoletti, M.; Pintore, G. A new approach to discriminate *Rosmarinus officinalis* L. plants with antioxidant activity, based on HPTLC fingerprint and targeted phenolic analysis combined with PCA. *Ind. Crops Prod.* **2016**, *94*, 665–672. [CrossRef]
26. Jemia, M.B.; Tundis, R.; Maggio, A.; Rosselli, S.; Senatore, F.; Menichini, F.; Bruno, M.; Kchouk, M.E.; Loizzo, M.R. NMR-based quantification of rosmarinic and carnosic acids, GC-MS profile and bioactivity relevant to neurodegenerative disorders of *Rosmarinus officinalis* L. extracts. *J. Funct. Foods* **2013**, *5*, 1873–1882. [CrossRef]
27. Moore, J.; Yousef, M.; Tsiani, E. Anticancer effects of rosemary (*Rosmarinus officinalis* L.) extract and rosemary extract polyphenols. *Nutrients* **2016**, *8*, 731. [CrossRef]
28. Bai, N.; He, K.; Roller, M.; Lai, C.S.; Shao, X.; Pan, M.H.; Ho, C.T. Flavonoids and phenolic compounds from *Rosmarinus officinalis*. *J. Agric. Food Chem.* **2010**, *58*, 5363–5367. [CrossRef]
29. Risaliti, L.; Kehagia, A.; Daoultzi, E.; Lazari, D.; Bergonzi, M.C.; Vergkizi-Nikolakaki, S.; Hadjipavlou-Litina, D.; Bilia, A.R. Liposomes loaded with *Salvia triloba* and *Rosmarinus officinalis* essential oils: *In vitro* assessment of antioxidant, antiinflammatory and antibacterial activities. *J. Drug Deliv. Sci. Technol.* **2019**, *51*, 493–498. [CrossRef]
30. Ainane, A.; Khammour, F.; Charaf, S.; Elabboubi, M.; Elkouali, M.; Talbi, M.; Benhima, R.; Cherroud, S.; Ainane, T. Chemical composition and insecticidal activity of five essential oils: *Cedrus atlantica*, *Citrus limonum*, *Rosmarinus officinalis*, *Syzygium aromaticum* and *Eucalyptus globules*. *Mater. Today Proc.* **2019**, *13*, 474–485. [CrossRef]
31. Ramadan, M.F.; Khider, M.; Abulreesh, H.H.; Assiri, A.M.A.; Elsanhoty, R.M.; Assaeedi, A.; Elbanna, K. Cold pressed rosemary (*Rosmarinus officinalis*) oil. *Cold Pressed Oils* **2020**, 683–694. [CrossRef]
32. Borrás-Linares, I.; Stojanović, Z.; Quirantes-Piné, R.; Arráez-Román, D.; Švarc-Gajić, J.; Fernández-Gutiérrez, A.; Segura-Carretero, A. *Rosmarinus officinalis* leaves as a natural source of bioactive compounds. *Int. J. Mol. Sci.* **2014**, *15*, 20585–20606. [CrossRef] [PubMed]
33. Okamura, N.; Haraguchi, H.; Hashimoto, K.; Yagi, A. Flavonoids in *Rosmarinus officinalis* leaves. *Phytochemistry* **1994**, *37*, 1463–1466. [CrossRef]
34. Zoral, M.A.; Futami, K.; Endo, M.; Maita, M.; Katagiri, T. Anthelmintic activity of *Rosmarinus officinalis* against *Dactylogyrus minutus* (Monogenea) infections in *Cyprinus carpio*. *Vet. Parasitol.* **2017**, *247*, 1–6. [CrossRef]
35. Ali, A.; Chua, B.L.; Chow, Y.H. An insight into the extraction and fractionation technologies of the essential oils and bioactive compounds in *Rosmarinus officinalis* L.: Past, present and future. *Trends Anal. Chem.* **2019**, *118*, 338–351. [CrossRef]
36. Csepregi, K.; Neugart, S.; Schreiner, M.; Hideg, É. Comparative evaluation of total antioxidant capacities of plant polyphenols. *Molecules* **2016**, *21*, 208. [CrossRef]
37. Mishra, K.; Ojha, H.; Chaudhury, N.K. Estimation of antiradical properties of antioxidants using DPPH- assay: A critical review and results. *Food Chem.* **2012**, *130*, 1036–1043. [CrossRef]
38. Szollosi, R.; Szollosi Varga, I. Total antioxidant power in some species of Labiatae (Adaptation of FRAP method). *Acta Biol. Szeged* **2002**, *46*, 125–127.
39. Olah, N.K.; Osser, G.; Câmpean, R.F.; Furtuna, F.R.; Benedec, D.; Filip, L.; Raita, O.; Hanganu, D. The study of polyphenolic compounds profile of some *Rosmarinus officinalis* L. extracts. *Pak. J. Pharm Sci.* **2016**, *29*, 2355–2361. [PubMed]
40. Gîrd, C.E.; Nencu, I.; Popescu, M.L.; Costea, T.; Duțu, L.E.; Balaci, T.D.; Olaru, O.T. Chemical, antioxidant and toxicity evaluation of rosemary leaves and its dry extract. *Farmacia* **2017**, *65*, 978–983.
41. Jacobo-Velázquez, D.A.; Cisneros-Zevallos, L. Correlations of antioxidant activity against phenolic content revisited: A new approach in data analysis for food and medicinal plants. *J. Food Sci.* **2009**, *74*, 107–113. [CrossRef]
42. Azzahra, L.F.; Fouzia, H.; Mohammed, L.; Noureddine, B. Antioxidant response of *Camellia sinensis* and *Rosmarinus officinalis* aqueous extracts toward H2O2 stressed mice. *J. Appl. Pharm Sci.* **2012**, *2*, 70–76. [CrossRef]
43. El-hawary, S.S.; Ezzat, S.M.; Elshibani, F. Gas chromatography-mass spectrometry analysis, hepatoprotective and antioxidant activities of the essential oils of four Libyan herbs. *J. Med. Plants Res.* **2013**, *7*, 1746–1753.
44. Boll, M.; Weber, L.W.D.; Becker, E.; Stampfl, A. Mechanism of carbon tetrachloride-induced hepatotoxicity. Hepatocellular damage by reactive carbon tetrachloride metabolites. *Z. Nat. Sect. C J. Biosci.* **2001**, *56*, 649–659. [CrossRef]
45. Stan, M.S.; Voicu, S.N.; Caruntu, S.; Nica, I.C.; Olah, N.K.; Burtescu, R.; Balta, C.; Rosu, M.; Herman, H.; Hermenean, A.; et al. Antioxidant and anti-inflammatory properties of a *Thuja occidentalis* mother tincture for the treatment of ulcerative colitis. *Antioxidants* **2019**, *8*, 416. [CrossRef]
46. Ielciu, I.; Frederich, M.; Hanganu, D.; Angenot, L.; Olah, N.K.; Ledoux, A.; Crisan, G.; Paltinean, R. Flavonoid analysis and antioxidant activities of the *Bryonia alba* L. aerial parts. *Antioxidants* **2019**, *8*, 108. [CrossRef]
47. Sevastre-Berghian, A.C.; Ielciu, I.; Mitre, A.O.; Filip, G.A.; Oniga, I.; Vlase, L.; Benedec, D.; Gheldiu, A.M.; Toma, V.A.; Mihart, B.; et al. Targeting oxidative stress reduction and inhibition of HDAC1, MECP2, and NF-kB pathways in rats with experimentally induced hyperglycemia by administration of *Thymus marshallianus* Willd. extracts. *Front. Pharmacol.* **2020**, *11*, 1–18. [CrossRef] [PubMed]
48. Badalica-Petrescu, M.; Dragan, S.; Ranga, F.; Fetea, F.; Socaciu, C. Comparative HPLC-DAD-ESI(+)MS fingerprint and quantification of phenolic and flavonoid composition of aqueous leaf extracts of *Cornus mas* and *Crataegus monogyna*, in relation to their cardiotonic potential. *Not. Bot. Horti Agrobot.* **2014**, *42*, 9–18. [CrossRef]

49. Hanganu, D.; Benedec, D.; Olah, N.K.; Ranga, F.; Mirel, S.; Tiperciuc, B.; Oniga, I. Research on enzyme inhibition potential and phenolic compounds from *Origanum vulgare* ssp. *vulgare*. *Farmacia* **2020**, *68*, 1075–1080. [CrossRef]
50. Mot, A.C.; Damian, G.; Sarbu, C.; Silaghi-Dumitrescu, R. Redox reactivity in propolis: Direct detection of free radicals in basic medium and interaction with hemoglobin. *Redox Rep.* **2009**, *14*, 267–274. [CrossRef] [PubMed]
51. Benzie, I.; Strain, J. The ferric reducing ability of plasma (FRAP) as a measure of "antioxidant power": The FRAP assay. *Anal. Biochem.* **1996**, *239*, 70–76. [CrossRef] [PubMed]
52. Oniga, I.; Puşcaş, C.; Silaghi-Dumitrescu, R.; Olah, N.K.; Sevastre, B.; Marica, R.; Marcus, I.; Sevastre-Berghian, A.C.; Benedec, D.; Pop, C.E.; et al. *Origanum vulgare* ssp. *vulgare*: Chemical composition and biological studies. *Molecules* **2018**, *23*, 2077. [CrossRef] [PubMed]
53. Alam, M.N.; Bristi, N.J.; Rafiquzzaman, M. Review on *in vivo* and *in vitro* methods evaluation of antioxidant activity. *Saudi Pharm J.* **2013**, *21*, 143–152. [CrossRef] [PubMed]
54. Organization for Economic Co-operation and Development (OECD/OCDE). *Test. No. 425: Acute Oral Toxicity: Up-and-Down Procedure*; OECD Guidelines for the Testing of Chemicals; OECD: Paris, France, 2008; pp. 1–27.

Article

Evaluation of the Antibacterial Activity and Efflux Pump Reversal of Thymol and Carvacrol against *Staphylococcus aureus* and Their Toxicity in *Drosophila melanogaster*

Zildene de Sousa Silveira [1,2], Nair Silva Macêdo [1,2], Joycy Francely Sampaio dos Santos [1], Thiago Sampaio de Freitas [3], Cristina Rodrigues dos Santos Barbosa [3], Dárcio Luiz de Sousa Júnior [1], Débora Feitosa Muniz [3], Lígia Claudia Castro de Oliveira [1], José Pinto Siqueira Júnior [4], Francisco Assis Bezerra da Cunha [1], Henrique Douglas Melo Coutinho [3,*], Valdir Queiroz Balbino [2] and Natália Martins [5,6,*]

1   Laboratory of Semi-Arid Bioprospecting (LABSEMA), Regional University of Cariri-URCA, Crato 63105-000, CE, Brazil; Zildenesousa15@gmail.com (Z.d.S.S.); naiirmacedo@gmail.com (N.S.M.); joycy.sampaio22@gmail.com (J.F.S.d.S.); darciolsjr@gmail.com (D.L.d.S.J.); ligiaclaudia@yahoo.com.br (L.C.C.d.O.); cunha.urca@gmail.com (F.A.B.d.C.)
2   Graduate Program in Biological Sciences-PPGCB, Federal University of Pernambuco-UFPE, Recife 50670-901, PE, Brazil; vqbalbino@gmail.com
3   Laboratory of Microbiology and Molecular Biology (LMBM), Regional University of Cariri-URCA, Crato 63105-000, CE, Brazil; thiagocrato@hotmail.com (T.S.d.F.); cristinase75@gmail.com (C.R.d.S.B.); deehmuniz78@gmail.com (D.F.M.)
4   Laboratory of Microorganism Genetics (LGM), Federal University of Paraiba-UFPB, João Pessoa 58051-900, PB, Brazil; siqueira@dbm.ufpb.br
5   Faculty of Medicine, University of Porto, Alameda Prof. Hernâni Monteiro, 4200-319 Porto, Portugal
6   Institute for Research and Innovation in Health (i3S), University of Porto, 4200-135 Porto, Portugal
*   Correspondence: hdmcoutinho@gmail.com (H.D.M.C.); ncmartins@med.up.pt (N.M.)

Academic Editors: Severina Pacifico and Simona Piccolella
Received: 9 April 2020; Accepted: 27 April 2020; Published: 30 April 2020

Abstract: The antibacterial activity and efflux pump reversal of thymol and carvacrol were investigated against the *Staphylococcus aureus* IS-58 strain in this study, as well as their toxicity against *Drosophila melanogaster*. The minimum inhibitory concentration (MIC) was determined using the broth microdilution method, while efflux pump inhibition was assessed by reduction of the antibiotic and ethidium bromide (EtBr) MICs. *D. melanogaster* toxicity was tested using the fumigation method. Both thymol and carvacrol presented antibacterial activities with MICs of 72 and 256 µg/mL, respectively. The association between thymol and tetracycline demonstrated synergism, while the association between carvacrol and tetracycline presented antagonism. The compound and EtBr combinations did not differ from controls. Thymol and carvacrol toxicity against *D. melanogaster* were evidenced with EC_{50} values of 17.96 and 16.97 µg/mL, respectively, with 48 h of exposure. In conclusion, the compounds presented promising antibacterial activity against the tested strain, although no efficacy was observed in terms of efflux pump inhibition.

Keywords: bacterial resistance; efflux pumps; terpenoids; thymol; carvacrol

---

1. Introduction

The prevalence of bacterial resistance to antibiotics, this being associated with increased mortality rates, has become a source of great concern for public health [1]. *Staphylococcus aureus* is a commensal microorganism associated with a wide variety of infections, since it has the capacity to acquire resistance to many classes of antibacterial agents, such as β-lactams, quinolones and macrolides [2,3].

There are several mechanisms by which *S. aureus* develops resistance to antimicrobials, including limited drug absorption, target modification, enzymatic inactivation and active efflux mechanisms [4]. The latter, also known as efflux pumps, are proteins integrated into the bacterial plasma membrane that reduce the intracellular concentration of antibiotics by extruding them from the cell [5]. Among these pumps, the TetK pump, belonging to the major facilitator superfamily (MFS), is present in *S. aureus* IS-58 strain. TetK powers its transport activity with energy derived from proton gradients and is responsible for resistance to the tetracycline class of antibiotics [6].

Given the above, the development of efflux pump inhibitors that act as competitive and non-competitive adjuvants to reduce antibiotic resistance has attracted the attention of researchers [7]. Natural bacterial resistance modifiers can facilitate the reintroduction of ineffective therapeutic antibiotics in the clinic, reducing the toxic risks of these drugs by acting as efflux system regulators or as efflux pump inhibitors (EPIs) when blocking their activity [8,9].

The compounds thymol and carvacrol are two phenolic terpenoids, geometric isomers, which can be found in the form of translucent crystals and a yellowish liquid, respectively, at room temperature. Both are obtained from essential oils, mainly from thyme (*Thymus vulgaris* L.) and oregano (*Origanum vulgare* L.) [10], where a number of pharmacological properties associated with these compounds have been previously described in the literature, including antifungal [11] and antibacterial activities [12,13]. Although some compounds are capable of acting as EPIs, their high eukaryotic cell toxicity prevents their development as EPIs [14]. Thus, studies assessing the toxicity of these substances are necessary. *Drosophila melanogaster* is a model organism in toxicological assays which aim to understand the genetic and molecular mechanisms of toxic substances since these are very sensitive to different concentrations of toxic substances [15].

Thus, the objective of this study was to assess the antibacterial activity and efflux pump reversal mechanisms of the isomers thymol and carvacrol against the *S. aureus* IS-58 bacterial strain and to evaluate their toxicity in a *D. melanogaster* model.

2. Results

2.1. Minimum Inhibitory Concentration (MIC)

The monoterpenes thymol and carvacrol demonstrated relevant direct antibacterial activity against the *S. aureus* IS-58 strain, with MIC values of 72 µg/mL and 256 µg/mL, respectively (Table 1), where the MIC value for thymol was more effective than that of the standard antibiotic tetracycline, with values varying between 128 and 114 µg/mL.

Table 1. Minimum inhibitory concentrations (MIC, µg/mL) of thymol, carvacrol and tetracycline against the *S. aureus* IS-58 strain.

Strain	MIC (µg/mL)		
	Thymol	Carvacrol	Tetracycline
S. aureus IS-58	72	256	128

2.2. Modulatory Effect over Antibiotic Activity and Ethidium Bromide

When thymol was combined at a sub-inhibitory concentration (MIC/8) with tetracycline, an interference of the antibiotic activity was observed, where a MIC reduction from 114 to 101 µg/mL was seen (Figure 1A). When the antibiotic was tested in association with standard inhibitors, the MIC values for chlorpromazine did not differ from the control, whereas in association with CCCP, a reduction in the antibiotic MIC was observed, indicating greater specificity of this inhibitor for the tested pump, with inhibition of the antibiotic resistance mechanism being observed.

With respect to efflux pump inhibition, the assays with ethidium bromide (EtBr) as a pump substrate found the association between thymol and EtBr did not differ from the control. Thus, the observed action of the compound, when in association with the antibiotic, suggests an activity over a resistance mechanism other than active efflux (Figure 1B).

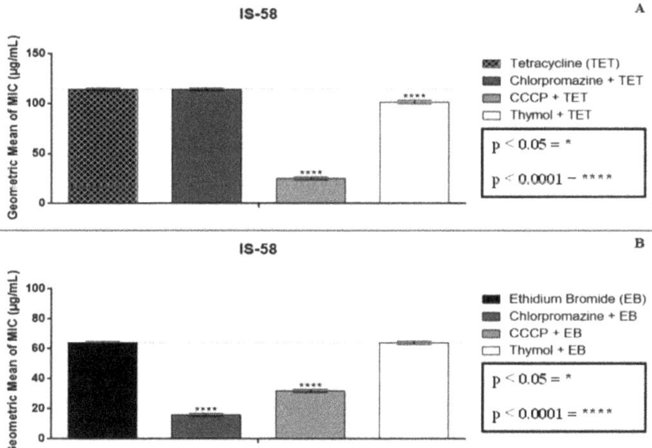

Figure 1. Effect of the association between thymol and tetracycline (**A**) and thymol and ethidium bromide (**B**) over *S. aureus* IS-58, expressing the TetK efflux protein. CCCP = carbonyl cyanide m-chlorophenylhydrazone; * p <0.05; **** p < 0.0001.

The data regarding the combined effect of carvacrol and tetracycline reported an antagonism, with the MIC value increasing from 128 to 203 µg/mL (Figure 2A). However, the results for the association of carvacrol with chlorpromazine did not differ from the control, while a synergism resulting from its association with carvacrol was observed for CCCP.

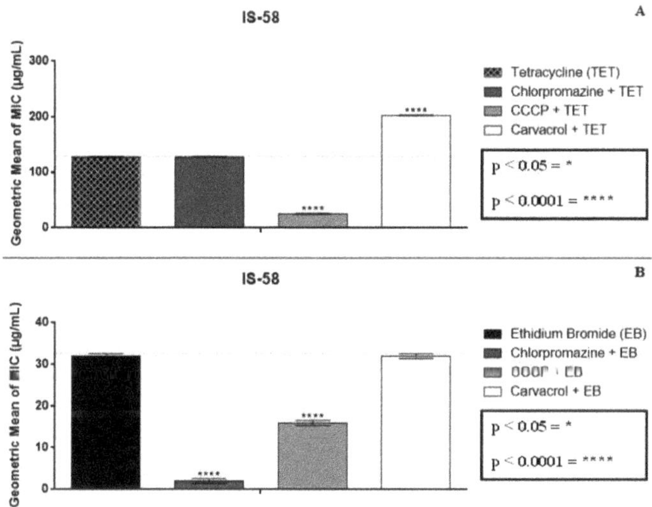

Figure 2. Effect of the association between carvacrol and tetracycline (**A**) and carvacrol and ethidium bromide (**B**) over *S. aureus* IS-58, expressing the TetK efflux protein. CCCP = carbonyl cyanide m-chlorophenylhydrazone; * p < 0.05; **** p < 0.0001.

When the EPI effect was evaluated using the MIC reduction of EtBr by carvacrol, the association was shown to not differ from the control, while standard inhibitors synergistically modulated this, demonstrating the presence of an efflux pump mechanism (Figure 2B).

The results observed in the graphs above indicate the presence of an efflux mechanism that is sensitive to the CCCP inhibitor when tested in association with the antibiotic against the *S. aureus* IS-58 strain. Furthermore, the association with EtBr demonstrated presence of an efflux pump mechanism sensitive to both standard inhibitors, which shows that the antibacterial activity exhibited by both thymol and carvacrol against *S. aureus* is not due to an EPI effect.

2.3. Drosophila Melanogaster Toxicity Assay and Negative Geotaxis

The monoterpenes thymol and carvacrol presented marked toxicity against *D. melanogaster* with EC_{50} values of 17.96 µg/mL and 16.97 µg/mL, respectively, within 48 h of exposure to the products. The mortality tests using thymol found the highest concentration tested, 31 µg/mL, and caused significant mortality compared to the control and the other concentrations from the first reading performed, which corresponded to 3 h of exposure to the compound (Figure 3A).

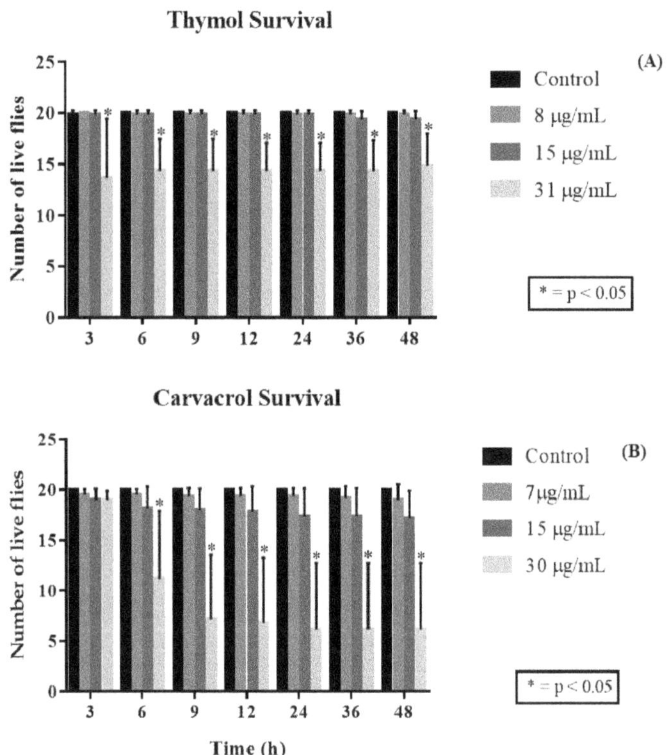

Figure 3. Toxic effect of different thymol (**A**) and carvacrol (**B**) concentrations on *D. melanogaster*.

In the 6 to 48-h readings, the concentration-dependent toxicity pattern remained constant, where the 8 and 15 µg/mL concentrations did not show significant differences compared to the control with respect to the number of dead flies. For carvacrol, a toxicity for this compound was found at 30 µg/mL, this being the most potent, since it began to differ statistically from the control after the first 6 h of exposure, becoming more effective throughout the exposure readings, showing a high mortality rate in the flies. The 7 and 15 µg/mL concentrations were not significant compared to the control (Figure 3B). Damage to locomotor capacity following exposure to thymol was determined by the negative geotaxis test, in which the 31 µg/mL concentration was found to cause damage to the *D. melanogaster* locomotor apparatus following 3 h of exposure to the compound, this being statistically significant compared to the control and the other concentrations (Figure 4A). The mobility of the flies was scarcely affected by the 8 and 15 µg/mL concentrations, with no statistically significant interferences in fly locomotion compared to the control group.

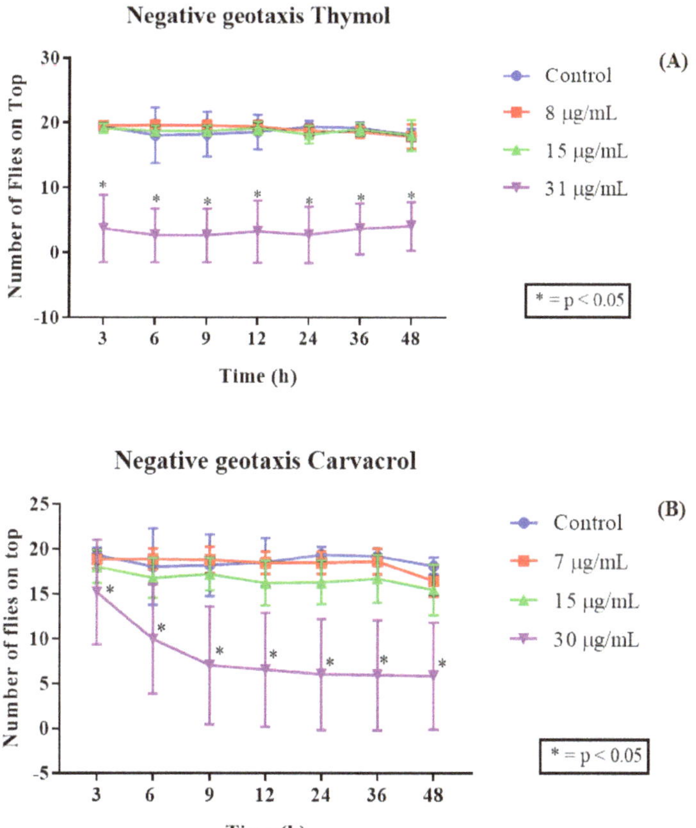

Figure 4. Toxic effect of varying thymol (**A**) and carvacrol (**B**) concentrations on the locomotor ability of *D. melanogaster*.

In the locomotor system damage assessment, a marked decline in the behavioral response of the flies was observed when these were exposed to a 30 µg/mL concentration, from the first 3 h of exposure to the compound. This effect was intensified in the following hours, such that by the 24-h reading the live flies showed great locomotor difficulties (Figure 4B).

3. Discussion

In this work, our results indicated that thymol and carvacrol exert relevant antibacterial activity, with MIC values of 72 and 256 µg/mL respectively, against the *S. aureus* IS-58 strain. These results are in accordance with that reported by Miladi et al. [16], in which thymol and carvacrol obtained MIC values of 64 and 256 µg/mL, respectively, against the *S. aureus* ATCC 25923 strain. Lambert et al. [17] also reported a greater antibacterial activity for thymol compared to carvacrol against *S. aureus* ATCC 6538.

These results may be justified by the hydrophobic nature and low solubility of thymol in the hydrophobic domain of the cytoplasmic membrane of bacterial cells. [18].

The structural differences in the hydroxyl group (OH) position in both thymol and carvacrol isomers did not affect the inhibitory effect against the assayed bacterial strain. By the same reason, the relative position of the hydroxyl group in the phenolic ring also failed to strongly influence the degree of antibacterial activity for thymol and carvacrol against *Bacillus cereus, S. aureus* and *Pseudomonas aeruginosa* [19]. However, the study by Dorman and Deans [20] reported that carvacrol and thymol act differently against gram-positive and gram-negative species. Our results evidenced an interference in the antibiotic activity when thymol and tetracycline were associated, resulting in a reduction of the MIC value of tetracycline from 114 to 101 µg/mL. However, Davies and Wright [21] stated in their study that when a compound is in association with an inhibitor, only a minimum of a 3-fold MIC reduction is acceptable as significant in terms of inhibiting resistance mechanisms, when carvacrol was associated with tetracycline an antagonism can be observed. These results are different from that observed by Cirino et al. [22], where thymol and carvacrol, used at sub-inhibitory concentrations (MIC/4), reduced the MIC value of tetracycline from 64 µg/mL to 32 µg/mL in both cases.

The TetK efflux pump is the main mechanism for bacterial resistance to tetracycline, being coded by the plasmid gene pt181. The main efflux protein protects the bacterial ribosome by extruding the antibiotic out of the bacterial cell [23]. Many studies have been conducted to face the bacterial resistance to antibiotics, mainly through using compounds that act as adjuvants to the antibiotic activity. For this reason, it is known that natural products and phytochemicals act synergistically with this objective [24–26]. In our assays, thymol and carvacrol had no effect as EPIs. Due to the fact that efflux pumps are the unique mechanism for EtBr extrusion, the MIC reduction of the EtBr indicates an EPI effect [27]. Thus, our results indicate that thymol and carvacrol act on other resistance mechanisms, regardless of the active efflux.

Given these results, we also investigated the thymol and carvacrol toxicity against *D. melanogaster*. Similar findings were also stated by Zhang et al. [28], who demonstrated that oxygenated monoterpenes, such as thymol and carvacrol, exhibited high toxicity against *D. melanogaster* while investigating the fumigant toxicity of monoterpenes against fruit flies. Negative geotaxis consists of flies' ability to move vertically, this being a common locomotion behavior associated with *D. melanogaster* [29]. The results found in this study corroborate data from the investigation by Karpouhtsis et al. [30], which reported an insecticidal and genotoxic activity for thymol against *D. melanogaster*. Previous studies also report a repellent activity for thymol against *Culex pipiens pallens* [31], as well as a toxic effect for its larvae [32].

The insecticidal potential of terpenoids has been associated with their low molecular weights, which makes them highly volatile, with these being often considered toxic or repellent against insects, where different functional groups and their locations seem to influence their biological effectiveness [33]. The toxicity mechanisms and locomotor capacity impairments produced by many compounds and/or essential oil may be associated with a decrease in AChE activity, which, considering its importance against neurotoxicity, functions as defense in stressful situations [34,35]. Another factor also reported in the literature is the increase in the biosynthesis of heat shock proteins, such as *hsp70*, which support the functional structure of important enzymes and proteins, as an insect's self-defense mechanism against stressors [36].

4. Materials and Methods

4.1. Bacterial Strain and Culture Media

The IS-58 *S. aureus* strain (gently furnished by Dr. Simon Gibbons, from the Imperial College, London, UK), with the PT181 plasmid carrying the gene for the tetracycline efflux protein TetK, was used. The culture media used in the tests were heart infusion agar (HIA, laboratories Difco Ltd.a., Campinas, Brazil) prepared according to the manufacturer and 10% brain heart infusion (BHI, laboratories Difco Ltd.a.) broth.

4.2. Substances

The antibiotic tetracycline, as well as thymol and carvacrol, were diluted in dimethyl sulfoxide (DMSO) and in sterile water to a final concentration of 1024 µg/mL. The DMSO proportion used was lower than 5%. Chlorpromazine and EtBr were dissolved in sterile distilled water, while carbonyl cyanide m-chlorophenylhydrazone (CCCP) was dissolved in methanol/water (1:3, *v/v*). All substances were diluted until reaching a concentration of 1024 µg/mL. The molecular structures of the compounds were obtained using the software ACD/ChemSketch (ACD/LABS, Toronto, ON, Canada) (Figure 5).

4.3. Determining the Minimum Inhibitory Concentration (MIC)

The MIC was determined for the isolated compounds thymol and carvacrol as per the broth microdilution method proposed by Javadpour et al. [37], with some adaptations. The inoculants were prepared 24 h after sowing the strains. Eppendorfs® were filled with 1440 µL of BHI and 160 µL of the inoculum. The plates were then filled with 100 µL of the final solution. Microdilution was performed with 100 µL of the products. Following 24 h of incubation, readings were taken by the addition of resazurin (7-hydroxy-3H-phenoxazine-3-one 10-oxide) [38]. The tests were performed in triplicates.

4.4. Evaluation of Efflux Pump Inhibition

Efflux pump inhibition was carried out using the methodology adapted from Coutinho et al. [39]. Eppendorfs® were filled with 160 µL of the inoculum, the sub-inhibitory concentration (MIC/8) of the compounds, and completed with BHI until reaching a volume of 1.6 mL. Microdilution was performed with 100 µL of the antibiotics, and readings were taken 24 h after incubation by adding resazurin [38]. The modulatory effect of the combination of the antibiotic, as well as EtBr, with the compounds' thymol and carvacrol was tested using a methodology adapted from Coutinho et al. [39] (Figure 5). For this, Eppendorfs® were filled with 160 µL of the inoculum with the compounds at sub-inhibitory concentrations (MIC/8) and completed with BHI until reaching a volume of 1.6 mL. A modulatory control was prepared with 160 µL of the inoculum and 1440 µL of BHI. Thereafter, the microdilution plates were filled, with rows G and H corresponding to the microbial growth controls. Sterility controls were performed on separate plates. Subsequently, microdilutions were performed with the antibiotic and EtBr (100 µL). After 24 h, readings were taken in the same manner as for the MIC tests. The tests were performed in triplicates.

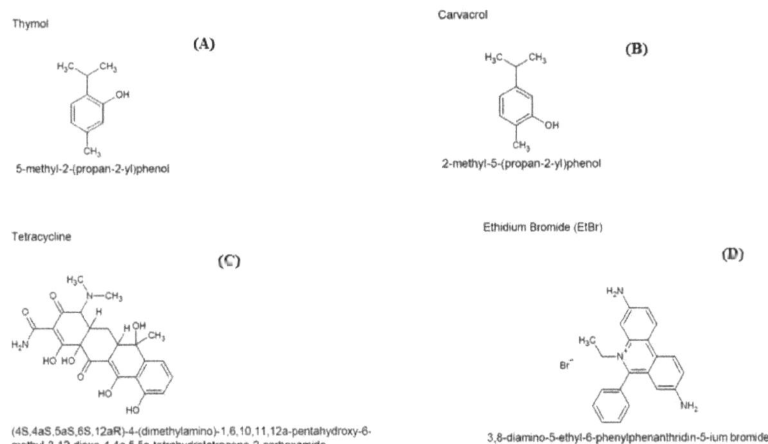

Figure 5. Chemical structures of thymol (**A**), carvacrol (**B**), tetracycline (**C**) and ethidium bromide (**D**).

4.5. Drosophila melanogaster Toxicity Assays

The fumigation bioassay method proposed by Cunha et al. [40] was used to assess toxicity. Adult flies (males and females), in multiples of 20, were placed in 130 mL flasks, previously prepared with 1 mL of 20% sucrose solution. The compound doses were impregnated in the glass cover on filter paper. The control received 20 µL of acetone, while the compounds were prepared at the concentrations of 200 µg/mL for thymol and 195.2 µg/mL for carvacrol, both diluted in acetone. Volumes of 20, 10 and 5 µL were taken from the stock solutions, resulting in the final concentrations of 31, 15 and 8 µg/mL in bottles with 130 mL of air for thymol and 30, 15 and 7 µg/mL in bottles with 130 mL of air for carvacrol, respectively.

The bioassays were conducted in a BOD-type greenhouse under controlled conditions. The tests were performed in triplicates and mortality rate readings were performed at 3, 6, 12, 24, 36 and 48 h [40].

4.6. Negative Geotaxis Assays

Damage to the locomotor system was determined as described by Coulom and Birman [41], after considering fly mortality. The negative geotaxis assay consists of counting the number of flies that rise above 3 cm in the experimental glass column during a 5 s time interval. The assays were repeated twice within a 1 min interval. The results were presented as the mean time (s) ± SE obtained from three independent experiments.

4.7. Statistical Analysis

A two-way analysis of variance (ANOVA) followed by Bonferroni's post hoc test was employed for the microbiological assays using GraphPad Prism 7.0 software (GraphPad Software, San Diego, CA, USA). Meanwhile, a two-way ANOVA followed by Tukey's multiple comparisons test was performed for the toxicity data analysis.

5. Conclusions

The monoterpenes thymol and carvacrol presented direct antibacterial activity against the *S. aureus* IS-58 strain, where the strain was shown to be more sensitive to thymol, with the isomeric difference being a possible factor in terms of antibacterial activity. However, despite the demonstrated antibiotic activity results, the compounds were ineffective at inhibiting the TetK efflux pump mechanism, thus indicating that the antibacterial activity of the compounds is not associated with this resistance mechanism. Thymol and carvacrol exerted a marked toxic activity against *D. melanogaster*.

Author Contributions: Conceptualization, F.A.B.d.C., N.M., H.D.M.C.; methodology, Z.d.S.S., J.F.S.d.S., T.S.d.F., C.R.d.S.B.; validation, N.S.M.; formal analysis, D.L.d.S.J., D.F.M.; resources, J.P.S.J., L.C.C.d.O.; writing—original draft preparation, Z.d.S.S, V.Q.B.; writing—review and editing, F.A.B.d.C., N.M., H.D.M.C.; supervision, F.A.B.d.C., N.M., H.D.M.C. All authors have read and agreed to the published version of the manuscript.

Funding: This study was funded by the Fundação Cearense de Apoio ao Desenvolvimento Científico e Tecnológico—FUNCAP (BP3-0139-00077.01.00/18).

Conflicts of Interest: The authors declare no conflicts of interest.

References

1. Porse, A.; Schou, T.S.; Munck, C.; Ellabaan, M.M.H.; Sommer, A.O. Biochemical mechanisms determine the functional compatibility of heterologous genes. *Nat. Commun.* **2018**, *9*, 522. [CrossRef] [PubMed]
2. Foster, T.J. Antibiotic resistance in *Staphylococcus aureus*. Current status and future prospects. *FEMS Microbiol. Rev.* **2017**, *41*, 430–449. [CrossRef] [PubMed]
3. Sun, J.; Deng, Z.; Yan, A. Bacterial multidrug efflux pumps: Mechanisms, physiology and pharmacological exploitations. *Biochem. Biophys. Res. Commun.* **2014**, *453*, 254–267. [CrossRef] [PubMed]
4. Yilmaz, E.Ş.; Aslantaş, Ö. Antimicrobial resistance and underlying mechanisms in *Staphylococcus aureus* isolates. *Asian Pac. J. Trop. Med.* **2017**, *10*, 1059–1064. [CrossRef]

5. Jang, S. Multidrug efflux pumps in *Staphylococcus aureus* and their clinical implications. *J. Microbiol.* **2016**, *54*, 1–8. [CrossRef]
6. Bhardwaj, A.K.; Mohanty, P. Bacterial efflux pumps involved in multidrug resistance and their inhibitors: Rejuvinating the antimicrobial chemotherapy. *Recent Pat. Anti-Infect. Drug Discov.* **2012**, *7*, 73–89. [CrossRef]
7. Mahmood, H.Y.; Jamshidi, S.; Sutton, M.J.; Rahman, K.M. Current advances in developing inhibitors of bacterial multidrug efflux pumps. *Curr. Med. Chem.* **2016**, *23*, 1062–1081. [CrossRef]
8. Abreu, A.C.; Mcbain, A.J.; Simoes, M. Plants as sources of new antimicrobials and resistance-modifying agents. *Nat. Prod. Rep.* **2012**, *29*, 1007–1021. [CrossRef]
9. Pagès, J.M.; Amaral, L. Mechanisms of drug efflux and strategies to combat them: Challenging the efflux pump of Gram-negative bacteria. *Biochim. Biophys. Acta (Baa)-Protein Struct. Mol. Enzym.* **2009**, *1794*, 826 833.
10. Kowalcze, M.; Jakubowska, M. Multivariate approach in voltammetric identification and simultaneous determination of eugenol, carvacrol, and thymol on boron-doped diamond electrode. *Mon. Chem. Chem. Mon.* **2019**, *150*, 991–1002. [CrossRef]
11. Braga, P.C.; Alfieri, M.; Culici, M.; Dal Sasso, M. Inhibitory activity of thymol against the formation and viability of Candida albicans hyphae. *Mycoses* **2007**, *50*, 502–506. [CrossRef] [PubMed]
12. Fadil, M.; Fikri-Benbrahim, K.; Rachiq, S.; Ihssane, B.; Lebrazi, S.; Chraibi, M.; Farah, A. Combined treatment of *Thymus vulgaris* L., *Rosmarinus officinalis* L. and *Myrtus communis* L. essential oils against Salmonella typhimurium: Optimization of antibacterial activity by mixture design methodology. *Eur. J. Pharm. Biopharm.* **2018**, *126*, 211–220. [CrossRef] [PubMed]
13. Vincenzi, M.; Stammati, A.; Vincenzi, A.; Silano, M. Constituents of aromatic plants: Carvacrol. *Fitoterapia* **2004**, *75*, 801–804. [CrossRef] [PubMed]
14. Limaverde, P.W.; Campina, F.F.; Cunha, F.A.; CRISPIM, F.D.; Figueredo, F.G.; Lima, L.F.; Tintino, S.R. Inhibition of the TetK efflux-pump by the essential oil of *Chenopodium ambrosioides* L. and α-terpinene against *Staphylococcus aureus* IS-58. *Food Chem. Toxicol.* **2017**, *109*, 957–961. [CrossRef]
15. Rand, M.D.; Montgomery, S.L.; Prince, L.; Vorojeikina, D. Developmental toxicity assays using the Drosophila model. *Curr. Protoc. Toxicol.* **2014**, *59*, 1–12. [CrossRef]
16. Miladi, H.; Zmantar, T.; Chaabouni, Y.; Fedhila, K.; Bakhrouf, A.; Mahdouani, K.; Chaieb, K. Antibacterial and efflux pump inhibitors of thymol and carvacrol against food-borne pathogens. *Microb. Pathog.* **2016**, *99*, 95–100. [CrossRef]
17. Lambert, R.J.W.; Skandamis, P.N.; Coote, P.J.; Nychas, G.J. A study of the minimum inhibitory concentration and mode of action of oregano essential oil, thymol and carvacrol. *J. Appl. Microbiol.* **2001**, *1*, 453–462. [CrossRef]
18. Trombetta, D.; Castelli, F.; Sarpietro, M.G.; Venuti, V.; Cristani, M.; Daniele, C.; Bisignano, G. Mechanisms of antibacterial action of three monoterpenes. *Antimicrob. Agents Chemother.* **2005**, *49*, 2474–2478. [CrossRef]
19. Obaidat, R.M.; Bader, A.; Al-Rajab, W.; Sheikha, G.A.; Obaidat, A.A. Preparation of mucoadhesive oral patches containing tetracycline hydrochloride and carvacrol for treatment of local mouth bacterial infections and candidiasis. *Sci. Pharm.* **2011**, *79*, 197–212. [CrossRef]
20. Dorman, H.J.D.; Deans, S.G. Antimicrobial agents from plants: Antibacterial activity of plant volatile oils. *J. Appl. Microbiol.* **2000**, *88*, 308–316. [CrossRef]
21. Davies, J.; Wright, G.D. Bacterial resistance to aminoglycoside antibiotics. *Trends Microbiol.* **1997**, *5*, 234–240. [CrossRef]
22. Cirino, I.C.S.; Menezes-Silva, S.M.P.; Silva, H.T.D.; Souza, E.L.; Siqueira-Júnior, J.P. The essential oil from *Origanum vulgare* L. and its individual constituents carvacrol and thymol enhance the effect of tetracycline against *Staphylococcus aureus*. *Chemotherapy* **2014**, *60*, 290–293. [CrossRef] [PubMed]
23. Khan, S.A.; Novick, R.P. Complete nucleotide sequence of pT181, a tetracycline-resistance plasmid from *Staphylococcus aureus*. *Plasmid* **1983**, *10*, 251–259. [CrossRef]
24. Andrade, L.M.S.; Oliveira, A.B.M.; Leal, A.L.A.B.; Alcântara, F.A.O.; Portela, A.L.; Lima Neto, J.D.S.; Barreto, H.M. Antimicrobial activity and inhibition of the NorA efflux pump of *Staphylococcus aureus* by extract and isolated compounds from *Arrabidaea brachypoda*. *Microb. Pathog.* **2020**, *40*, 103935. [CrossRef]
25. Nafis, A.; Kasrati, A.; Jamali, C.A.; Custódio, L.; Vitalini, S.; Iriti, M.; Hassani, L. A Comparative Study of the in Vitro Antimicrobial and Synergistic Effect of Essential Oils from *Laurus nobilis* L. and *Prunus armeniaca* L. from Morocco with Antimicrobial Drugs: New Approach for Health Promoting Products. *Antibiotics* **2020**, *9*, 140. [CrossRef] [PubMed]

26. Liu, Q.; Niu, H.; Zhang, W.; Mu, H.; Sun, C.; Duan, J. Synergy among thymol, eugenol, berberine, cinnamaldehyde and streptomycin against planktonic and biofilm-associated food-borne pathogens. *Lett. Appl. Microbiol.* **2015**, *60*, 421–430. [CrossRef]

27. Kaatz, G.W.; Moudgal, V.V.; Seo, S.M.; Kristiansen, J.E. Fenotiazinas e tioxantenos inibem a atividade da bomba de efluxo de múltiplas drogas no *Staphylococcus aureus*. *Antimicrob. E Quim.* **2003**, *47*, 719–726.

28. Zhang, Z.; Yang, T.; Zhang, Y.; Wang, L.; Xie, Y. Fumigant toxicity of monoterpenes against fruitfly, *Drosophila melanogaster*. *Ind. Crop. Prod.* **2016**, *81*, 147–151. [CrossRef]

29. Rhodenizer, D.; Martin, I.; Bhandari, P.; Pletcher, S.D.; Grotewiel, M. Genetic and environmental factors impact age-related impairment of negative geotaxis in Drosophila by altering age-dependent climbing speed. *Exp. Gerontol.* **2008**, *43*, 739–748. [CrossRef]

30. Karpouhtsis, I.; Pardali, E.; Feggou, E.; KokkinI, S.; Scouras, Z.G.; Mavragani-Tsipidou, P. Insecticidal and genotoxic activities of oregano essential oils. *J. Agric. Food Chem.* **1998**, *46*, 1111–1115. [CrossRef]

31. Zahran, H.E.D.M.; Abdelgaleil, A.S. Insecticidal and developmental inhibitory properties of monoterpenes on Culex pipiens L. (Diptera: Culicidae). *J. Asia-Pac. Entomol.* **2011**, *14*, 46–51. [CrossRef]

32. Park, B.S.; Choi, W.S.; Kim, J.H.; Kim, K.H.; Lee, S.E. Monoterpenes from thyme (*Thymus vulgaris*) as potential mosquito repellents. *J. Am. Mosq. Control Assoc.* **2005**, *21*, 80–84. [CrossRef]

33. Moreira, X.; Abdala-Roberts, L.; Nell, C.S.; Vázquez-González, C.; Pratt, J.D.; Keefover-Ring, K.; Mooney, K.A. Sexual and genotypic variation in terpene quantitative and qualitative profiles in the dioecious shrub *Baccharis salicifolia*. *Sci. Rep.* **2019**, *9*, 1–10. [CrossRef]

34. Calisi, A.; Lionetto, M.G.; Schettino, T. Biomarker response in the earthworm Lumbricus terrestris exposed to chemical pollutants. *Sci. Total Environ.* **2011**, *409*, 4456–4464. [CrossRef] [PubMed]

35. Hu, X.; Fu, W.; Yang, X.; Mu, Y.; Gu, W.; Zhang, M. Effects of cadmium on fecundity and defence ability of *Drosophila melanogaster*. *Ecotoxicol. Environ. Saf.* **2019**, *171*, 871–877. [CrossRef] [PubMed]

36. Basile, A.; Sorbo, S.; Lentini, M.; Conte, B.; Esposito, S. Water pollution causes ultrastructural and functional damages in *Pellia neesiana* (Gottsche) Limpr. *J. Trace Elem. Med. Biol.* **2017**, *43*, 80–86. [CrossRef] [PubMed]

37. Javadpour, M.M.; Juban, M.M.; Lo, W.C.J.; Bishop, S.M.; Alberty, J.B.; Cowell, S.M.; Mclaughlin, M.L. De novo antimicrobial peptides with low mammalian cell toxicity. *J. Med. Chem.* **1996**, *39*, 3107–3113. [CrossRef]

38. CLSI. *Performance Standards for Antimicrobial Susceptibility Testing: Twentythird Informational Supplement*; Clinical and Laboratory Standards Institute: Wayne, PA, USA, 2013.

39. Coutinho, H.D.M.; Costa, J.G.; Lima, E.O.; Falcão-Silva, V.S.; Siqueira-Júnior, J.P. Enhancement of the antibiotic activity against a multiresistant Escherichia coli by *Mentha arvensis* L. and chlorpromazine. *Chemotherapy* **2008**, *54*, 328–330. [CrossRef]

40. Cunha, F.A.B.; Wallau, G.L.; Pinho, A.I.; Nunes, M.E.M.; Leite, N.F.; Tintino, R.S.; Franco, J.L. Eugenia uniflora leaves essential oil induces toxicity in *Drosophila melanogaster*: Involvement of oxidative stress mechanisms. *Toxicol. Res.* **2015**, *4*, 526–534. [CrossRef]

41. Coulom, H.; Birman, S. Chronic exposure to rotenone models sporadic Parkinson's disease in *Drosophila melanogaster*. *J. Neurosci.* **2004**, *24*, 10993–10998. [CrossRef]

Sample Availability: Samples of the all compounds assayed are available from the authors.

 *molecules*

Article

In Vitro and In Vivo Anticancer Activity of the Most Cytotoxic Fraction of Pistachio Hull Extract in Breast Cancer

Maryam Seifaddinipour [1], Reyhaneh Farghadani [2], Farideh Namvar [3,*], Jamaludin Bin Mohamad [1,*] and Nur Airina Muhamad [1,*]

[1] Institute of Biological Sciences, Faculty of Science, University of Malaya, Kuala Lumpur 50603, Malaysia; m.seyfadini@gmail.com
[2] Department of Molecular Medicine, Faculty of Medicine, University of Malaya, Kuala Lumpur 50603, Malaysia; r_farghadani@yahoo.com
[3] Faculty of Medicine, Mashhad Branch, Islamic Azad University, Mashhad 917568, Iran
* Correspondence: farideh.namvar@gmail.com (F.N.); jamal@um.edu.my (J.B.M.); nurairina@um.edu.my (N.A.M.); Tel.: +98-9151252011 (F.N.); +60-1128208747 (J.B.M.)

Academic Editors: Severina Pacifico and Simona Piccolella
Received: 17 January 2020; Accepted: 11 March 2020; Published: 13 April 2020

Abstract: Pistacia (*Pistacia vera*) hulls (PV) is a health product that has been determined to contain bioactive phytochemicals which have fundamental importance for biomedical use. In this study, PV ethyl acetate extraction (PV-EA) fractions were evaluated with the use of an MTT assay to find the most cytotoxic fraction, which was found to be F13b1/PV-EA. After that, HPTLC was used for identify the most active compounds. The antioxidant activity was analyzed with DPPH and ABTS tests. Apoptosis induction in MCF-7 cells by F13b1/PV-EA was validated via flow cytometry analysis and a distinctive nuclear staining method. The representation of genes like *Caspase 3*, *Caspase 8*, *Bax*, *Bcl-2*, *CAT* and *SOD* was assessed via a reverse transcription (RT_PCR) method. Inhabitation of Tubo breast cancer cell development was examined in the BALB-neuT mouse with histopathology observations. The most abundant active components available in our extract were gallic acid and the flavonoid quercetin. The F13b1/PV-EA has antiradical activity evidence by its inhibition of ABTS and DPPH free radicals. F13b1/PV-EA displayed against MCF-7 a suppressive effect with an IC_{50} value of 15.2 ± 1.35 μg/mL. Also, the expression of *CAT*, *SOD*, *Caspase 3*, *Caspase 8* and *Bax* increased and the expression of *Bcl-2* decreased. F13b1/PV-EA dose-dependently inhibited tumor development in cancer-induced mice. Thus, this finding introduces F13b1/PV-EA as an effectual apoptosis and antitumor active agent against breast cancer.

Keywords: pistacia (*Pistacia vera*) hulls; breast cancer; anticancer

---

1. Introduction

The pistachio hull refers to the epicarp which has a reddish/yellow color during development and when it ripens, it is a rosy and light yellow [1,2]. Usually, collected pistachio nuts are encased in this shell which is removed by a dehulling process. During the pistachio dehulling process many types of by-products are generated that are currently considered as an agriculture waste and to a lesser extent, are used as fodder by local livestock farmers. Hulls may also be used as an herbal medicine for stomach pains and the prevention of diarrhea and to improve hemorrhoids. Pistachio hull has caught the attention of researchers in recent years due to its natural phenolic and antioxidant compounds. Recent literature has proven that pistachio hull extracts have antioxidant, antimicrobial and antimutagenicity activities. Several reports have validated and established the pharmacological activities and medicinal properties of pistachio hull [1,3,4]. In a report by Tomaino the antioxidant activity of the polyphenols extracts from natural shelled pistachios (NP) was determined. In the rats treated with NP a remarkable decrease was observed for CAR-induced histological paw damage, nitrotyrosine formation and neutrophil infiltration. These results demonstrated that the polyphenols display antioxidant properties in lower doses [1].

In Goli [4] report pistachio hulls were extracted with three different solvents (water, methanol and ethyl acetate) and its total phenol content were determined using the Folin–Ciocalteu method. Additionally, the effect of water and methanolic extracts on the stability of soybean oil that was heated to 60 °C was ascertained. The pistachio hull extract (PHE) slowed down the process of oil deterioration at 60 °C with a concentration-dependent increase between 0.02–0.06%. The 0.06% PHE showed a similar activity pattern to butylhydroxyanisole (BHA) and butylated hydroxytoluene (BHT) of. Thus, pistachio hulls, which at the moment are mainly considered as agricultural waste, contain antioxidants that may be compatible for adding them to food products [4].

It should be noted that the content of antioxidant compounds could vary depending on the extraction procedures adopted. Indeed, for pistachio it has been demonstrated by Garavand et al. [5], who studied and measured the phytochemical substances and radical scavenging activity of pistachio hull extracts, obtained using diverse solvents (water, ethanol, and butanol). Their results showed that ultrasound-assisted aqueous extraction of the hull using ultrasound power (35 kHz) was more effective in increasing the phytochemicals content than a sonochemical ultrasonication method (130 kHz). The amount of vanillic acid, *p*-coumaric acid, naringenin, and catechin in the ultrasound-assisted extracts increased as demonstrated by high-performance liquid chromatography-mass spectrometry. The content of phenolics and antioxidant properties of the aqueous extract decreased remarkably after post-extraction sonication. Contrariwise, the amount of phenolics and flavonoids improved with microwave-assisted extraction in a power-dependent trend [5]. Grace et al. [6] described the presence of anacardic acids, fatty acids, carotenoids, tocopherols and phytosterols as the main components in pistachio hulls. Quercetin-3-*O*-glucoside together with smaller concentrations of quercetin, myricetin and luteolin flavonoids were found in a polar (P) extract. Gallotannins and other phenolic compounds esterified with a gallic acid moiety were characterized in the P extract. Release of nitric oxide (NO) and reactive oxygen species (ROS) were inhibited by the P extract in lipopolysaccharide-stimulated RAW 264.7 macrophage cells. In addition, in the macrophages the non-mitochondrial oxidative burst associated with inflammatory response were reduced by the P extract [6].

Bulló, et al. [7] also investigated the anti-inflammatory properties of polyphenol extracts from natural raw shelled pistachios (NP). For the determination of the amount of protection offered by NP against lipopolysaccharide (LPS)-induced inflammation, the monocyte/macrophage cell line J774 was utilized. The in vitro study illustrated that pre-treatment with NP decreased the TNF-α and IL-1β production and degradation of IκB-α, although not significantly. These results show that, at lower doses, the polyphenols present in pistachios possess anti-inflammatory properties [7].

The hulls of pistachio have been shown to have in vitro antioxidant and in vivo photoprotective effects [2], and also exhibit antimicrobial and antimutagenicity [8] as well as enzyme inhibitory and also possess radical scavenging activities [4].

Some phytochemical assessments have revealed the presence of wide ranging levels of phenolic and flavonoids compounds such as gallic acid, catechin, cyanidin-3-*O*-galactoside, eriodictyol-7-*O*-glucoside and epicatechin in the skin of pistachio, which is even 10 times richer than the seeds [9].

In our previous study, we have demonstrated a promising cytotoxic effect and anti-angiogenesis potential of the ethyl acetate extract from pistachio (*Pistacia vera*) hulls (PV-EA) against MCF-7 breast cancer cells [10]. Therefore, in the current study, we have taken this research a step further and investigated the anticancer activity of the most cytotoxic fraction of PV-EA through the utilization of in vitro and in vivo models of breast cancer.

2. Results

2.1. Separation of the Bioactive Compound

Dried hulls of *Pistacia vera* were extracted with ethyl acetate. The ethyl acetate extract (16 g) was fractionated in three steps by column chromatography on silica gel 60, which yielded some fractions in each step. After doing an MTT assay and choosing the most cytotoxic fraction in every step, we continued with the next step until the isolation and purification of final fraction (F13b1/PV-EA) that was about 10 mg with an IC$_{50}$ 15.2 µg/mL (Figure 1). Preparative HPTLC using 100% methanol as the mobile phase and silica gel as the stationary phase with 10 concentrations or tracks was done. The image and spectrum of spots were scanned at two wavelengths (254 and 320 nm). All spectra were the same and in an identical region (Figures 2–4). Chemical profiling of F13b1/PV-EA was investigated by the use of HPTLC again. After comparing retention times, first with blended standards (gallic acid, cyanidin and the flavonoid quercetin and a second time only with gallic acid and the purified compound from pistachio it was found that gallic acid and quercetin were present in the F13b1/PV-EA fraction (Figure 5).

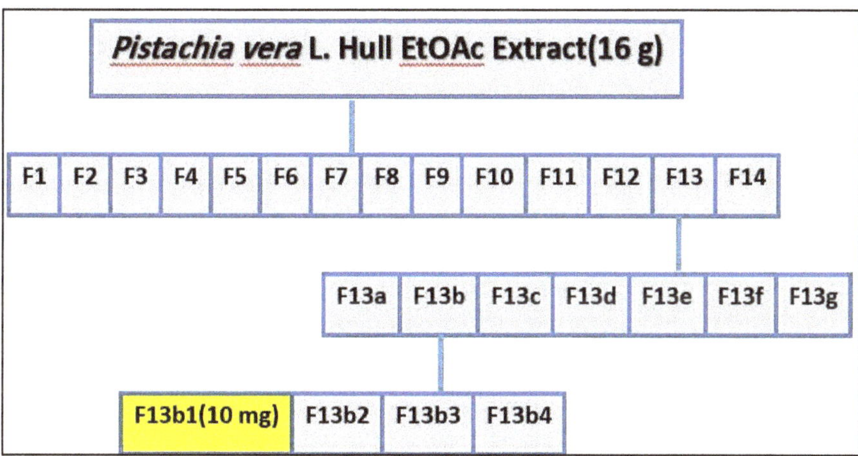

Figure 1. Flow chart from the three steps of bioassay guided fractionation of F13b/PVLH-EAE.

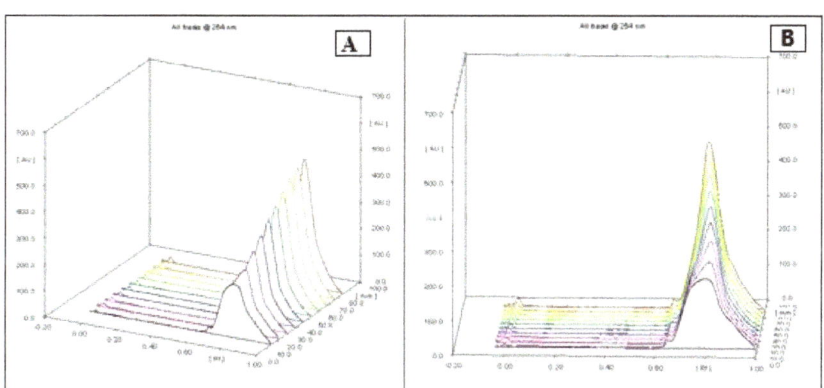

Figure 2. The image and spectrum of all 10 HPTLC spots scanned at 254 nm wavelength on 2 sides (**A**) and (**B**).

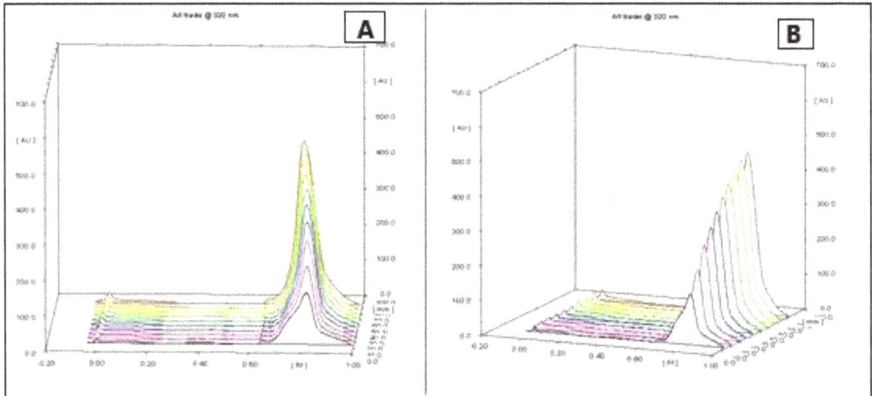

Figure 3. The image and spectrum of all 10 HPTLC spots scanned at 320 nm wavelength on 2 sides (**A**) and (**B**).

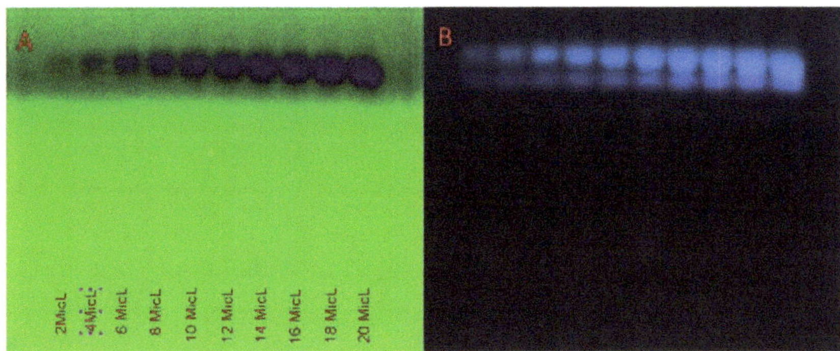

Figure 4. The image and spectrum of all 10 spots scanned at (**A**) 254 nm and (**B**) 320 nm wavelengths.

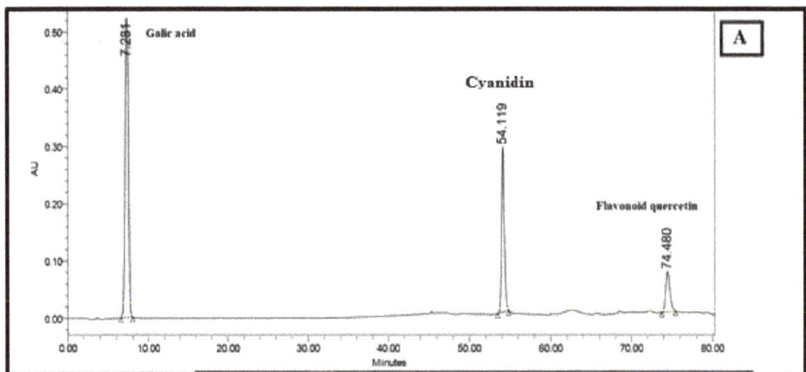

Figure 5. *Cont.*

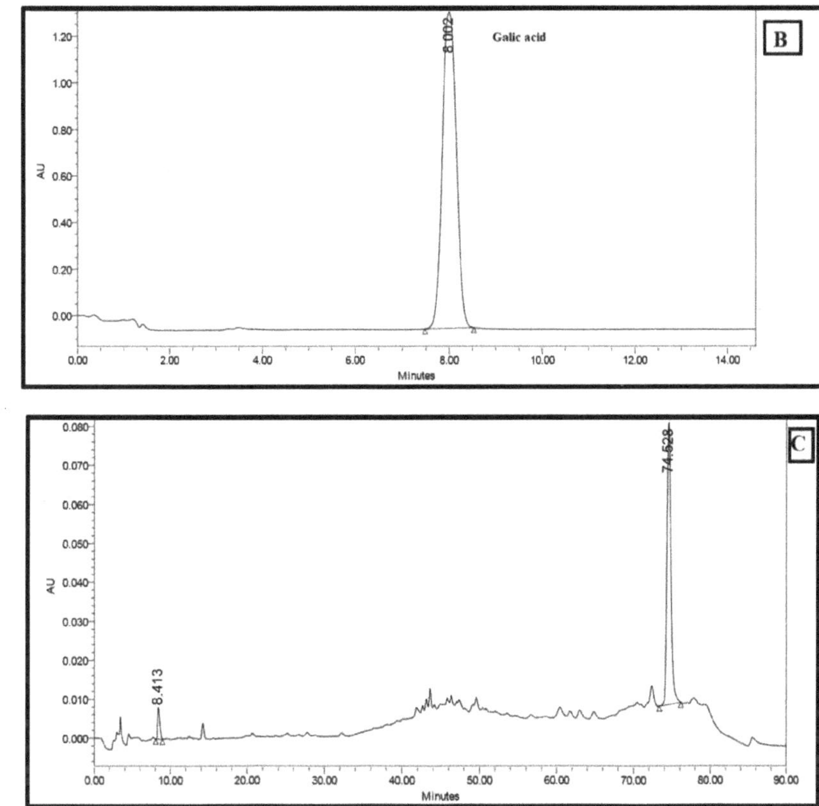

Figure 5. HPTLC analysis. (**A**) Injection of blended standard and, (**B**) Injection of gallic acid standard, (**C**) injection of F13b1/PV-EA to the column.

2.2. Cytotoxic effect of F13b1/PV-EA toward MCF-7 Cells

An evaluation of the cytotoxic properties of F13b1/PV-EA in the MCF-7 cell line was performed using the prescribed MTT assay.

Different concentrations ranging from 7.8 to 250 µg/mL of the compound were used and the amount of formazan formed was specified and detected after 24, 48 and 72 h of incubation. Figure 6 display that F13b1/PV-EA resulted in dose-dependent and time-dependent decline in cell viability with increasing concentration and treatment period. The results suggest that cell growth was prevented when the cells were incubated in the presence of the compound.

2.3. Apoptotic Morphological Variations

Figure 7 shows the results acquired after performing the AO/PI tests. From the data, it can be seen that the compound has dose-dependent effects on cell viability and induces apoptotic morphological variations in treated cells. The results show reduced viability as more apoptotic cells (red in color) were seen at all three concentrations of treatment. In addition, Hoechst 33342 staining (Figure 8), also revealed that the F13b1/PV-EA stimulates apoptotic morphological variations. The cells underwent amazing nuclear changes when treated. However, in the untreated group, the cells were uniformly stained by the fluorescence Hoechst dye indicating the nuclei of the cells were virgin. However, with increasing concentration level of the compound, there was an increase of intensity captured on fluorescence signals and luminous points where the cells expressed apoptotic morphological variations.

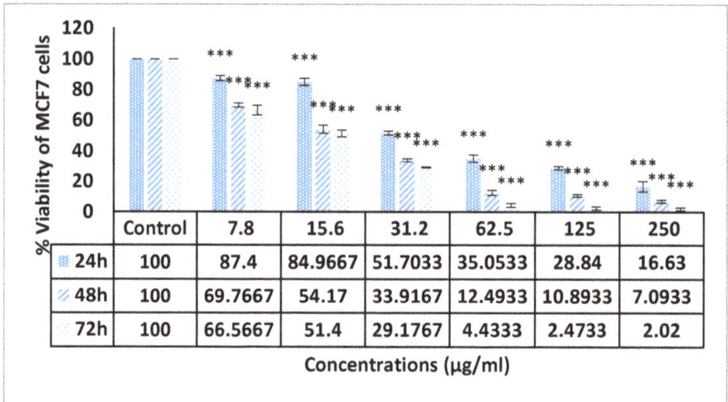

	Control	7.8	15.6	31.2	62.5	125	250
24h	100	87.4	84.9667	51.7033	35.0533	28.84	16.63
48h	100	69.7667	54.17	33.9167	12.4933	10.8933	7.0933
72h	100	66.5667	51.4	29.1767	4.4333	2.4733	2.02

Figure 6. Shows the growth inhibition effects of F13b1/PV-EA on MCF-7 cells noted at different intervals (24, 48 and 72 h) and concentrations. (*** p value < 0.001). All of the in vitro experiments were done in triplicate.

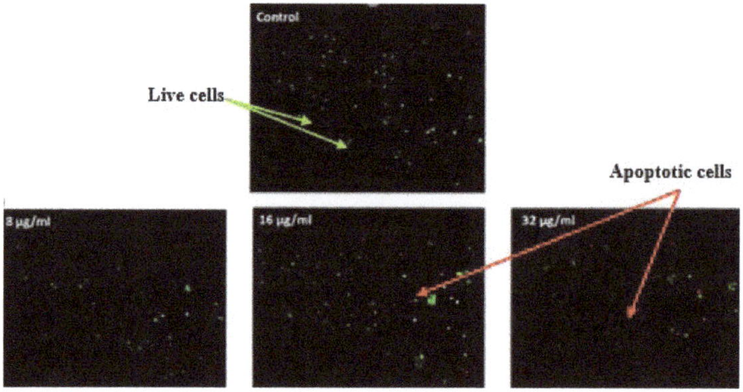

Figure 7. Fluorescent images of MCF-7 cells dyed by AO/PI. Untreated cells (×200) and treated with three concentration of F13b1/PV-EA for 48 h (×200).

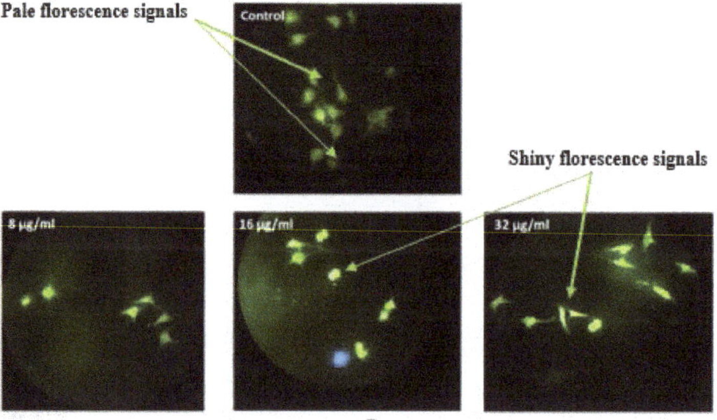

Figure 8. Fluorescent images of Hoechst 33342 stained MCF-7 cells. Untreated cells (×200) and cells treated with 8, 16 and 32 µg/mL of F13b1/PV-EA for 48 h (×200).

2.4. Flow Cytometer Analysis

By utilizing PI staining, we tried to establish whether MCF-7 cells treated with F13b1/PV-EA underwent apoptosis accompanied by alteration in the cell cycle, and the distribution index was also noted. This was in tandem with growth in the Sub-G1 population with increasing concentrations as shown in Figure 9. As depicted in the mentioned figure, high concentration treatment with the compound (32 µg/mL) led to a growth in the percentage of Sub-G1 phase up to 62.1% ± 0.41 when compared to the control cells which were at 2.8% ± 0.86%, thus indicating a change in arrested cells towards a Sub-G1 population which is known as apoptotic cells. The population of cells that possesses sub-diploid DNA content is a clear indication of DNA fragmentation happening at the time of apoptosis.

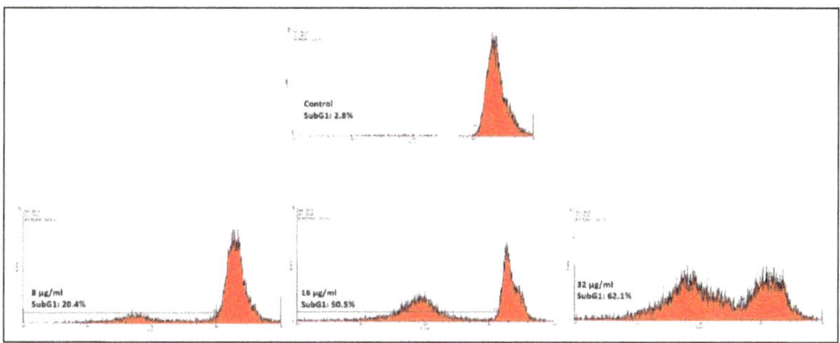

Figure 9. MCF-7 cell cycle analysis of untreated cells (×200) and cells treated with 8, 16 and 32 µg/mL of F13b1/PV-EA for 48-h interval (×200).

2.5. RT-PCR Evaluation

The link of some genes containing *Caspase 3*, *Caspase 8*, *Bax*, *Bcl-2*, *CAT* and *SOD* with apoptosis induced by F13b1/PV-EA were observed using RT-PCR.

As explained in Figures 10 and 11, *Caspase 3*, *Caspase 8*, *CAT*, *Bax* and *SOD* gene expressions increased, respectively, when compared to the control (gene expression in cancer cells without any treatment). Further investigation revealed that the compound treatment lowered the expression level of *Bcl-2* over time. These results show that F13b1/PV-EA could stimulate apoptosis by shifting the regulation of apoptotic genes exclusively through the up-regulation of *Bax* and down-regulation of *BCL-2*.

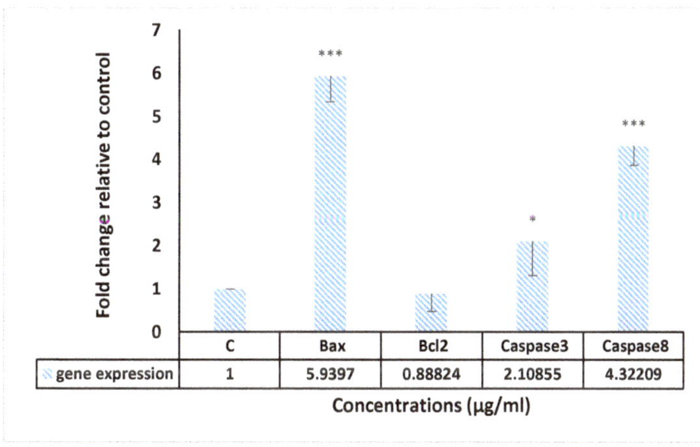

Figure 10. The *Bax*, *Bcl-2*, *Caspase 3* and *Caspase 8* genes expression of MCF-7 cells treated with 15 µg/mL of F13b1/PV-EA for 24 h. (* $p < 0.05$, *** $p < 0.001$).

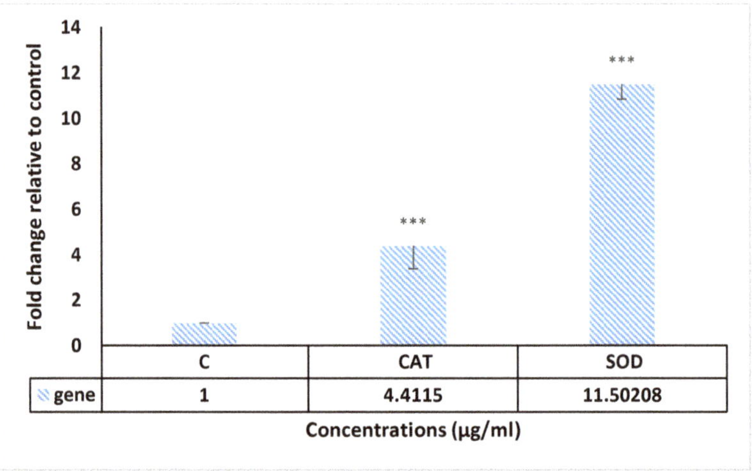

Figure 11. The *CAT* and *SOD* genes expression of MCF-7 cells treated with 15 µg/mL of F13b1/PV-EA for 24 h. (*** p value < 0.001).

2.6. Analysis of Radical Scavenging Effect

Examination to evaluate the antioxidant activity of F13b1/PV-EA, ABTS and DPPH free radical scavenging activity was performed. Figure 12 shows that F13b1/PV-EA demonstrated antiradical activity by inhibiting ABTS radical with IC_{50} values less than 125 µg/mL. F13b1/PV-EA displayed a dose-dependent activity and the ABTS scavenging effect has measured at 63% at a concentration of 125 µg/mL. In addition, pure compound displayed a dose-dependent activity and the DPPH scavenging effect was 38.8% at a concentration of 1000 µM. F13b1/PV-EA thus displayed a moderate inhibitory effect on DPPH free radicals (Figure 13).

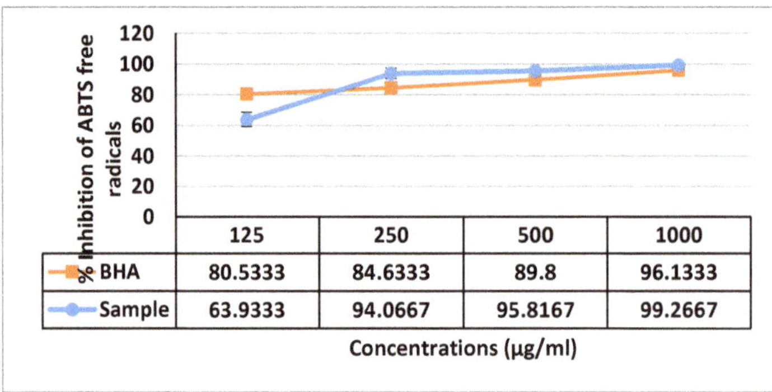

Figure 12. Inhibition activity of F13b1/PV-EA and comparison of a BHA group with treated samples. Data are expressed as mean ± standard division.

2.7. Animal Study

2.7.1. LD_{50} Tests (Lethal Dose 50 Test)

LD_{50} is the amount of a material that results in loss of life of 50% (one half) of animals in an experiment. The LD_{50} is one way to measure the short-term poisoning potential (acute toxicity) of a material. In this study three concentrations of F13b1/PV-EA were tested (12.5, 25 and 50 µg/mL) and in the concentration of 50 µg/mL, 50% of the mice died. This result indicated that a concentration of

50 µg/mL or higher was poisonous to mice and it was best for our main animal experiments to use concentrations of less than 50 µg/mL, which meant only using 12.5 and 25 µg/mL.

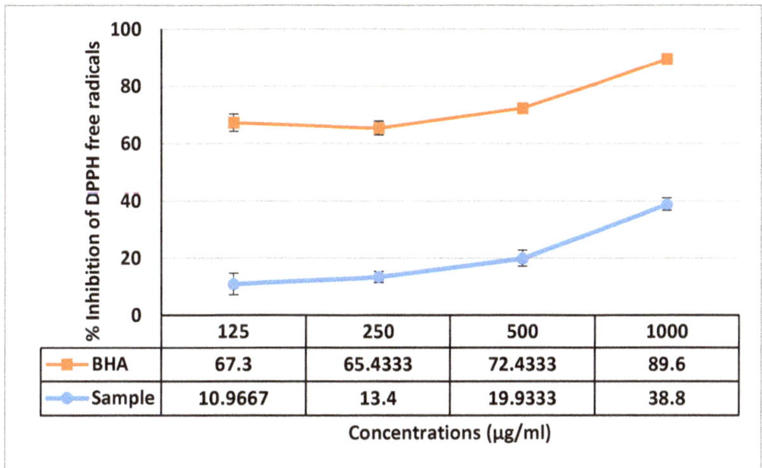

Figure 13. Shows F13b1/PV-EA radical inhibition activity and comparison of BHA group with treated samples. Data are expressed as mean ± standard division.

2.7.2. Average Tumor Volume

To investigate the effects of F13b1/PV-EA in the inducement of apoptosis in Tubo breast cancer cells, mice were treated with two different concentrations of the compound (12.5 and 25 µg/mL), tamoxifen was used as a standard drug and one group without any treatment. At the end of the experiment, the mice were euthanized, tumors were excised from the mice and weighted. The tumor volumes (Table 1) were measured according to the formula below:

$$\text{tumor volume} = A \times B^2 \times 0.5$$

where A: length, B: width.

Table 1. The tumor volume of Tubo cancer cells in the four treatment groups.

Group	Tumor Volumes
Negative control group	10.7 ± 1.2 cm^3
Positive control group	0.95 ± 0.7 cm^3
Experimental group A(12.5 µg/mL)	2.5 ± 0.8 cm^3
Experimental group B (25 µg/mL)	0.8 ± 0.7 cm^3

Statistical analysis showed that the total tumor volume in all treatment groups was smaller than that of the control group.

2.7.3. Histological Analysis

A histopathological examination by H & E staining was done to confirm the effect of treatment by isolated compounds present in *Pistacia vera* hull extract F13b1/PV-EA. Five different sections from each H & E slide were monitored at 100× magnification (Figures 14–17) and mean score was calculated from these five sections (Table 2). The scoring method is described in Table 4 in the Materials and Methods section.

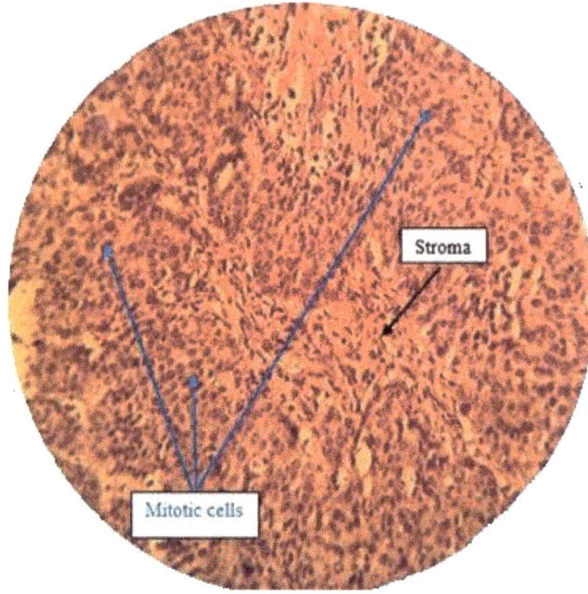

Figure 14. Negative control group with a high density of mitotic cells with dense nuclei (100×).

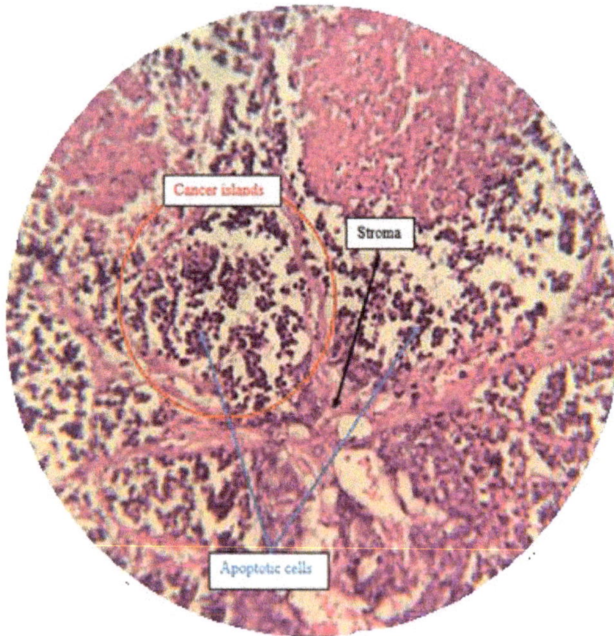

Figure 15. Standard drug control group with a high density of apoptotic cells with disintegrated nuclei in the cancer islands (100×).

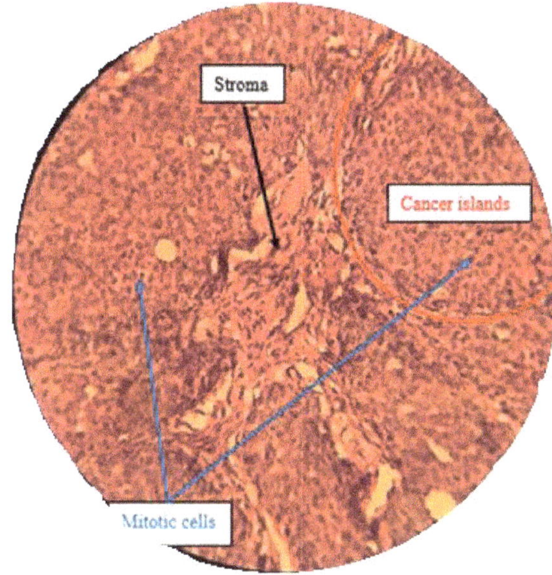

Figure 16. Low dose treatment group with a high density of mitotic cells with dense nuclei (100×).

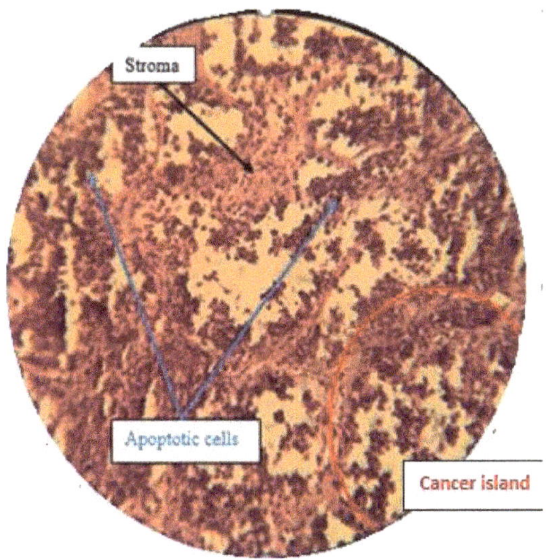

Figure 17. High dose treatment group with a high density of apoptotic cells with disintegrated nuclei in the cancer islands (100×).

Table 2. Apoptotic index, and mitotic index in animal treated with F13b1/PV-EA.

Experimental Group	Apoptotic Index	Mitotic Index
Negative control	1 (2 ± 1)	1 (15 ± 3)
Positive control	2 (9 ± 0.5)	1 (4 ± 0.8)
Low-dose	2 (5 ± 0.1)	1 (9 ± 1)
High-dose	3 (11 ± 0.9)	1 (5 ± 0.7)

3. Discussion

One of the unique characteristics of cancer is the capability of malignant cells to elude apoptosis [11,12]. Therefore, an all-inclusive perception of the apoptotic signaling pathways that are involved is of crucial importance for the discovery and development of target selective therapeutics. Mouse models are useful tool for carcinogenic study. They will greatly enrich the understanding of pathogenesis and molecular mechanisms for cancer. [13,14]. According to our previous study [10], the molecular and cellular response exerted by the PV-EA on the MCF-7 is noteworthy, thus it is vital to ascertain and determine the bioactive constituents that are present. Thus, the Isolation and identification of the bioactive compounds present were performed accordingly to help identify which compound(s) play a role in the safe and effective use for therapeutic purposes.

In this study, we managed to purify 14 fractions from the ethyl acetate extract of this plant in three steps and they were characterized using diverse spectroscopic analyses with subsequent confirmation using the HPTLC method. Chemical profiling of F13b1/PV-EA showed the presence of gallic acid and quercetin.

Phytochemical investigations conducted previously on the hulls of ripe pistachio have led to the identification of structurally varied secondary metabolites [15]. Barreca, et al. [9] in their research identified 20 derivatives from extracts of hull of pistachio, the most plentiful being gallic acid, followed by 4-hydroxybenzoic acid, protocatechuic acid, naringin, eriodictyol-7-*O*-glucoside, isorhamnetin-7-*O*-glucoside, quercetin-3-*O*-rutinoside, isorhamnetin-3-*O*-glucoside and catechin. The key difference between the red and green hulls was the presence of anthocyanins in the green hulls. For the first time, differently galloylated hydrolysable tannins, anthocyanins, and minor anacardic acids were identified. Thus pistachio hulls have structurally varied and potentially bioactive phenolic compounds [9].

One of the phenolic compounds is gallic acid (GA), chemically known as 3,4,5-trihydroxybenzoic acid [16]. Gallic acid is structured in such a way that it has phenolic groups that are a source of activated hydrogen atoms so that generated radicals can be delocalized over the phenolic moieties [17]. Another polyphenolic flavonoid compound is quercetin (3,3′,4′,5,7-pentahydroxyflavone) that is ubiquitous in plants and foods of plant origin. The most notable property of quercetin is its ability to act as an antioxidant. Quercetin seems to be a strong flavonoid for defending the body against reactive oxygen species, which is very important in cancer therapy [15].

Young et al. [18] have examined polyphenols as potential inhibitors of UGDP-glucose dehydrogenase (UGDH) activity. Gallic acid and quercetin decreased the specific activity of UGDH and inhibited the proliferation of MCF-7 human breast cancer cells. Western blot analysis showed that gallic acid and quercetin did not affect UGDH protein expression, suggesting that UGDH activity is inhibited by polyphenols at the post-translational level. Kinetics studies using human UGDH revealed that gallic acid was a non-competitive inhibitor with respect to UDP-glucose and NAD^+. In contrast, quercetin showed a competitive inhibition and a mixed-type inhibition with respect to UDP-glucoseand NAD^+, respectively. These results indicate that gallic acid and quercetin are effective inhibitors of UGDH that exert strong antiproliferative activity in breast cancer cells.

By evaluating the cytotoxic properties of F13b1/PV-EA on MCF-7 cell line, it was observed that there was decreased cell viability in tandem with increasing concentration and time of treatment. Multiple papers about gallic acid and its pharmacological activities have been published. Gallic acid has shown some activities that include in the following: angiogenesis, repression of cell viability and reproduction in human glioma cells, prevention of the propagation of HeLa cervical cancer cells, inhibition of ribonucleotide reductase, induction of apoptosis in humoral cell lines, prevention of lymphocyte duplication and cyclooxygenases in human HL-60 promyelocytic leukemia cells, stimulation inactivating phosphorylation via ATM-Chk2 activation and anti-oxidant activity [11,18,19].

The effects of three different doses of F13b1/PV-EA on MCF-7 was evaluated through acridine orange/propidium iodide (AO/PI) staining and fluorescence microscopy. The tests confirmed that the compound has a dose-dependent effect on cell viability and induces apoptotic morphological changes in the treated cells. The results show reduced viability with the presence of higher levels of apoptotic cells for all three treatment concentrations.

To get a better understanding of the efficacy of the bioactive compound on the nucleus, treated cells were stained with Hoechst stain. The cells was seen to have gone through major nuclear changes upon treatment. In the control, the cells were uniformly stained via the fluorescent Hoechst stain, indicating the nuclei of the cells were intact. However, with increasing concentrations of the compound, there was a growth of severity seen captured on the fluorescence signals and luminous points where the cells exhibited apoptotic morphological changes.

By utilizing PI staining, it was established whether the MCF-7 cells treated with F13b1/PV-EA underwent apoptosis and if it was accompanied by notable alterations in the cell cycle, and the distribution index was also investigated via PI staining. This was accompanied by growth in the Sub-G1 population with increasing concentrations. This cells population possessed a sub-diploid DNA content which is indicative of DNA fragmentation occurring during apoptosis. F13b1/PV-EA was thus shown to be able to bring about apoptosis by changing the regulation of apoptotic genes, particularly through up-regulation of *Bax* and down-regulation of *BCL-2*.

Caspase 3, *Caspase 8*, *CAT*, *Bax* and *SOD* genes expression increased compared to control (gene expression in cancer cells without any treatment). Further investigations revealed that compound treatment eventually decreased the expression level of Bcl2.

The cytotoxicity and anti-cancer effects of hydro-alcoholic extracts of pistachio shell on HepG2 and L929 cells was elucidated by Harandi et al. [20]. Cell viability of HepG2 and L929 was decreased after 24 and 48 h of treatment with IC_{50} 1500 and 1000 µg/mL for HepG2 and 2000 and 1500 µg/mL for L929. *Bax* and *P53* genes were shown to be up-regulated and *Bcl-2* gene was displayed to be down-regulated after treatment.

ROS are stabilized by reactions which in turn cause cellular damage and the formation of carcinogenic DNA adducts. The consumption of antioxidants has been shown to reduce the risks of getting cancer. Our compound displayed a dose dependent activity on ABTS and a slow inhibitory effect on DPPH free radicals.

Hashemi, et al. [21] investigated the antioxidant activity of a *Pistacia atlantica* extract. The antioxidant activity of the extract was 4.6 ± 0.66 µg/mL, while it was 25.41 ± 1.89 µg/mL for butylated hydroxytoluene (BHT). The total phenol, flavonoid and flavanol contents were 269 mg GAE/g, 40.7 mg RUT/g and 88.12 mg RUT/g, respectively. In recent times, identifying and an emphasis on chemical agents and natural products with the capability of preventing human cancer has been an important objective in preclinical cell culture and animal efficacy testing models. In clinical chemoprevention tests, according to the toxicity screening experiments, only the most active factors have potential as human chemopreventives.

Our animal experiment study results showed that the total tumor volume of all treatment groups were smaller than that those in the control group. According to several papers, quercetin has anti-proliferative effects to cancer, enhances the efficacy of chemotherapeutic factors, in vivo lymphocyte tyrosine kinase prevention and anti-tumor activity. LD_{50} tests in rats showed that injection of boldine remarkable decreased breast cancer tumor size and at dose of 100 mg/kg body weight was well tolerated.

Ferulago angulata leaf hexane extract (FALHE) is capable of inducing apoptosis on MCF-7 cells. An in vivo study showed that FALHE reduce the tumor size from 2031 ± 281 mm^3 to 432 ± 201 mm^3 after treatment. Acute toxicity tests revealed an absence of toxic effects of the two compounds on mice [22].

This study explained the potential use of red *Pistacia vera* hull as chemopreventive drug, as it can be exploited as a new lead compound for prodrug therapy. Since MCF7 is an estrogen-receptor negative human breast cancer, it's good for potential future examination of our extract and its active compound in another estrogen receptor-responsive cell line like MDA-231 [23]. Also as extracts of plants may activate the immune system of the hosts and kill cancer cells, one can consider testing the concentrations

using micro syringe. The plates were saturated for 20 min in a twin trough glass chamber with the mobile phase of methanol (100%). The plates were placed in the mobile phase and dried. A densitometric scanning of plates were performed at 254 nm and 320 nm using a Camag TLC scanner III operated in reflectance–absorbance mode. To examine the chemical profiling of F13b1/PV-EA, analysis was carried out using a C_{18} column on a Breeze2 system. First, 20 μL of blended standard (gallic acid, cyaniding and flavonoid quercetin) was injected to the column with water and acetonitrile solvent. Subsequently, 20 μL of standard of gallic acid and finally 20 μL of compound or F13b1/PV-EA was injected to the column. For comparing the three injections' properties and identification of the chemical profile of our compound, after each injection, the retention time (RT) or the amount of time that a compound spends on the column from injection to detection, was calculated.

4.7. Flow Cytometry Analysis

In the next step, three different concentrations (8, 16 and 32 μg/mL) of F13b1/PV-EA was added for 2 days to MCF-7 cells that were seeded (5×10^5 cells/well) in a 35 mm dish for 24 h. Then nuclear fractions from the cells were obtained according to the kit's propidium iodide staining protocol. The intensity of the fluorescence was detected using a FAC Scan flow cytometer (BD Biosciences, San Jose, CA, USA) and analyzed via Cell Quest software.

4.8. Acridine Orange/Propidium Iodide Staining (AO/PI)

After a 24-h incubation of the 1×10^6 cells/well of MCF-7 cells in a 6-well plate, the cells were treated with 8, 16 and 32 μg/mL of F13b1/PV-EA for 48 h. Then detached cells were dyed with AO/PI stain according to manufacturer's protocol and examined using fluorescence microscope.

4.9. Hoechst 33342 Staining

Using a functional vital dye the classical morphological criteria, the quantification and determination of cell death notation was carried out. The MCF-7 cells were treated using three different concentrations (8, 16 and 32 μg/mL) of F13b1/PV-EA for 48 h. Hoechst 33342, which is a specific stain used for AT-rich regions of double-stranded DNA was utilized. The cells were incubated for 15 min with Hoechst 33342 dye (5 μg/mL in PBS)) and subsequently visualized using a BHZ, RFCA microscope (Olympus, Tokyo, Japan) equipped with a fluorescent light source with an excitation wavelength of 330 nm and a barrier filter of 420 nm.

4.10. Gene Expression Assay

By employing the RT-PCR, the gene expression of *Bax*, *Bcl2*, *Caspase 3*, *Caspase 8*, *CAT* and *SOD* were analyzed. RNA was extracted from MCF-7 cells (3×10^6 cells/well) that were treated with 15 μg/mL of F13b1/PV-EA for 24 h, using the manufacturer's instructions for RNA extraction. The mRNA was transcribed in reverse to cDNA by adhering to the manufacturer's protocol using the Advantage RT-PCR kit. cDNA was amplified via a real time and sybr green kit. Table 3 shows the specific primers used for amplifying the cDNA.

4.11. DPPH Radical Scavenging Assay

The free radical scavenging activity of F13b1/PV-EA was assessed based on its effect trapping 2,2-diphenyl-1-picrylhydrazyl (DPPH) free radicals. 0.1 mM methanolic DPPH solution was mixed with the varying concentrations of F13b1/PV-EA (125, 250, 500 and 1000 μg/mL), in an equal volume. After 30 min of incubation, the absorbance of the samples was read at 517 nm. In the control group water and BHA was used as a standard compound. The percentage of inhabitation of DPPH free radical was calculated according to the formula below:

Percentage of inhibition of DPPH free radical = Absorbance of control − Absorbance of sample/Absorbance of control × 100

Table 3. Primer sequence for amplifying cDNA.

BCL-2	5 CATGTGTGTGGAGAGCGTCAA 3 F 5 CAGATAGGCACCCAGGGTGA 3 R
BAX	5 TTTGCTTCAGGGTTTCATCCA 3 F 5 CTCCATGTTACTGTCCAGTTCGT 3 R
Caspase 3	5 GTGGAACTGACGATGATATGGC 3 R 5 CGCAAAGTGACTGGATGAACC 3 R
Caspase 8	5 CTGGGAAGGATCGACGACGAT 3 F 5 CATGTCCTGCATTTTGATGG 3 R
CAT	5 CTTCCCGCTTGAATGTGAAG 3 F 5 CCGATTACATAAACCCATCA 3 R
SOD	5 GCTCCTAAGCCGCTTACGGTT 3F 5 CACGCCATCGGCATTGGCAAT 3R

4.12. ABTS Radical Scavenging Assay

In brief 1 mL of various concentrations of F13b1/PV-EA (125, 250, 500 and 1000 µg/mL), were mixed with 1 mL of ABTS·+ working solution. After incubation period of 1 h in room temperature in the dark, absorbance was read at 734 nm. To prepare the ABTS·+ stock solution, 7 mM of ABTS and 2.45 mM of potassium persulfate were mixed, incubated at room temperature for 12–16 h and finally ABTS·+ stock solution was diluted with distilled water to gain 0.70 ± 0.02. Percentage of inhibition of ABTS free radical was calculated according to formula in previous part.

4.13. Experimental Animals

The experiments on the animal were divided into 2 parts. First, the lethal dose 50% (LD_{50}) test and after that in vivo anti-tumor assessment was carried out. For LD_{50} testing and to determine the acute toxicity of the proposed compounds, 18 healthy male Balb/C mice (25 ± 5 g, five-weeks-old) were provided by the animal house of the University of Malaya Animal Experimental Unit (AEU), in clean, sterile and polyvinyl cages. The mice were maintained under standard conditions, temperature of 22–26 °C, 45–50% relative humidity with water, food and sterile diet under pathogen-free environment and maintained on a 12 h light/dark photo period. For the second part or in vivo anti-tumor assessment, 24 healthy female Balb/C mice were purchased from Pasteur Institute of Mashhad (Iran). The mice were maintained under conditions that were mentioned above. The animal studies were performed after approval of the protocol by the FOM Institutional Animal Care and Use Committee, University of Malaya (FOM, IACUC, ethic No.: 2016-190405/IBS/R/MS).

4.13.1. Lethal Dose 50% (LD_{50}) Test

LD_{50} is the amount of the substance (usually per body weight) required to kill 50% of the test population within a specific time. In order to observe the overall effect of our compound on a living subject, 18 male Balb/C mice were divided into three groups of animals in each group. Every group was treated with one concentration of F13b1/PV-EA, 1 time every week until 2 weeks, for a total of two times. Three doses (12.5, 25 and 50 µg/mL) of pure compound was dissolved in 10% tween 20 and given orally by gavage to the mice. For purpose of calculating the consumable dose (dosage compound for each mouse) we used the formula below:

(Concentration of the compound) × (mouse body weight) × 6 (the rate of metabolism mice to the human) = consumable dose

Animals were monitored for 30 min, 2, 4, 8, 24 and 48 h for up to two weeks after the dosing. Any signs to the animals involving toxicity and/or mortality and behavior changes were observed keenly and recorded throughout the experimental period for 14 days.

4.13.2. In Vivo Anti-Tumor Assessment

To construct an allograft breast carcinoma model, Tubo cancer cell lines were grown and harvested under appropriate conditions. Tubo cells are a cloned cell line established in vitro from a BALB-neuT mouse mammary carcinoma. 1.5×10^6 cells were suspended in 0.2 mL PBS and were injected this subcutaneously into the right flank of BALB/c mice ($n = 6$). After the tumor inoculation (approximately 7 days later) the animals were randomly divided into four groups of six mice:

Group 1: Negative control (just received 10% tween 20 orally) once daily for 2 weeks.

Group 2: Standard drug control group as a positive control (Tubo induced + tamoxifen 10 mg/kg dissolved in 10% tween 20 via oral administration) once daily for 2 weeks.

Group 3: Experimental group A (Tubo induced + low dose of compound dissolved in 10% Tween 20 via oral administration) once daily for 2 weeks.

Group 4: Experimental group B (Tubo induced + high dose of compound dissolved in 10% Tween 20 via oral administration) once daily for 2 weeks.

For calculating the consumable dose of F13b1/PV-EA in the experimental groups, we used two concentrations (12.5 and 25 µg/mL) based on the LD_{50} result. The mice were euthanized at the end of 14 days treatment period and the tumors were then excised and weighted (Figure 20). The tumor volumes were measured according to the formula below:

$$\text{tumor volumes} = A \times B^2 \times 0.5$$

where A: length, B: width

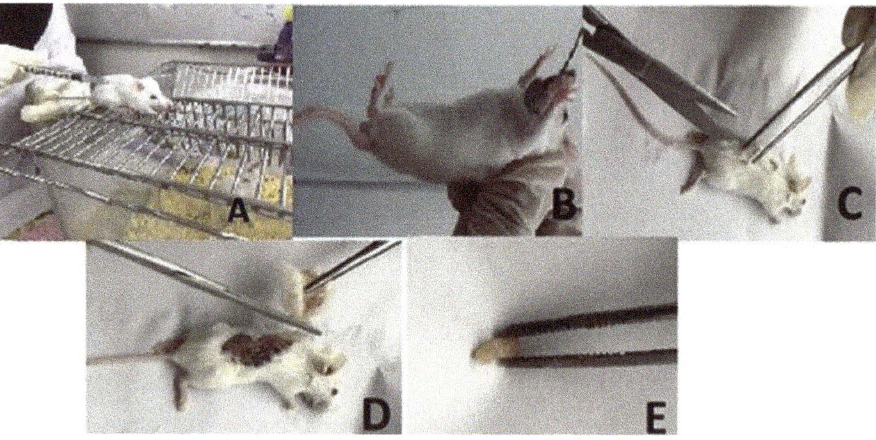

Figure 20. In vivo anti-tumor assessment in mouse. (**A**) breast cancer injection; (**B**) oral gavage treatment; (**C,D**) tumor isolation and (**E**) tumor mass.

H & E staining was performed in order to analyze the histopathological samples.

4.13.3. Histological Analysis and Scoring Method

After separating tumor from the body, it was instantly was fixed in 10% formalin overnight, embedded in paraffin, cut into 4 µm sections and stained with hematoxylin-eosin (H & E). Tissue scoring was conducted according to Elston and Ellis method [25], with some slight modifications. Randomly five different sections from each H & E slide view at 100× magnification and mean score is calculated from these five sections and the scoring is referred to in Table 4. Only structures which could not be misinterpreted as anything except mitotic figures were taken into account. For apoptosis only hyperchromatin or pyknotic nuclei with hollow signs were counted:

$$\text{Mitotic index} = \text{The numbers of mitotic cells/the number of total cells}$$

$$\text{Apoptotic index} = \text{The numbers of apoptotic cells/the number of total cells}$$

Table 4. Scoring method for apoptosis and mitotic indexes.

Apoptotic Index		Mitotic Index	
Apoptotic Count	**Score**	**Mitotic Count**	**Score**
0–4	1	<25	1
5–10	2	25–50	2
11–15	3	51–100	3
>15	4	>100	4

Adapted from Elston and Ellis [25], with some slight modifications.

4.14. Statistical Analysis

For statistical assessment *t*-test was applied for non-dose responses. Dose response experiments were studied using the Omnibus test followed by Rodger's method. All values are declared as mean ± S.D. Probability values * $p < 0.05$ was considered as statistically significant.

5. Conclusions

To conclude, the results from the present study provide an understanding of the cytotoxic effect of *Pistacia vera* red hull. Cytotoxicity studies of *Pistacia vera* red hull ethyl acetate (PV-EA) extract on different cancer cell lines and subsequent chemical purification of bioactive PV-EA have led to the isolation of gallic acid and quercetin. Purified compound (F13b1/PV-EA) was shown to possess cytotoxic effects against MCF-7 cells. This finding may have an impact on the future of cancer treatment by providing another positive avenue for the discovery of a combined anticancer drug treatment that may promote induced apoptosis.

Author Contributions: M.S. carried out the tests analyzed the data and wrote the manuscript; R.F. analyzed the data and wrote the manuscript; F.N. designed and analyzed the data and wrote the final draft of the paper and proof read the manuscript; J.B.M. designing the experimental work and proof reading the manuscript; N.A.M. designing the experiments and analyzed data. All authors have read and agreed to the published version of the manuscript.

Funding: This paper was supported by a grant from the University Malaya (GPF006B -2018).

Conflicts of Interest: The authors declare no conflict of interest.

References

1. Tomaino, A., Martorana, M., Arcoraci, T., Monteleone, D., Giovinazzo, C.; Saija, A. Biochimie Antioxidant activity and phenolic profile of pistachio (*Pistacia vera* L., variety Bronte) seeds and skins. *Biochimie* **2010**, *92*, 1115–1122. [CrossRef]
2. Martorana, M.; Arcoraci, T.; Rizza, L.; Cristani, M.; Paolo, F.; Saija, A.; Tomaino, A. In vitro antioxidant and in vivo photoprotective effect of pistachio (*Pistacia vera* L., variety Bronte) seed and skin extracts. *Fitoterapia* **2013**, *85*, 41–48. [CrossRef]
3. Kennedy-Hagan, K.; Painter, J.E.; Honselman, C.; Halvorson, A.; Rhodes, K.; Skwir, K. The effect of pistachio shells as a visual cue in reducing caloric consumption. *Appetite* **2011**, *57*, 418–420. [CrossRef]

4. Goli, A.H.; Barzegar, M.; Sahari, M.A. Antioxidant activity and total phenolic compounds of pistachio (*Pistachia vera*) hull extracts. *Food Chem.* **2005**, *92*, 521–525. [CrossRef]

5. Garavand, F.; Madadlou, A.; Moini, S. Determination of phenolic profile and antioxidant activity of pistachio hull using HPLC-DAD-ESI-MS as affected by ultrasound and microwave. *Int. J. Food Prop.* **2015**, *2912*. [CrossRef]

6. Grace, M.H.; Esposito, D.; Timmers, M.A.; Xiong, J.; Yousef, G.; Komarnytsky, S.; Lila, M.A. Chemical composition, antioxidant and anti-inflammatory properties of pistachio hull extracts. *Food Chem.* **2016**, *210*, 85–95. [CrossRef]

7. Bulló, M.; Juanola-Falgarona, M.; Hernández-Alonso, P.; Salas-Salvadó, J. Nutrition attributes and health effects of pistachio nuts. *Br. J. Nutr.* **2015**, *113*, S79–S93. [CrossRef]

8. Rajaei, A.; Barzegar, M.; Mobarez, A.M.; Sahari, M.A.; Esfahani, Z.H. Antioxidant, anti-microbial and antimutagenicity activities of pistachio (*Pistachia vera*) green hull extract. *Food Chem. Toxicol.* **2010**, *48*, 107–112. [CrossRef]

9. Barreca, D.; Laganà, G.; Leuzzi, U.; Smeriglio, A.; Trombetta, D. Evaluation of the nutraceutical, antioxidant and cytoprotective properties of ripe pistachio (*Pistacia vera* L., variety Bronte) hulls. *Food Chem.* **2016**, *196*, 493–502. [CrossRef]

10. Seifaddinipour, M.; Farghadani, R.; Namvar, F.; Mohamad, J.; Abdul Kadir, H. Cytotoxic effects and anti-angiogenesis potential of pistachio (*Pistacia vera* L.) hulls against MCF-7 human breast cancer cells. *Molecules* **2018**, *23*, 110. [CrossRef]

11. He, M. Mechanisms of antiprostate cancer by gum mastic: NF-κB signal as target. *Acta Pharmacol. Sin.* **2007**, *28*, 446–452. [CrossRef]

12. Mense, S.M.; Hei, T.K.; Ganju, R.K.; Bhat, H.K. Phytoestrogens and breast cancer prevention: Possible mechanisms of action. *Environ. Health Perspect.* **2008**, *116*, 426–433. [CrossRef]

13. Jiang, Y.; Yu, Y. Transgenic and gene knockout mice in gastric cancer research. *Oncotarget* **2017**, *8*, 36–96. [CrossRef]

14. Dunn, B.K.; Umar, A.; Richmond, E. Introduction: Cancer chemoprevention and its context. *Semin. Oncol.* **2016**, *43*, 19–21. [CrossRef]

15. Wang, L.; Chen, C.; Su, A.; Zhang, Y.; Yuan, J.; Ju, X. Structural characterization of phenolic compounds and antioxidant activity of the phenolic-rich fraction from defatted adlay (Coix lachryma-jobi L. var. ma-yuen Stapf) seed meal. *Food Chem.* **2016**, *196*, 509–517. [CrossRef]

16. Shay, J.; Elbaz, H.A.; Lee, I.; Zielske, S.P.; Malek, M.H.; Hüttemann, M. Molecular Mechanisms and Therapeutic Effects of Epicatechin and Other Polyphenols in Cancer, Inflammation, Diabetes, and Neurodegeneration. *Oxid. Med. Cell. Longev.* **2015**, *2015*, 181–260. [CrossRef]

17. Müllauer, L.; Gruber, P.; Sebinger, D.; Buch, J.; Wohlfart, S.; Chott, A. Mutations in apoptosis genes: A pathogenetic factor for human disease. *Mutat. Res.* **2001**, *488*, 211–231. [CrossRef]

18. Young, E.; Huh, J.; Choi, M.; Young, S. Inhibitory effects of gallic acid and quercetin on UDP-glucose dehydrogenase activity. *FEBS Lett.* **2008**, *582*, 3793–3797.

19. Sourani, Z.; Pourgheysari, B.; Beshkar, P.; Shirzad, H.; Shirzad, M. Gallic acid inhibits proliferation and induces apoptosis in lymphoblastic leukemia cell line (C121). *IJBMS* **2016**, *41*, 525.

20. Sarkhail, P.; Salimi, M.; Sarkheil, P.; Kandelous, H.M. Anti-melanogenic activity and cytotoxicity of *Pistacia vera* hull on human melanoma SKMEL-3 cells. *Acta Med. Iran.* **2017**, *55*, 422–428. [CrossRef]

21. Harandi, H.; Majd, A.; Falahati-Pour, S.K.; Mahmoodi, M. Anti-cancer effects of hydro-alcoholic extract of pericarp of pistachio fruits. *Asian Pac. J. Trop. Biomed.* **2018**, *8*, 598–603.

22. Hashemi, L.; Asadi-Samani, M.; Moradi, M.-T.; Alidadi, S. Anticancer activity and phenolic compounds of Pistacia atlantica extract. *Int. J. Pharm.* **2017**, *7*, 26–31. [CrossRef]

23. Karimian, H.; Zorofchian, S.; Fadaeinasab, M.; Golbabapour, S.H.; Razavi, M.; Hajrezaie, M.; Arya, A. Ferulago angulata activates intrinsic pathway of apoptosis in MCF-7 cells associated with G1 cell cycle arrest via involvement of p21/p27. *Drug Des. Dev. Ther.* **2014**, *8*, 1481. [CrossRef] [PubMed]

24. Farghadani, R.; Rajarajeswaran, J.; Hashim, N.B.M.; Abdulla, M.A.; Muniandy, S.A. Novel β-diiminato manganese III complex as the promising anticancer agent induces G 0/G 1 cell cycle arrest and triggers apoptosis via mitochondrial-dependent pathways in MCF-7 and MDA-MB-231 human breast cancer cells. *RSC Adv.* **2017**, *7*, 24387–24398. [CrossRef]
25. Elston, C.W.; Ellis, I.O. Pathological prognostic factors in breast cancer. I. The value of histological grade in breast cancer: Experience from a large study with long-term follow-up. *Histopathology* **2002**, *41*, 151. [CrossRef]

Sample Availability: Samples of the compounds are available from the authors.

Article

Study of the Quality Parameters and the Antioxidant Capacity for the FTIR-Chemometric Differentiation of *Pistacia Vera* Oils

Lydia Valasi [1], Dimitra Arvanitaki [1], Angeliki Mitropoulou [1], Maria Georgiadou [2] and Christos S. Pappas [1,*]

[1]   Laboratory of Chemistry, Department of Food Science & Human Nutrition, Agricultural University of Athens, Iera Odos 75, 11855 Athens, Greece; lydia.valasi@aua.gr (L.V.); dimitrarvntk@gmail.com (D.A.); aggelikimitropoulou8@gmail.com (A.M.)

[2]   Laboratory of Food Process Engineering, Department of Food Science & Human Nutrition, Agricultural University of Athens, Iera Odos 75, 11855 Athens, Greece; m.georgiadou@aua.gr

*   Correspondence: chrispap@aua.gr; Tel.: +30-2105294262

Academic Editors: Severina Pacifico and Simona Piccolella
Received: 29 February 2020; Accepted: 30 March 2020; Published: 1 April 2020

Abstract: The aim of this work was to characterize the pistachio oil of the Greek variety, "Aegina", evaluate its various quality indices, and investigate the potential use of FTIR as a tool to discriminate different oil qualities. For this purpose, the antioxidant capacity, the tocopherol content and the oxidation and degradation of fatty acids, as described by k, Δk, R-values, and free acidity were evaluated using 45 samples from eight different areas of production and two subsequent years of harvesting. The antioxidant capacity was estimated using 2,2'-azinobis(3-ethylbenzothiazoline-6-sulfonic acid diammonium salt (ABTS) and 2,2-diphenyl-1-(2,4,6-trinitrophenyl)hydrazine (DPPH) assays, and the tocopherol content was quantified through HPLC analysis. FTIR spectra were recorded for all samples and multivariate analysis was applied. The results showed significant differences between the oil samples of different harvesting years, which were successfully discriminated by a representative FTIR spectral region based on R-value, total antioxidant capacity, and scavenging capacity, through ABTS. A similar approach could not be confirmed for the other quality parameters, such as the free acidity and the tocopherol content. This research highlighted the possibility of developing a simple, rapid, economic, and environment friendly method for the discrimination of pistachio oils according to their quality profile, through FTIR spectroscopy and multivariate analysis.

Keywords: *Pistacia vera*; antioxidant; quality; tocopherol; FTIR; discriminant analysis

---

1. Introduction

Pistacia, a genus of the *Anacardiaceae* family, includes at least eleven species, among them *Pistacia vera* L. is the only edible commercial species [1]. The pistachio nut is an important agricultural commodity for a number of countries. Iran, United States, Turkey, Syria, Greece, Italy, and Spain are the main pistachio producers [2]. Pistachio nut can be considered to be a functional food, and has recently been ranked among the first 50 food products with the highest antioxidant potential [3]. The Dietary Guidelines recommend that consuming nuts (almonds, hazelnuts, walnuts, pistachios, pecans, and peanuts) as a part of a daily diet has a beneficial effect on human health [4].

The increasing consumption and demand for novel edible oils has led to a market expansion for the plant-derived oils, which receive particular attention due to their attractive sensory characteristics and their high nutritional properties [5]. The pistachio (*Pistacia vera* L.) oil content ranges from 50% to

60% (dry weight) in kernels, depending mainly on the cultivar, crop year, and geographic location. Even though no specific standards for pistachio oil have been set by the Codex Alimentarius on Fats and Oils, it is claimed to be a niche product [6,7]. It has gained attention due to its special organoleptic characteristics [8] and its richness in some nutrients and health promoting compounds that exhibit high antioxidant capacity [9,10].

The most abundant components in pistachio oil are the fatty acids profiled as (mono- and poly-unsaturated, saturated, and as esters of triglycerides) [5]. The structure of fatty acids might change due to oxidation caused by bad agricultural practices, inappropriate harvesting time, and storage conditions, which can be detected by ultraviolet-visible (UV–Vis) spectroscopic measurements. Absorbencies at 232, 268, 270, and 274 nm are correlated with the state of oxidation, through the detection of secondary oxidation compounds, and possible adulteration with refined oils. Delta (Δ) k and R-value, resulting from the k_{232}, k_{268}, k_{270}, and k_{274} values, are quick indicators of quality assessment of pistachio oil discriminating between high and poor quality oil and are correlated with possible adulteration [11]. Furthermore, the hydrolysis of the oil results in the formation of free fatty acid (FFA) and glycerol residues, indicating that higher quality oils exhibit very low FFA percentage and acidity [12]. As a result, free acidity is claimed to be an early indicator of the potential storage stability of the product. Additionally, it should be mentioned that the formation of oxidized compounds in pistachio oil is blocked by its own natural preservatives that carry out an important antioxidant activity. Pistachio oil contains numerous phenolic compounds that increase its shelf life, and prevent or reduce the damage to cells, caused by free radicals. The evaluation of the total antioxidant capacity is certainly a very useful property of the pistachio oil, ensuring the preservation of the most important health benefits and sensory characteristics. Additionally, tocopherols' content is associated with health benefits, as confirmed by clinical evidence [13]. Tocopherols are the main antioxidant components that are equally active to vitamin E, thus, are considered to be important biofunctional compounds of the human diet. Due to their non-polar nature, their presence in oils is profound [7]. The major tocopherol isomer in pistachio oil is γ-tocopherol, the most prevalent form of vitamin E [14].

Published works have shown that Fourier transform infrared (FTIR) spectroscopy combined with statistical methods for the discrimination of wines, according to the variety and vintage year, can be used as a rapid, accurate, simple, environment friendly, and economical approach [15,16]. The chemical composition of pistachio oil might be influenced by several factors like variety, climatic and agronomic conditions (weather, soil), the growing season, and agricultural practices [17,18]. Very limited data are available in the literature for the quality profile of the pistachio oil that is extracted from the main Greek pistachio variety, "Aegina".

The present work focused on the characterization of the pistachio oil of the variety "Aegina", evaluated various quality indices, and investigated the potential use of FTIR as a tool to discriminate different oil qualities.

2. Results and Discussion

All pistachio oil samples were stored in a freezer (−20 °C), in order to maintain their initial quality until analysis. Storage at low temperatures prevents from increasing or reducing the concentrations of oil components and helps to maintain the oil's primary quality [19].

2.1. Oil Extraction

The oil yield ranged between 59.7%–68.2% *w*/*w*, with an average of 62.5 ± 2.6% *w*/*w* for the samples of 2017 and between 52.5%–64.8% *w*/*w* for the samples of 2018, with a mean value of 59.6 ± 2.8% *w*/*w*, respectively. No statistically significant differences were observed in the oil yield between the samples obtained in the two years.

2.2. Evaluation of Antioxidant Capacity

Total antioxidant capacity (TAC) and scavenging capacity of the analyzed pistachio oil samples, as measured by the DPPH and ABTS assays are shown in Table 1.

Table 1. Total antioxidant capacity (mM Trolox equivalents per mL pistachio oil) (TAC) and scavenging capacity (%) of pistachio oils of different regions, as determined through the 2,2-diphenyl-1-(2,4,6-trinitrophenyl)hydrazine (DPPH) and 2,2′-azinobis(3-ethylbenzothiazoline-6-sulfonic acid diammonium salt (ABTS) assays.

Samples No	Origin	Year of Harvest	TAC		Scavenging Capacity	
			DPPH	ABTS	DPPH	ABTS
1	AEGINA	2017	6.05	7.10	37.69	68.14
2	AEGINA	2017	6.45	8.63	41.76	82.85
3	AEGINA	2017	6.35	8.60	44.29	84.95
4	AEGINA	2017	3.11	9.45	32.17	91.79
5	AEGINA	2017	3.17	4.92	32.44	45.28
6	AEGINA	2017	5.62	7.44	41.44	74.74
7	AEGINA	2017	3.82	5.46	34.94	48.50
8	MEGARA	2017	5.95	9.32	39.68	90.03
9	MEGARA	2017	5.36	9.01	37.82	85.15
10	MEGARA	2017	3.21	9.44	28.75	90.84
11	MEGARA	2017	6.70	5.02	39.02	42.53
12	PHTHIOTIS	2017	3.59	10.02	31.51	97.25
13	PHTHIOTIS	2017	6.45	9.02	39.42	88.70
14	PHTHIOTIS	2017	4.96	7.85	37.07	73.33
15	TRIZINA	2017	4.68	8.73	35.33	81.85
16	AEGINA	2018	2.12	9.63	39.30	82.26
17	AEGINA	2018	7.92	8.23	37.97	76.33
18	AEGINA	2018	7.10	8.91	42.70	83.65
19	AEGINA	2018	6.49	8.58	40.40	84.02
20	AEGINA	2018	1.99	7.99	40.52	74.89
21	AEGINA	2018	8.44	8.61	47.87	82.05
22	AEGINA	2018	4.83	8.38	34.05	77.05
23	AEGINA	2018	5.86	9.46	40.19	90.34
24	AEGINA	2018	5.30	7.71	34.84	66.41
25	AEGINA	2018	7.47	8.27	38.43	75.91
26	AEGINA	2018	6.70	8.72	41.40	86.10
27	AEGINA	2018	5.69	7.96	36.21	73.62
28	EVIA	2018	5.51	7.07	36.83	64.38
29	EVIA	2018	3.80	9.49	28.96	91.06
30	EVIA	2018	6.67	9.22	40.32	88.08
31	MEGARA	2018	0.28	9.61	32.01	91.97
32	MEGARA	2018	6.47	8.58	24.12	81.59
33	MEGARA	2018	7.35	8.98	38.53	85.76
34	MEGARA	2018	5.34	8.76	35.45	80.68
35	MEGARA	2018	7.45	9.63	43.20	94.15
36	TRIZINA	2018	7.65	8.47	44.10	77.16
37	PHTHIOTIS	2018	5.84	9.80	38.83	93.99
38	PHTHIOTIS	2018	6.04	6.78	40.30	61.75
39	PHTHIOTIS	2018	0.10	9.22	31.87	87.01
40	PHTHIOTIS	2018	1.35	9.83	36.60	94.10
41	PHTHIOTIS	2018	1.40	8.28	36.50	74.18
42	PHTHIOTIS	2018	5.82	10.00	36.09	96.60
43	VOLOS	2018	10.28	9.95	52.29	96.31
44	THIVA	2018	7.34	9.87	43.04	95.39
45	AVLONAS	2018	8.43	9.16	45.60	88.19

The Trolox calibration curve equations used for transforming absorbance inhibition values (AI) to Trolox equivalents (TE, mM) for the DPPH and ABTS assays, were Equations (1) and (2), respectively.

$$AI_{DPPH} = (-0.388) \times TE + 0.797, R^2 = 0.971 \tag{1}$$

$$AI_{ABTS} = (-0.675) \times TE + 0.695, R^2 = 0.995 \tag{2}$$

Antioxidant capacity, as determined by the DPPH assay, ranged between 3.11–6.70 mM with a mean TAC value of 5.03 ± 1.3 mM, and between 0.10–10 with a mean TAC value of 5.57 ± 2.56 mM, for the samples of 2017 and 2018, respectively. Following this, the scavenging capacity ranged between 28.75%–44.29% and 28.96%–52.29% for the samples of 2017 and 2018, with mean values of 36.89 ± 4.33% and 38.62 ± 5.55%, respectively.

TAC values, as determined by the ABTS assay, ranged between 4.92–10.02 mM and 7.07–10.00 mM for the samples of 2017 and 2018, respectively, resulting in mean TAC values of 8.00 ± 1.68 mM and 8.84 ± 0.84 mM. The scavenging capacity, as measured by the ABTS assay, ranged between 42.53%–97.25%, with an average of 76.39 ± 17.75% for the 2017 samples and between 61.75%–96.6% for the 2018 samples, with a mean value of 83.17 ± 9.56%.

Concerning mean TAC and mean scavenging capacity values, results from the ABTS assay were significantly and consistently higher than those from DPPH in both years of harvest. This was due to the applicability of DPPH to hydrophobic systems. Specifically, DPPH was discolored in the presence of compounds that were capable of either transferring an electron or donating hydrogen (lipophilic components). On the other hand, ABTS was freely soluble in both organic and aqueous solvents, thus, it could be used to screen both hydrophilic and lipophilic antioxidants, exhibiting a better estimation of the overall antioxidant capacity of the foods [20,21]. Consequently, the results that were obtained from the ABTS assay were only considered for further statistical analysis. TAC and scavenging capacity, as estimated by ABTS were statistically different between 2017 and 2018. The differences were statistically significant.

2.3. UV-Vis Spectroscopic Assessment

The quality indices associated with the k_{232}, k_{268}, k_{270}, k_{274}, Δk, and R values were evaluated in 45 pistachio oil samples and the results are displayed in Table 2. The European Quality Standard of Commission Regulation (EEC) No 2568/91 (Annex IX of the Regulation) has set the standard values for extra virgin olive oil (EVOO), as described in Table 2. Considering that an official protocol to predict the quality indicators of the pistachio oil or other nut oils based on the Δk and R-value has not been established, the existing limits were used for the evaluation of pistachio oil samples.

The quality of the oil was assessed by the UV–Vis absorption screening, which identifies changes in the structure of fatty acids due to oxidation. A low absorption in this region is indicative of high-quality oil, whereas old, refined, and generally poor-quality oils show a greater level of absorption in this region, implying high degree of oxidation. The absorbance at 232 nm is caused by hydroperoxides (primary stage of oxidation) and conjugated dienes (intermediate stage of oxidation). The absorbance at 270 nm was caused by carbonylic compounds (secondary stage of oxidation) and conjugated trienes (technological treatments). In the oils, due to oxygen fixation in linolenic and linoleic acids' double bond position, hydroperoxides arise. The double bond provokes the formation of conjugate diene systems between the carbon atoms. This kind of conjugate systems presents a maximum absorption at 232 nm. During more advanced oxidation states, the products are generated with conjugate diene systems of carbon–oxygen. The maximum absorption in this case ranges between 260–280 nm [11].

Table 2. Ultraviolet–visible (UV–Vis) spectroscopy, acid values (AV), and free fatty acid (FFA) of pistachio oil samples against the extra virgin olive oil (EVOO) corresponding values.

Samples No	k_{232}	k_{268}	k_{270}	k_{274}	Δk	R	AV [1] (as oleic acid)	% FFA [1]
EVOO	≤2.50	≤0.22	≤0.22	≤0.22	≤0.01	≤11.36	≤4.000	≤0.350
1	0.126	0.009	0.009	0.010	0.000	14.073	6.615 ± 0.000	3.327 ± 0.000
2	0.129	0.015	0.015	0.015	0.000	8.547	3.186 ± 0.325	1.603 ± 0.163
3	0.146	0.010	0.011	0.011	0.000	13.759	2.249 ± 0.000	1.131 ± 0.000
4	0.148	0.011	0.011	0.011	0.000	13.550	1.676 ± 0.019	0.843 ± 0.010
5	0.145	0.017	0.018	0.018	0.000	8.155	1.676 ± 0.019	0.843 ± 0.010
6	0.172	0.020	0.020	0.021	0.000	8.430	1.676 ± 0.019	0.843 ± 0.010
7	0.164	0.010	0.011	0.011	0.000	15.537	3.373 ± 0.000	1.697 ± 0.000
8	0.156	0.013	0.013	0.014	0.000	11.672	1.687 ± 0.000	0.848 ± 0.000
9	0.166	0.020	0.020	0.021	0.000	8.308	0.532 ± 0.001	0.268 ± 0.000
10	0.132	0.014	0.014	0.015	0.000	9.277	1.102 ± 0.000	0.555 ± 0.000
11	0.173	0.027	0.027	0.027	0.000	12.281	2.509 ± 0.336	1.262 ± 0.169
12	0.159	0.018	0.019	0.019	0.000	8.550	1.687 ± 0.000	0.848 ± 0.000
13	0.142	0.009	0.009	0.009	0.000	15.987	1.124 ± 0.000	0.566 ± 0.000
14	0.157	0.013	0.013	0.013	0.000	11.996	1.102 ± 0.000	0.555 ± 0.000
15	0.173	0.025	0.025	0.025	0.000	6.854	1.676 ± 0.019	0.843 ± 0.010
16	0.070	0.011	0.011	0.012	0.000	6.190	0.821 ± 0.007	0.413 ± 0.003
17	0.075	0.016	0.016	0.016	0.000	4.699	0.805 ± 0.269	0.405 ± 0.135
18	0.069	0.011	0.011	0.011	0.000	6.248	1.171 ± 0.166	0.589 ± 0.083
19	0.071	0.011	0.011	0.012	0.000	6.179	0.994 ± 0.161	0.500 ± 0.081
20	0.052	0.007	0.008	0.008	0.000	6.806	0.982 ± 0.164	0.494 ± 0.082
21	0.068	0.011	0.012	0.012	0.000	5.842	1.087 ± 0.009	0.547 ± 0.004
22	0.056	0.012	0.012	0.013	0.000	4.524	1.362 ± 0.270	0.685 ± 0.136
23	0.073	0.014	0.014	0.014	0.000	5.254	0.989 ± 0.148	0.498 ± 0.074
24	0.059	0.020	0.020	0.020	0.000	2.879	8.116 ± 0.191	4.082 ± 0.096
25	0.071	0.011	0.011	0.012	0.000	6.202	0.807 ± 0.014	0.406 ± 0.007
26	0.053	0.008	0.008	0.008	0.000	6.933	1.279 ± 0.163	0.643 ± 0.082
27	0.074	0.015	0.015	0.016	0.000	4.831	0.995 ± 0.166	0.500 ± 0.083
28	0.068	0.012	0.012	0.012	0.000	5.647	1.079 ± 0.015	0.543 ± 0.008
29	0.070	0.013	0.014	0.014	0.000	5.197	0.800 ± 0.251	0.402 ± 0.126
30	0.075	0.019	0.019	0.018	0.000	4.067	0.709 ± 0.614	0.357 ± 0.309
31	0.073	0.015	0.015	0.016	0.000	4.865	1.004 ± 0.156	0.505 ± 0.078
32	0.070	0.011	0.011	0.011	0.000	6.431	1.078 ± 0.023	0.542 ± 0.011
33	0.054	0.011	0.012	0.012	0.000	4.666	0.978 ± 0.151	0.492 ± 0.076
34	0.069	0.014	0.014	0.014	0.000	4.871	0.819 ± 0.277	0.412 ± 0.139
35	0.072	0.014	0.014	0.015	0.000	5.050	1.007 ± 0.157	0.506 ± 0.079
36	0.072	0.017	0.017	0.018	0.000	4.228	0.982 ± 0.161	0.494 ± 0.081
37	0.070	0.012	0.012	0.013	0.000	5.764	0.898 ± 0.151	0.452 ± 0.076
38	0.054	0.009	0.009	0.009	0.000	6.118	1.004 ± 0.157	0.505 ± 0.079
39	0.053	0.010	0.011	0.011	0.000	5.009	0.808 ± 0.016	0.406 ± 0.008
40	0.072	0.013	0.014	0.014	0.000	5.316	0.632 ± 0.155	0.318 ± 0.078
41	0.053	0.011	0.011	0.011	0.000	4.969	1.009 ± 0.160	0.507 ± 0.080
42	0.068	0.012	0.012	0.012	0.000	5.690	0.913 ± 0.164	0.459 ± 0.083
43	0.070	0.012	0.012	0.013	0.000	5.771	0.821 ± 0.004	0.413 ± 0.002
44	0.070	0.010	0.010	0.011	0.000	6.848	0.821 ± 0.007	0.413 ± 0.003
45	0.053	0.010	0.010	0.010	0.000	5.476	0.911 ± 0.166	0.458 ± 0.083

[1] mean ± SD ($n = 3$).

The mean k values for each harvesting year were 0.15 ± 0.02 (2017), 0.07 ± 0.01 (2018), 0.01 ± 0.01 (2017), 0.01 ± 0.00 (2018), 0.02 ± 0.01 (2017), 0.01 ± 0.00 (2018), 0.02 ± 0.01 (2017), and 0.01 ± 0.00 (2018) for k_{232}, k_{268}, k_{270}, and k_{274}, respectively. Δk was 0.00 ± 0.00 for all samples, regardless of the year of harvest or the origin. The mean R-values were 11.13 ± 3.00 and 5.42 ± 0.91, for the samples of 2017 and 2018, respectively.

There were no significant differences between the years of harvest, based on k and Δk, as all measurements ranged into the high-quality limits. As for the R-value, the 2017 samples were systematically higher than 2018 samples. Specifically, 82% of the samples, which mostly originated from 2018 harvest complied with the EVOO standard, except for the R-value of the remaining 18% of the total samples, which belonged to the 2017 harvest (samples 1, 3, 4, 7, 8, 11, 13, and 14).

2.4. Acid (AV) and FFA Values

AV value is a measure of the number of carboxylic acid groups. It is used as an indicator for edibility of oil and is expressed in milligrams per gram. However, FFA are expressed as a percentage of oleic acid. According to the Codex Standard for Edible Fats and Oils, acid value of oil suitable for edible purposes should not exceed 4 mg/g.

FFA has been reported to play a very important role in the aroma and flavor. FFA also contributes to the organoleptic quality of foods, when present in adequate concentration. FFA content is an index of lipase activity and an indicator of freshness, storage time, and stability of many fat-rich foods. It is well-known that FFAs are more susceptible to lipid oxidation, leading to rancidity and production of off-odor, compared to intact fatty acids in triglycerides. It is considered to be an early indicator of the storage stability of the oil [22], with a supreme limit that is less than 0.35% [11].

As presented in Table 2, the AV ranged from 0.53 to 6.61 and from 0.63 to 8.12 mg/g for 2017 and 2018, respectively. The mean AV (1.19 ± 1.32 mg/g oil) of 2018 exhibited lower values than the mean AV (2.12 ± 1.46 mg/g oil) of 2017. The FFA content ranged from 0.27% to 3.33% and from 0.32% to 4.08% for 2017 and 2018, respectively. Similarly, the mean FFA (0.60 ± 0.66%) of 2018 was lower than the mean FFA (1.07 ± 0.73) of 2017, leading to the conclusion that the 2018 harvesting exhibited a superior antioxidant capacity. Based on the standard for edible oils, only two samples (No. 1 and No. 24) from the area of Aegina showed values out of the acceptable levels of AV, whereas only two samples (No. 9 and No. 40) exhibited acceptable levels of FFA content.

2.5. Tocopherol Analysis

The limit of detection (LOD) for tocopherol analysis was 0.15 μg/mL. The tocopherol calibration curves used for the qualitative separation of samples were (Equations (3)–(6)):

$$\text{Area (mV x s)} = 8.0556 \times C_{\alpha\text{-T}} \ (\mu g/mL) + 0.531, R^2 = 0.999 \tag{3}$$

$$\text{Area (mV x s)} = 10.724 \times C_{\beta\text{-T}} \ (\mu g/mL) + 1.9311, R^2 = 0.998 \tag{4}$$

$$\text{Area (mV x s)} = 13.786 \times C_{\gamma\text{-T}} \ (\mu g/mL) + 2.5428, R^2 = 0.978 \tag{5}$$

$$\text{Area (mV x s)} = 14.617 \times C_{\delta\text{-T}} \ (\mu g/mL) + 2.1461, R^2 = 0.997 \tag{6}$$

As for recovery evaluation, the amount of vitamin E isomers added to the samples corresponded to 98.47%, 77.86%, 47.44%, and 110.37% (Equation (14)) of the expected α-T, β-T, γ-T, and δ-T, and the intraday analytical precision was 3.08%, 5.99%, 4.89%, and 2.75% (Equation (15)), respectively.

Figure 1 illustrates the separation of the most important vitamin E isomers, as determined with the HPLC method, using fluorescence detection. The retention times for the α-, β-, γ-, and δ-tocopherols were 8, 10, 12, and 16 min, approximately. The concentration of each tocopherol for all 45 pistachio oil samples is presented in Table 3. The results were obtained and corrected on the basis of recovery and repeatability of the method, as determined by the coefficient of variation (CV). The tocopherol contents of pistachio oils expressed as 10^2 μg/mL pistachio oil, ranged from 0.53 (No. 25) to 5.90 (No. 43), 0.33 (No. 13) to 2.25 (No. 29), 97.56 (No. 36) to 235.06 (No. 6), and 0.84 (No. 12) to 2.31 (No. 20) for α-, β-, γ-, and δ-tocopherol, respectively. The above minimum and maximum values corresponded to 13.25–147.50, 8.25–56.25, 2439.00–5876.50, and 21.00–57.75 mg/kg of pistachio oil for α-, β-, γ-, and δ-tocopherol, respectively. It is important to mention that γ-tocopherol is coeluted with β-tocotrienol, as a result, the calculated content of γ-tocopherol includes both isomers. The data indicate that the main form in all samples was γ-tocopherol (by coelution with β-tocotrienol), whereas the β-tocopherol content was limited. These results are in agreement with Martinez et al. (2016) [23]. The minimum and maximum values of each tocopherol presented in the samples were compared to the standard for vegetable oils provided by the Codex Alimentarius Commission on Fats and Oils (Table 4). With regards to pistachio oils, the quantity of α- and δ-tocopherol, as measured in the present work for the variety "Aegina", ranges within the limits that have been set by the standard. However, no values were described for the β-tocopherol and β-tocotrienol, in contrast to the present study.

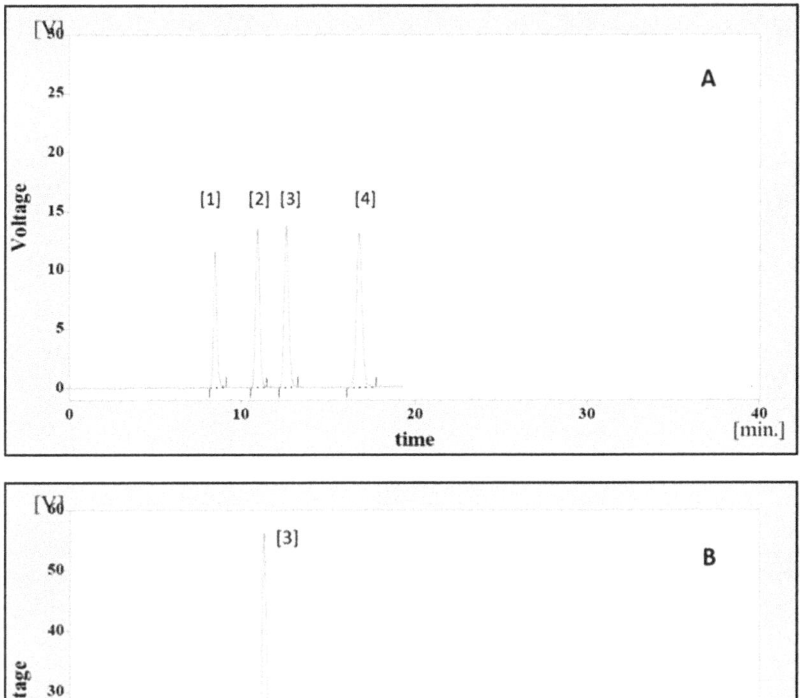

Figure 1. Chromatogram of a working standard mixture (**A**) and of a pistachio oil sample (**B**), determined through high performance liquid chromatography (HPLC)-fluorescence. Peaks: 1, α-tocopherol; 2, β-tocopherol; 3, γ-tocopherol, and β-tocotrienol; 4, δ-tocopherol.

Table 3. Tocopherol (T) content ($\times 10^2$ μg/mL pistachio oil) obtained by Soxhlet extraction and repeatability assessment.

Samples No	Concentration [1]				Repeatability (CV %, $n = 3$)			
	α-T	β-T	γ-T [2]	δ-T	α-T	β-T	γ-T	δ-T
1	1.57 ± 0.16	1.07 ± 0.24	186.77 ± 3.65	1.79 ± 0.04	10.52	22.26	1.95	2.13
2	1.32 ± 0.10	1.14 ± 0.19	202.72 ± 12.77	1.73 ± 0.13	7.45	16.60	6.30	7.35
3	0.99 ± 0.05	0.63 ± 0.09	191.13 ± 4.38	1.57 ± 0.02	4.94	13.88	2.29	1.62
4	1.73 ± 0.15	0.39 ± 0.05	222.09 ± 10.95	1.59 ± 0.05	8.54	12.38	4.93	3.09
5	1.86 ± 0.30	0.45 ± 0.24	187.09 ± 3.09	1.80 ± 0.15	16.21	52.60	1.65	8.10
6	3.12 ± 0.29	1.13 ± 0.46	235.06 ± 7.77	2.11 ± 0.07	9.27	40.99	3.31	3.42
7	2.13 ± 0.12	1.10 ± 0.10	210.49 ± 18.33	1.77 ± 0.14	5.86	8.88	8.71	7.83
8	1.53 ± 0.16	1.06 ± 0.04	174.01 ± 6.78	1.37 ± 0.09	10.31	4.14	3.89	6.62
9	0.59 ± 0.17	0.97 ± 0.17	173.56 ± 11.78	1.41 ± 0.11	29.47	17.01	6.79	8.17
10	1.24 ± 0.07	0.38 ± 0.03	173.29 ± 5.46	1.06 ± 0.09	5.72	7.18	3.15	8.38
11	1.71 ± 0.27	0.98 ± 0.12	170.07 ± 12.59	1.78	15.60	12.51	7.41	0.10
12	1.32 ± 0.18	0.83 ± 0.31	134.59 ± 4.15	0.84 ± 0.22	13.46	37.48	3.08	26.52
13	1.80 ± 0.13	0.33 ± 0.20	199.49 ± 7.96	1.60 ± 0.06	7.49	60.45	3.99	3.48
14	3.37 ± 0.44	0.55 ± 0.15	231.93 ± 9.84	1.17 ± 0.19	12.96	26.87	4.24	16.08
15	1.87 ± 0.17	0.48 ± 0.07	199.88 ± 6.87	1.58 ± 0.12	9.16	14.98	3.44	7.35
16	2.29 ± 0.06	1.25 ± 0.06	193.13 ± 5.12	2.01 ± 0.14	2.75	4.69	2.65	7.11
17	1.59 ± 0.08	0.86 ± 0.11	184.07 ± 9.59	1.81 ± 0.01	5.20	13.06	5.21	0.39
18	4.10 ± 0.08	1.68 ± 0.11	223.82 ± 2.45	2.20 ± 0.09	2.04	6.45	1.09	4.24
19	2.83 ± 0.20	1.26 ± 0.11	201.78 ± 9.48	1.89 ± 0.22	7.16	8.61	4.70	11.67
20	2.78 ± 0.35	0.87 ± 0.15	204.80 ± 2.12	2.31 ± 0.28	12.65	17.70	1.04	12.12
21	3.22 ± 0.67	0.77 ± 0.07	195.44 ± 5.02	1.97 ± 0.04	20.91	8.57	2.57	1.91
22	1.72 ± 0.14	1.33 ± 0.03	152.10 ± 2.14	1.73 ± 0.13	8.33	2.01	1.41	7.36
23	1.73 ± 0.02	0.65 ± 0.12	157.93 ± 4.27	1.79 ± 0.03	1.04	18.85	2.70	1.79
24	2.15 ± 0.20	0.83 ± 0.11	213.88 ± 3.55	2.12 ± 0.04	9.29	13.35	1.66	1.99
25	0.53 ± 0.03	1.01 ± 0.28	114.45 ± 39.60	1.43 ± 0.43	6.62	27.72	34.60	30.02
26	2.56 ± 0.17	1.79 ± 0.34	196.23 ± 2.86	1.99 ± 0.10	6.73	19.18	1.46	5.07
27	1.88 ± 0.25	0.96 ± 0.08	203.79 ± 2.77	2.13 ± 0.08	13.14	8.17	1.36	3.78
28	4.39 ± 0.15	1.79 ± 0.11	216.32 ± 7.47	1.88 ± 0.07	3.35	6.38	3.45	3.52
29	3.07 ± 0.28	2.25 ± 0.14	174.04 ± 4.57	1.79 ± 0.11	9.03	6.40	2.63	6.02
30	3.51 ± 0.22	1.37 ± 0.16	178.60 ± 6.09	1.73 ± 0.13	6.36	11.48	3.41	7.28
31	2.93 ± 0.12	0.66 ± 0.07	160.22 ± 2.28	1.41 ± 0.17	4.15	11.13	1.42	12.27
32	1.98 ± 0.20	1.75 ± 0.21	191.41 ± 7.90	2.17 ± 0.31	10.17	12.13	4.13	14.22
33	0.70 ± 0.11	1.11 ± 0.08	119.08 ± 4.32	1.29 ± 0.03	15.51	7.03	3.62	2.64
34	2.26 ± 0.41	0.45 ± 0.00	156.43 ± 1.80	1.33 ± 0.04	18.36	0.56	1.15	3.26
35	2.23 ± 0.48	1.58 ± 0.16	203.53 ± 5.74	1.96 ± 0.24	21.50	9.91	2.82	12.24
36	1.56 ± 0.22	1.07 ± 0.32	97.56 ± 8.15	1.58 ± 0.03	14.41	30.46	8.36	2.11
37	3.28 ± 0.07	0.62 ± 0.02	157.49 ± 5.24	1.20 ± 0.12	2.22	3.61	3.33	10.15
38	4.15 ± 0.04	0.78 ± 0.07	185.36 ± 11.25	1.50 ± 0.04	1.03	9.21	6.07	2.41
39	2.11 ± 0.19	1.82 ± 0.21	164.57 ± 23.70	1.96 ± 0.74	9.01	11.51	14.40	37.91
40	1.47 ± 0.13	1.96 ± 0.06	169.36 ± 7.54	1.79 ± 0.10	8.67	3.24	4.45	5.86
41	2.47 ± 0.43	0.71 ± 0.09	158.31 ± 14.51	1.46 ± 0.32	17.35	12.85	9.16	21.66
42	4.20 ± 0.07	2.22 ± 0.65	170.70 ± 2.42	1.90 ± 0.02	1.68	29.47	1.42	1.20
43	5.90 ± 0.22	1.56 ± 0.37	225.78 ± 5.51	1.99 ± 0.05	3.69	23.81	2.44	2.76
44	1.98 ± 0.19	0.70 ± 0.02	157.50 ± 4.63	1.59 ± 0.05	9.69	2.29	2.94	2.99
45	2.58 ± 0.11	ND [3]	216.02 ± 3.06	1.98 ± 1.12	4.47	-	1.42	56.55

[1] expressed as mean ± SD; [2] γ-T is co-eluted with β-tocotrienol; [3] ND = Not Detected.

Table 4. Limits (min–max) of tocopherol content (mg/kg dried sample) for different vegetable oils. according to the Codex Alimentarius Commission on Fats and Oils Standard.

Oils	α-T	β-T	γ-T	δ-T	Total
almond	20–545	ND [1]–10	5–104	ND–5	20–600
hazelnut	100–420	6–12	18–194	ND–10	200–600
walnut	ND–170	ND–110	120–400	ND–60	309–455
pistachio	10–330	ND	0–100	ND–50	100–600
flax/linseed	2–20	ND	100–712	3–14	150–905
avocado	50–450	ND	10–20	ND–10	50–450

[1] ND – Not Detected.

The mean values of each harvesting year were 1.74 ± 0.72 (2017), 2.60 ± 1.15 (2018), 0.77 ± 0.31 (2017), 1.19 ± 0.55 (2018), 192.81 ± 26.41 (2017), 178.12 ± 32.01 (2018), 1.55 ± 0.33 (2017), and 1.80 ± 0.29 (2018), expressed as 10^2 µg/mL pistachio oil for α-, β-, γ-, and δ-tocopherol, respectively. The aforementioned values corresponded to a total tocopherol content of 196.87 for the 2017 harvest and 183.71 for 2018 (expressed as 10^2 µg/mL pistachio oil).

2.6. FTIR Spectroscopy Study

Figure 2 shows two representative FTIR spectra of a pistachio oil sample with its basic peaks marked. The presented spectra depict samples of common origin but of different harvesting year. It is interesting to note that both spectra are optically very similar and, thus, the use of discriminant analysis is necessary. Each peak corresponds to a certain wavenumber that is attributed to specific vibrations and chemical structures of components from pistachio oil (Table 5).

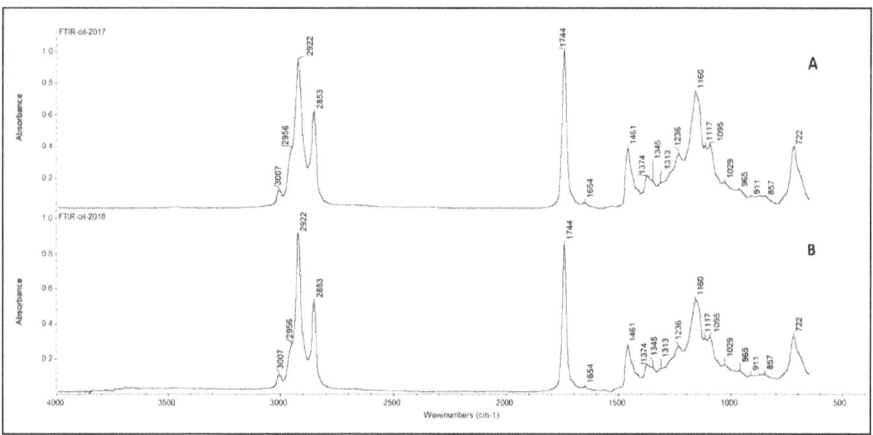

Figure 2. Representative FTIR spectra of pistachio oil samples from the same origin, but from different years of harvest, 2017 (**A**) and 2018 (**B**).

Table 5. Peak correspondence of the pistachio oil FTIR spectra.

Wavenumber (cm^{-1})	Function Group	Abbreviations	Reference
3007	C-H symmetric stretching vibration of -CH$_3$	v_s(CH$_3$)	[24–26]
2956	C-H asymmetric stretching vibration of -CH$_3$	v_{as}(CH$_3$)	[27]
2922	C-H asymmetric stretching vibration of -CH$_2$-	v_{as}(CH$_2$)	[24–26,28–30]
2853	C-H symmetric stretching vibration of -CH$_2$-	v_s(CH$_2$)	[24,26–28,30]
1744	C=O stretching vibration	v(C=O)	[24,25,28,29]
1654	>C=C< cis-olefinic stretching vibration	v(C=C)	[24]
1461	C-H in-plane bending vibration of -CH$_2$- (scissoring)	δ_s(CH$_2$)	[24,26,30,31]
1374	C-H symmetric bending vibration of -CH$_3$	δ(CH$_3$)	[25,27,28,30]
1345, 1313	-CH$_2$- out-of-plane bending vibration (wagging)	ω(CH$_2$)	[27]
1236, 1160, 1117	C-O asymmetric stretching vibration	v_{as}(C-O)	[25,27,28,30,31]
1095, 1029	in-phase-C-C stretching vibration	v(C-C)	[27,30]
965	C-H in-plane bending vibration (scissoring)	δ_s(C=C=C)	[27,28]
911, 857	-CH$_2$- plane vibration	γ(CH$_2$)	[27,28]
722	-CH=CH- cis-stretching vibration	v(C=C)	[24,28,29]

Pistachios are rich in lipids (48%–63%), with a balanced content of mono- (56%–77%) and poly-unsaturated (10%–31%) fatty acids, protein (18%–22%), and dietary fibers (8%–12%). Moreover, they present a high content of bioactive compounds, such as tocopherols, phytosterols, and phenolic compounds [32]. Main lipid acids absorb in the same spectral region (3007–772 cm^{-1}) as phenols, tocopherols, and sterols [24,26,28,33,34].

2.7. Statistical Analysis

In case the number of samples for each test group exceeded 30, Levene's test (*t*-test) was applied without testing whether the data were normally distributed or not. Additionally, a normality test was applied in order to accept or reject the null hypothesis that each test group was statistically different from a normal distribution. Kolmogorov–Smirnov's and Shapiro–Wilk's normality tests evaluated if the groups followed normal distribution (*p*-value > 0.05), with the Shapiro–Wilk's result exhibiting higher validity, as it comes from a more conservative test. If the data did not follow a normal distribution, the appropriate normalization was made to fix the skewness and kurtosis values, at the accepted levels.

When the normality test was confirmed, Levene's test was used to assess the equality of variances. Levene's test checked the null hypothesis that the test group variances were equal (homogeneity of variance or homoscedasticity). If the resulting *p*-value of Levene's test was less than the required significance level (typically 0.05), the obtained differences in sample variances were unlikely to have occurred, based on random sampling from a population with equal variances, so the test group were significantly different.

MetaboAnalyst checked data integrity and continued on to data filtering. The purpose of the data filtering was to identify and remove variables that were unlikely to be of use when modeling the data. This step is strongly recommended for datasets with a large number of variables, many of which are from baseline noises. Based on the total number of variables, 10% of data were filtered, logarithmically transformed, and auto-scaled (mean-centered and divided by the SD of each variable).

The total number of samples (45 pistachio oil samples) was differentiated according to their year of harvest.

2.7.1. Discrimination Based on Antioxidant Capacity

Levene's test of SPSS was used to assess the equality of ABTS variances for two years of harvest, 2017 and 2018. Levene's test tested the null hypothesis that the ABTS variances of 2017 and 2018 were equal. *P*-value less than 0.05 rejected the null hypothesis and proved that the TAC (Wilks' Lambda = 0.895) and the scavenging capacity (Wilks' Lambda = 0.939) of ABTS were different in 2017 and 2018. From cross-validated grouped cases, 71.10% were classified correctly according to their antioxidant capacity and year of harvest.

2.7.2. Discrimination Based on R-Value Study

Based on R-values, 86.70% of cross-validated grouped cases were correctly classified to their year of harvest and the *P*-value (<0.001) proved the accuracy and robustness of the forecasting model, using SPSS. Therefore, the R-value of the samples was exploited to classify the samples according to the year of production (Wilks' Lambda = 0.315).

2.7.3. Discrimination Based on Acid Value and Free Fatty Acid

Levene's test examined the null hypothesis that the AV of 2017 and 2018 harvest were equal and the same assumption was made for the FFA content. The results (p-value > 0.05) failed to reject the null hypothesis and indicated that AV and FFA were not significantly different between the two years of harvest. Discrimination analysis displayed a percentage of correct classification at 71.10% (cross-validated grouped cases).

2.7.4. Discrimination Based on the Tocopherol Analysis

SPSS could not discriminate between the years of harvest of the 45 pistachio oil samples, according to their total tocopherol content. Levene's test for equality of variances between the 2017 and 2018 harvest exhibited a p-value > 0.05 and a 61.40% cross-validation level.

2.7.5. Discrimination Based on FTIR Spectroscopy Study

The spectral regions 3030–2795 and 1805–650 cm^{-1} were selected for the discriminant analysis, i.e., the regions where the peaks were observed (Figure 2). Applying the principal component analysis, the initial set of variables was reduced to a number of hidden variables of principal components (PC). The scree plot (Figure 3) revealed that the greatest impact on the variance of the analysed spectra for the pistachio oil samples was related to the first two principal components. Figures 4 and 5 present the score and loading plot for the principal components (PC) in the principal components analysis (PCA) model. The pistachio oil samples were clearly classified into two groups (2017 and 2018 harvest year). As depicted in Figure 6, MetaboAnalyst could correctly classify 100% of the cross-validated grouped cases, according to their chemical composition and year of harvest with R^2 = 0.992 and Q^2 = 0.987, which indicate a high predictive accuracy. *P*-values less than 0.05 proved that the FTIR method could be used as an accurate rapid screening tool for the differentiation of pistachio oils by their year of harvest.

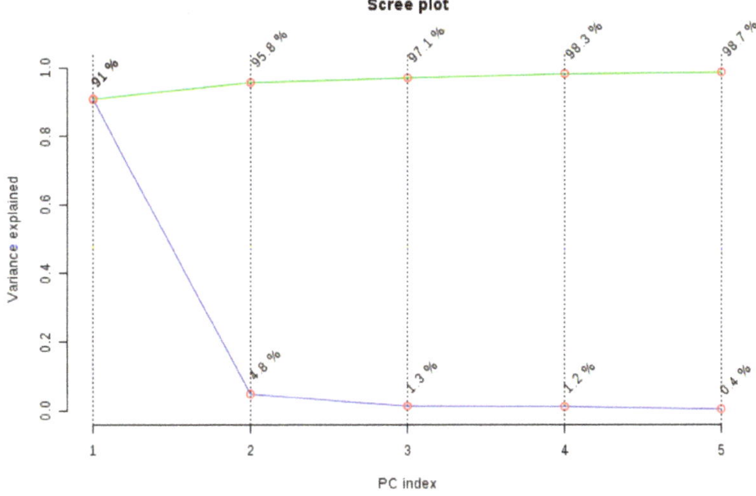

Figure 3. Plot of explained variance for principal components analysis (PCA) of the FTIR spectra.

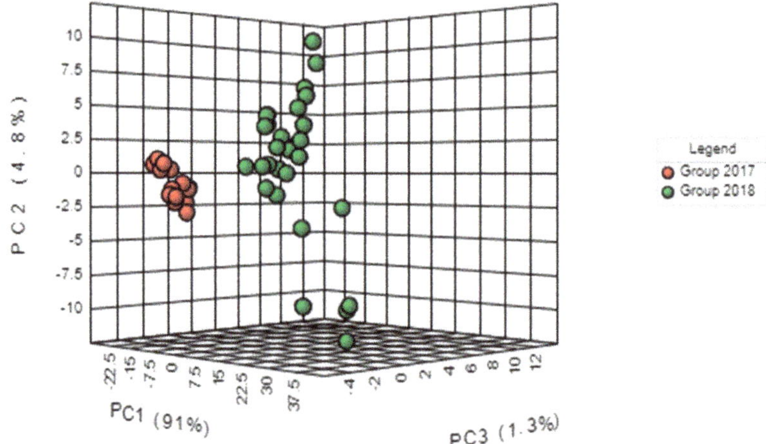

Figure 4. 3D Score plot of principal components analysis (PCA).

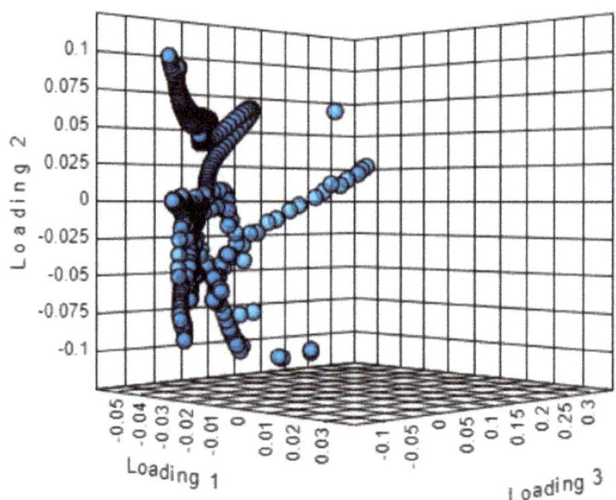

Figure 5. 3D loading plot of principal components analysis (PCA).

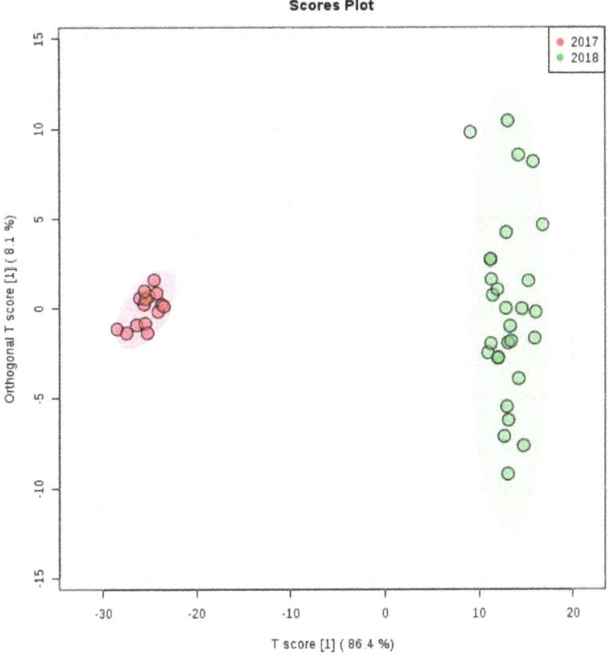

Figure 6. Orthogonal Partial Least Squares–Discrimination (orthoPLS–DA) using MetaboAnalyst.

2.7.6. Statistical Models Comparison

The evaluation of the total antioxidant capacity of pistachio oil samples through an ABTS assay showed that TAC and their scavenging capacity could be statistically differentiated, among the years of harvesting, as was also observed with the results based on R-value. It is worth noticing that in the case of AV, FFA, and HPLC-fluorescence analysis, there were no statistically significant differences between 2017 and 2018, at a 95% confidence level. However, FTIR spectroscopy combined with the statistical methods represent an appropriate rapid technique to discriminate pistachio oils of different quality, based on their antioxidant profile.

3. Materials and Methods

3.1. Samples

A total of 45 pistachio samples of the Greek variety 'Aegina' were provided by pistachio farmers from eight different regions of Greece (Aegina, Megara, Phthiotis, Evia, Volos, Trizina, Thiva, and Avlona) during the 2017 and 2018 harvest seasons. Due to alternate bearing, the number of samples of 2017 was less than the succeeding year. The pistachios were sound and had the typical characteristics of the variety. They were dried under the sun or mechanically at moisture level 5%–7%, after dehulling at farm level. In the laboratory, each sample was shelled and finely ground in an IKA M 20 (IKA, Königswinter, Germany) laboratory mill, at a maximum rotational speed 20,000 rpm, followed by particle size separation using sieves (500 μm < size < 800 μm). After preparation, all samples were put in sealed bags, protected from light, and stored in the freezer (−20 °C) until analysis.

3.2. Reagents

Petroleum ether, 2,2-diphenyl-1-(2,4,6-trinitrophenyl)hydrazine (DPPH), 2,2′-azinobis(3-ethylbenzothiazoline-6-sulfonic acid diammonium salt (ABTS), potassium persulfate ($K_2S_2O_8$), potassium hydrogen phthalate (KHP), sodium hydroxide pellets (NaOH), 6-hydroxy-2,5,7,8-tetramethylchroman-2-carboxylic acid (Trolox), ethyl acetate, tetrahydrofuran (THF), n-heptane, cyclohexane 99.8%, methanol (MeOH), and ethanol (EtOH) were purchased from Sigma-Aldrich (Steinhein, Germany). (+)-α-, (+)-β-, (+)-γ-, and (+)-δ-tocopherol standards of 99.99% purity were obtained from Merck (Darmstadt, Germany). Distilled water and phenolphthalein indicator solution were also used. All compounds and solvents were of analytical grade.

3.3. Oil Extraction

Pistachio oil was extracted from 4 g of kernel flour with 250 mL of petroleum ether in a Soxhlet apparatus for 6 h, according to the AOAC Official Method 948.22. After evaporation of the solvent under reduced pressure, the oil was weighed to measure the lipids' mass and was kept in a freezer (–20 °C) to maintain its initial quality, until analysis. The extraction was carried out in triplicates and the mean value with the standard deviation was calculated.

3.4. Evaluation of the Antioxidant Capacity

3.4.1. DPPH Assay

DPPH radical-scavenging capacity was determined according to Minioti and Georgiou (2010) [17], with some modifications using a JASCO V-550 spectrophotometer (JASCO Corporation, Tokyo, Japan). Briefly, 100 μL of pistachio oil were mixed with 4 mL of DPPH working solution (8.1×10^{-5} M working solution of the DPPH radical in ethyl acetate). The reaction mixture was vigorously stirred for a few seconds and kept in a dark place for 30 min, at room temperature. Absorbencies were measured at 515 nm against a blank (100 μL of ethyl acetate instead of pistachio oil). Pistachio oil antioxidants scavenged the DPPH radical, resulting in decolorization of its purple solution. Analyses were performed in triplicates. The scavenging capacity was calculated using Equation (7):

$$\text{Scavenging capacity} = [(A_{515} \text{ of control} - A_{515} \text{ of sample})/A_{515} \text{ of control}] \times 100 \qquad (7)$$

A calibration curve (0.08–1 mM) was constructed using Trolox as the external standard and the obtained values were expressed as mmol/L of Trolox equivalents per mL of oil.

3.4.2. ABTS Assay

The ABTS assay was slightly modified, based on the methods of Rajaei et al. (2010) [35] and Torres-Martinez et al. (2017) [36], using an Agilent 8453 spectrophotometer. In brief, 96 mg of ABTS with distilled water were diluted in a 25 mL volumetric flask and 440 μL of $K_2O_8S_2$ solution (0.14 M in distilled water) were added. The mixture was maintained for 18 h, protected by light, at room temperature for stabilization of the ABTS oxidation. Prior to further use, the ABTS$^+$ solution was diluted with EtOH, at an absorbance value of 0.7 ± 0.005 (working solution). Antioxidant capacity was evaluated by measuring the scavenging effect of 100 μL of pistachio oil, mixed with 2 mL of ABTS$^+$ working solution, followed by shaking and incubation in the dark, for 6 min at room temperature. The decrease in absorbance was then measured at 734 nm against a control solution (100 μL of EtOH). All measurements were performed in triplicates. The scavenging capacity was calculated using Equation (8):

$$\text{Scavenging capacity} = [(A_{734} \text{ of control} - A_{734} \text{ of sample})/A_{734} \text{ of control}] \times 100 \qquad (8)$$

Trolox was used as a reference compound for the calibration curve with a concentration range of 0.05–1 mM and a total antioxidant capacity, expressed as mmol/L of Trolox equivalents per mL of oil, was calculated and reported as mean ± SD.

3.5. Quality Assessment of Pistachio Oil

3.5.1. UV–Vis Assessment

The Agilent Cary 60 UV–Vis spectrophotometer (Agilent Technologies, Mississauga, ON, Canada) and rectangular quartz cuvettes with an optical length of 1 cm were used according to EEC No 2568/91 (Annex IX of the Regulation). Pistachio oil samples (45 in total) were diluted in cyclohexane. A total of 0.1 g of pistachio oil was weighed accurately into a 10 mL graduated flask, filled up to the mark with the solvent, and homogenized. The resulting solution (10 g/L) was perfectly clear. If opalescence or turbidity was present, it was filtered through the paper. All samples were measured in cuvettes, running a solvent blank as a reference. Absorption measurements for purity determination were made at 232, 268, 270, and 274 nm in triplicates, and the average was used for the determination of pistachio oil purity. K values were calculated according to Equation (9):

$$k = A/(C \times s) \tag{9}$$

where A is the absorbance at the specified nanometer; C is the concentration in grams per liter; and s is the cuvette thickness in centimeter. Delta (Δ) k and R-value were evaluated using Equations (10) and (11):

$$\text{Delta } (\Delta)\, k = k_{270} - [(k_{268} + k_{274})/2] \tag{10}$$

$$\text{R-value was calculated} = k_{232}/k_{270} \tag{11}$$

3.5.2. Determination of AV and FFA

The AV and FFA content were determined in triplicates, according to Otemuyiwa and Adewusi (2013) [22]. In brief, titration of pistachio oil (1 g) dissolved in 5 mL EtOH was applied, using a 0.1 M NaOH solution as the standard reagent to a phenolphthalein endpoint (when the addition of a single drop of alkali produces a slight but definite color change that persists for at least 15 s). The AV value was expressed as oleic acid, according to Equation (12). All determinations were performed in triplicates. The acid value was calculated according to Equation (13):

$$AV = (56 \times C \times V)/m \tag{12}$$

where V is the titration volume (mL) of the standard volumetric NaOH solution used; C is the concentration (M) of the standard volumetric NaOH solution used; and m is the mass (g) of the pistachio oil sample. The percentage of FFAs in the pistachio oil was calculated using Equation (13):

$$\% \text{ FFA} = 0.503 \times AV \tag{13}$$

Then, the AV and FFA mean values and the corresponding SDs were calculated.

3.6. Tocopherol Analysis

3.6.1. Apparatus and Chromatographic Conditions

The chromatographic analysis was carried out in an analytical HPLC unit equipped with a JASCO PU 980 pump, with a 100 µL injection loop, a JASCO FP920 fluorescence detector (Co. Ltd., Tokyo, Japan) supported by Clarity Lite software (DataApex, Prague, Czech Republic) for data processing, and an ODS Hypersyl column (4.6 × 250 mm, 5 µm particle size, Thermo Fisher Scientific Inc., Waltham, MA, USA).

The determination of the α-, β-, γ-, and δ-tocopherol (T) content using HPLC, followed the ISO 9936:2006 standard. The mobile phase consisted of the THF/n-heptane (4:96 *v/v*) at a flow rate of 1.0 mL/min and the injection volume was 10 µL. The effluent was detected in a fluorescence detector, with an excitation filter at 295 nm and an emission wavelength at 330 nm. The system was operated at ambient temperatures. The tocopherol compounds were identified by chromatographic comparisons of the retention times of the analytes in a standard solution and quantified by the respective calibration curves. The results were obtained from triple measurements, and the mean values and corresponding SDs were calculated.

3.6.2. Standard Solutions

Stock standard solutions, α-T (96.53 µg/mL), β-T (85.74 µg/mL), γ-T (87.40 µg/mL), and δ-T (81.14 µg/mL) in n-heptane were prepared and stored in the dark at −20 °C. Combined working standard mixtures, with concentrations in the expected sample ranges, were prepared daily from the stock standard solutions, by diluting appropriate volumes of stock solutions with n-heptane. Then, a calibration curve for each tocopherol was constructed.

3.6.3. Validation Method

Calibration and linearity: Calibration curves were prepared using standard solution of vitamin E isomers at nine concentrations (C), ranging from 0.15–20 µg/mL.

Recovery: Extraction recoveries were evaluated by adding known amounts of isomers (+)-α-T, (+)-β-T, (+)-γ-T, and (+)-δ-T to the pistachio oil samples. The amounts added were of low, medium and high tocopherol content (0.2, 10 and 20 µg/mL). Recovery was calculated by Equation (14):

$$\text{Recovery} = (\text{C of spiked sample}/(\text{C of sample} + \text{C of standard added})) \times 100 \tag{14}$$

Analytical precision: Interday precision was determined by analyzing two concentrations (15 and 20 µg/mL) of standard vitamin E isomers in three replicates on three different days. The following equation was used:

$$\text{Precision, \%} = (\text{SD/Mean C}) \times 100 \tag{15}$$

The repeatability of the tocopherols' measurements in pistachio oil was calculated by Equation (15).

3.7. FTIR Spectra Recording

The FTIR spectra of the pistachio oil samples were recorded in triplicates on a Thermo Nicolet 6700 FTIR spectrophotometer (Thermo Electron Corporation, Madison, WI, USA) equipped with a deuterated triglycine sulfate (DTGS) detector. The spectra were in an attenuated total reflection (ATR) mode with a Horizontal ATR accessory from Spectra-Tech Inc. (Stamford, CT, USA). The accessory was equipped with a ZnSe-ATR crystal of a trapezoid shape (800 × 10 × 4 mm). The crystal provided an angle of incidence of 45° and was enclosed in a stainless-steel cuvette. For spectra recording, an aliquot of 200 µL of pistachio oil or tocopherol standard mixture was poured on the ATR crystal and allowed to dry, forming a uniform film. Spectra were recorded with a resolution of 4 cm^{-1} and 100 scans. The speed of the interferometer moving mirror was 0.6329 mm/s. Background spectrum was collected using only ATR crystal, prior to spectrum recording of each sample.

FTIR spectra were smoothed using the Savitsky–Golay algorithm (5-point moving second-degree polynomial) and the baseline was corrected using the 'automatic baseline correct function' (second-degree polynomial, twenty iterations). Then, the average spectrum of each sample was measured and normalized (absorbance maximum value of 1). Each average spectrum was extracted and saved as a csv file for their use in discriminant analysis. Spectral data collection and processing was carried out using the OMNIC ver. 8.2.0.387 software (Thermo Fisher Scientific Inc., Waltham, MA, USA).

3.8. Statistical Analysis

Discriminant analysis was performed using IBM SPSS Statistics 22 (ver. 8.0.0.245) (SPSS Inc., Chicago, IL, USA) and the MetaboAnalyst 4.0 software (McGill University, Montreal, QC, Canada) for a comprehensive and integrative data analysis.

4. Conclusions

The results of this work showed that the pistachio oil samples of the variety "Aegina" were within the limits set by the specific standards in terms of high quality. The oil yields of the samples from the two harvest seasons (2017, 2018) were found to be similar, while statistically significant differences were evident for the antioxidant capacity and the R-value between pistachio oil samples, from different years of harvesting. These differences might be attributed to agroclimatic factors, such as different agricultural practices, average temperature, and rainfall from year to year. The FTIR spectroscopy succeeded to classify pistachio oil samples according to the differences which are related to quality parameters, particularly described by the antioxidant capacity, and the R-value. The developed method is fast, accurate, non-destructive, with no excessive sample preparation, and has the additional advantage of not requiring the use of large quantities of solvents, being especially suitable for the screening of large number of samples. Furthermore, the present results provide evidence that the FTIR method could be a promising discriminating tool against fraud related to plant-derived oils, through the use of quality parameters as indicators.

Author Contributions: Conceptualization, C.S.P., L.V., and M.G.; methodology, C.S.P and M.G.; formal analysis, L.V., D.A., and A.M.; investigation, L.V., D.A., and A.M.; data curation, L.V., D.A., and A.M.; writing—original draft preparation, L.V. and M.G.; writing—review and editing, L.V., M.G., and C.S.P. All authors have read and agreed to the published version of the manuscript.

Funding: This research received no external funding.

Acknowledgments: This research work was supported by the Hellenic Foundation for Research and Innovation (HFRI) and the General Secretariat for Research and Technology (GSRT), under the HFRI PhD Fellowship grant (GA. no. 1313).

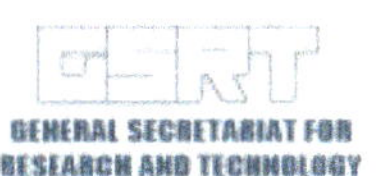

Conflicts of Interest: The authors declare no conflict of interest.

References

1. Gentile, C.; Tesoriere, L.; Butera, D.; Fazzari, M.; Monastero, M.; Allegra, M.; Livrea, M.A. Antioxidant Activity of Sicilian Pistachio (*Pistacia vera* L. Var. Bronte) Nut Extract and Its Bioactive Components. *J. Agric. Food Chem.* **2007**, *55*, 643–648. [CrossRef] [PubMed]

2. Rodríguez-Bencomo, J.J.; Kelebek, S.; Ahmet, S.S.; Rodríguez-Alcalá, L.M.; Fontecha, J.; Selli, S. Characterization of the Aroma-Active, Phenolic, and Lipid Profiles of the Pistachio (*Pistacia vera* L.) Nut as Affected by the Single and Double Roasting Process. *J. Agric. Food Chem.* **2015**, *63*, 7830–7839.

3. Halvorsen, B.L.; Carlsen, M.H.; Phillips, K.M.; Bøhn, S.K.; Holte, K.; Jacobs, D.R., Jr.; Blomhoff, R. Content of redox-active compounds (ie, antioxidants) in foods consumed in the United States. *Am. J. Clin. Nutr.* **2006**, *84*, 95–135. [CrossRef] [PubMed]

4. John, J.A.; Shahidi, F. Phenolic compounds and antioxidant activity of Brazil nut (*Bertholletia excelsa*). *J Funct. Foods* **2010**, *2*, 196–209. [CrossRef]

5. Desamparados, S.; Ojeda-Amador, R.; Fregapane, G. Virgin Pistachio (*Pistachia vera* L.) Oil. In *Fruit Oils: Chemistry and Functionality*; Ramadan, M., Ed.; Springer: Cham, Switzerland, 2019; p. 183.

6. Tsantili, E.; Takidelli, C.; Christopoulos, M.V.; Lambrinea, E.; Rouskas, D.; Roussos, P.A. Physical, compositional and sensory differences in nuts among pistachio (*Pistachia vera* L.) varieties. *Sci. Hortic.* **2010**, *125*, 562–568. [CrossRef]

7. Ojeda-Amador, R.M.; Fregapane, G.; Desamparados, S.M. Composition and properties of virgin pistachio oils and their by-products from different cultivars. *Food Chem.* **2018**, *240*, 123–130. [CrossRef]

8. Aceña, L.; Vera, L.; Guasch, J.; Busto, O.; Mestres, M. Comparative study of two extraction techniques to obtain representative aroma extracts for being analysed by gas chromatography-olfactometry: Application to roasted pistachio aroma. *J. Chromatogr. A* **2010**, *1217*, 7781–7787. [CrossRef]

9. Kay, C.D.; Gebauer, S.K.; West, S.G.; Kris-Etherton, P.M. Pistachios Increase Serum Antioxidants and Lower Serum Oxidized-LDL in Hypercholesterolemic Adults. *J. Nutr.* **2010**, *140*, 1093–1098. [CrossRef]

10. Martorana, M.; Arcoraci, T.; Rizza, L.; Cristani, M.; Bonina, F.P.; Saija, A.; Trombetta, D.; Tomaino, A. In vitro antioxidant and in vivo photoprotective effect of pistachio (*Pistacia vera* L., variety Bronte) seed and skin extracts. *Fitoterapia* **2013**, *85*, 41–48. [CrossRef]

11. Houshia, O.J.; Zaid, O.; Shqair, H.; Zaid, M.; Fashafsheh, N.; Bzoor, R. Effect of Olive Oil Adulteration on Peroxide Value, Delta-K and on the Acidity Nabali-Baladi Olive Oil Quality. *Adv. Life Sci.* **2014**, *4*, 235–244.

12. El-Abassy, R.M.; Donfack, P.; Materny, A. Rapid Determination of Free Fatty Acid in Extra Virgin Olive Oil by Raman Spectroscopy and Multivariate Analysis. *J. Am. Oil Chem. Soc.* **2009**, *86*, 507–511. [CrossRef]

13. Servili, M.; Montedoro, G. Contribution of phenolic compounds to virgin olive oil. *Eur. J. Lipid Sci. Technol.* **2002**, *104*, 602–613. [CrossRef]

14. Juhaimi, F.A.; Ozcan, M.M.; Ghafoor, K.; Babiker, E.E.; Hussain, S. Comparison of cold-pressing and soxhlet extraction systems for bioactive compounds, antioxidant properties, polyphenols, fatty acids and tocopherols in eight nut oils. *J. Food Sci. Technol.* **2018**, *55*, 3163–3173. [CrossRef] [PubMed]

15. Geana, E.I.; Ciucure, C.T.; Apetrei, C.; Artem, V. Application of Spectroscopic UV-Vis and FT-IR Screening Techniques Coupled with Multivariate Statistical Analysis for Red Wine Authentication: Varietal and Vintage Year Discrimination. *Molecules* **2019**, *24*, 4166. [CrossRef] [PubMed]

16. Pérez-Ràfols, C.; Subirats, X.; Serrano, N.; Díaz-Cruz, J.M. New discrimination tools for harvest year and varieties of white wines based on hydrophilic interaction liquid chromatography with amperometric detection. *Talanta* **2019**, *201*, 104–110. [CrossRef]

17. Minioti, K.S.; Georgiou, C.A. Comparison of different tests used in mapping the Greek virgin olive oil production for the determination of its total antioxidant capacity. *Grasas Y Aceites* **2010**, *61*, 45–51.

18. Yu, J.; Wang, H.; Zhan, J.; Huang, W. Review of recent UV-Vis and infrared spectroscopy researches on wine detection and discrimination. *Appl. Spectrosc.* **2017**, *53*, 65–86. [CrossRef]

19. Rowshan, V.; Bahmanzadegan, A.; Saharkhiz, M.J. Influence of storage conditions on the essential oil composition of Thymus daenensis Celak. *Ind. Crop. Prod.* **2013**, *49*, 97–101. [CrossRef]

20. Kim, D.O.; Lee, K.W.; Lee, H.J.; Lee, C.Y. Vitamin C equivalent antioxidant capacity (VCEAC) of phenolic phytochemicals. *J. Agric. Food Chem.* **2002**, *50*, 3713–3717. [CrossRef]

21. Floegel, A.; Kim, D.O.; Chung, S.J.; Koo, S.I.; Chun, O.K. Comparison of ABTS/DPPH assays to measure antioxidant capacity in popular antioxidant-rich US foods. *J. Food Compos. Anal.* **2011**, *24*, 1043–1048. [CrossRef]

22. Otemuyiwa, I.O.; Adewusi, S.R.A. Fatty Acid, Carotenoid and Tocopherol Content of Some Fast Foods from a Nigerian Eatery. *Food Nutr. Res.* **2013**, *1*, 82–86.

23. Martínez, M.L.; Fabani, M.P.; Baroni, M.V.; Huaman, R.N.M.; Ighani, M.; Maestri, D.M.; Wunderlin, D.; Tapia, A.; Feresin, G.E. Argentinian pistachio oil and flour: A potential novel approach of pistachio nut utilization. *J. Food Sci. Technol.* **2016**, *53*, 2260–2269. [CrossRef] [PubMed]

24. Uncu, O.; Ozen, B.; Tokatli, F. Use of FTIR and UV–visible spectroscopy in determination of chemical characteristics of olive oils. *Talanta* **2019**, *201*, 65–73. [CrossRef] [PubMed]

25. Luna, A.S.; Silva, A.P.; Ferré, J.; Boqué, R. Classification of edible oils and modeling of their physico-chemical properties by chemometric methods using mid-IR spectroscopy. *Spectrochim. Acta A* **2013**, *100*, 109–114. [CrossRef]

26. Gericke, A.; Hühnerfuss, H. Investigation of Z- and E-Unsaturated Fatty Acids, Fatty Acid Esters, and Fatty Alcohols at the Air/Water Interface by Infrared Spectroscopy. *Langmuir* **1995**, *11*, 225–230. [CrossRef]

27. Dymińska, L.; Calik, M.; Albegar, A.M.M.; Zając, A.; Kostyń, K.; Lorenc, J.; Hanuza, J. Quantitative determination of the iodine values of unsaturated plant oils using infrared and Raman spectroscopy methods. *Int. J. Food Prop.* **2017**, *20*, 2003–2015. [CrossRef]

28. Rohman, A.; CheMan, Y.B. Fourier transform infrared (FTIR) spectroscopy for analysis of extra virgin olive oil adulterated with palm oil. *Food Res. Int.* **2010**, *43*, 886–892. [CrossRef]

29. Moros, J.; Roth, M.; Garrigues, S.; Guardia, M. Preliminary studies about thermal degradation of edible oils through attenuated total reflectance mid-infrared spectrometry. *Food Chem.* **2009**, *114*, 1529–1536. [CrossRef]

30. Christy, A.A.; Egeberg, P.K. Quantitative determination of saturated and unsaturated fatty acids in edible oils by infrared spectroscopy and chemometrics. *Chemom. Intell. Lab. Syst.* **2006**, *82*, 130–136. [CrossRef]

31. Gutiérrez, L.F.; Quiñones-Segura, Y.; Sanchez-Reinoso, Z.; Leonardo, D.D.; Abril, J.I. Physicochemical properties of oils extracted from γ-irradiated Sacha Inchi (*Plukenetiavolubilis* L.) seeds. *Food Chem.* **2017**, *237*, 581–587.

32. Bullo, M.; Juanola-Falgarona, M.; Hernandez-Alonso, P.; Salas-Salvado, J. Nutrition attributes and health effects of pistachio nuts. *Br. J. Nutr.* **2015**, *113*, 879–893. [CrossRef] [PubMed]

33. Ahmed, M.K.; Daun, J.K.; Przybylski, R. FT-IR based methodology for quantitation of total tocopherols, tocotrienols and plastochromanol-8 in vegetable oils. *J. Food Compos. Anal.* **2005**, *18*, 359–364. [CrossRef]

34. Rosenkrantz, H. Infra-red absorption spectra of tocopherols and related structure. *J. Biol. Chem.* **1948**, *173*, 439–447. [PubMed]

35. Rajaei, A.; Barzegar, M.; Mobarez, A.M.; Sahari, M.A.; Esfahani, Z.H. Antioxidant, anti-microbial and antimutagenicity activities of pistachio (*Pistachia vera*) green hull extract. *Food Chem. Toxicol.* **2010**, *48*, 107–112. [CrossRef]

36. Torres-Martínez, R.; García-Rodríguez, Y.M.; Ríos-Chávez, P.; Saavedra-Molina, A.; López-Meza, J.E.; Ochoa-Zarzosa, A.; Garciglia, R.S. Antioxidant Activity of the Essential Oil and its Major Terpenes of *Satureja macrostema* (Moc. and Sessé ex Benth.) Briq. *Pharm. Mag.* **2018**, *13*, 875–880.

Sample Availability: Samples of the compounds are available from the authors.

 *molecules*

Article

Hempseed Lignanamides Rich-Fraction: Chemical Investigation and Cytotoxicity towards U-87 Glioblastoma Cells

Ersilia Nigro [1,2,†], Giuseppina Crescente [1,†], Marialuisa Formato [1,†], Maria Tommasina Pecoraro [1], Marta Mallardo [1,2], Simona Piccolella [1], Aurora Daniele [1,2] and Severina Pacifico [1,*]

[1] Department of Environmental Biological and Pharmaceutical Sciences and Technologies, University of Campania "Luigi Vanvitelli", Via Vivaldi 43, I-81100 Caserta, Italy; nigro@ceinge.unina.it (E.N.); giuseppina.crescente@unicampania.it (G.C.); marialuisa.formato@unicampania.it (M.F.); mariatommasina.pecoraro@unicampania.it (M.T.P.); marta.mallardo@unicampania.it (M.M.); simona.piccolella@unicampania.it (S.P.); aurora.daniele@unicampania.it (A.D.)

[2] CEINGE-Advanced Biotechnologies, Scarl, 80131 Napoli, Italy

[*] Correspondence: severina.pacifico@unicampania.it; Tel.: +39-0823-274578

[†] These authors equally contributed to the work.

Academic Editor: Stefano Dall'Acqua
Received: 18 February 2020; Accepted: 25 February 2020; Published: 26 February 2020

Abstract: The weak but noteworthy presence of (poly)phenols in hemp seeds has been long overshadowed by the essential polyunsaturated fatty acids and digestible proteins, considered responsible for their high nutritional benefits. Instead, lignanamides and their biosynthetic precursors, phenylamides, seem to display interesting and diverse biological activities only partially clarified in the last decades. Herein, negative mode HR-MS/MS techniques were applied to the chemical investigation of a (poly)phenol-rich fraction, obtained from hemp seeds after extraction/fractionation steps. This extract contained phenylpropanoid amides and their random oxidative coupling derivatives, lignanamides, which were the most abundant compounds and showed a high chemical diversity, deeply unraveled through high resolution tandem mass spectrometry (HR-MS/MS) tools. The effect of different doses of the lignanamides-rich extract (LnHS) on U-87 glioblastoma cell line and non-tumorigenic human fibroblasts was evaluated. Thus, cell proliferation, genomic DNA damage, colony forming and wound repair capabilities were assessed, as well as LnHS outcome on the expression levels of pro-inflammatory cytokines. LnHS significantly inhibited U-87 cancer cell proliferation, but not that of fibroblasts, and was able to reduce U-87 cell migration, inducing further DNA damage. No modification in cytokines' expression level was found. Data acquired suggested that LnHS acted in U-87 cells by inducing the apoptosis machinery and suppressing the autophagic cell death.

Keywords: *Cannabis sativa* L.; phenylamides; lignanamides; hemp seeds; high resolution tandem mass spectrometry; U-87 glioblastoma cells; cytotoxicity

---

1. Introduction

Plant foods, thanks to the functionality of their bioactive secondary metabolites, are considered to be both safe and able to promote good health [1], explaining some important targeted effects in humans, and preventing affluence diseases, such as cardiovascular diseases and cancers [2]. Bioactive compounds are also in ancient crops, whose actual recovery further renewed the phytochemical research into the discovery of beneficial substances, which could be present, after food processing, in the daily meals or be exceptionally lost in produced waste and by-products [3]. This is the case of lignans, chemically characterized by a phenylpropanoid core, which are reported to exert numerous biological effects in mammals, including antitumor and antioxidant activities [4]. They act as phytoestrogens and are converted by intestinal microflora into mammalian lignans or enterolignan compounds [5].

Lignans are abundant in dietary sources like whole-grain cereals and legumes. The preventive benefits of some edible seeds, and the increasing intake of no-conventional foods as chia, quinoa, flax, canola and pumpkin seeds could be ascribed to their richness in these compounds [6]. For instance, flaxseed (*Linum usitatissimum* L.) is reported to contain about 75–800 times more lignans than cereal grains, legumes, fruits and vegetables, together with antioxidant flavonols and hydroxycinnamic acids [7], and lariciresinol was one of the main constituents of pumpkin seeds (*Cucurbita pepo* L.) [8].

Hemp seeds (non-drug type of *Cannabis sativa* L.) also contain, beyond proteins, and polyunsaturated fatty acids, bioactive lignan derivatives, known as lignanamides [9–12]. These latter could be found in hemp fruits together with their biosynthetic precursors, namely phenylamides, whose presence is functional in another seed, such as oat seed, which produces avenanthramides (AVAs) with important anti-inflammatory and antiproliferative effects [13]. Until a few years ago, phenylamides and lignanamides seemed to constitute a small group of natural products, whose distribution in plant kingdom was thought to be limited to plants of the Cannabaceae and Solanaceae family. Indeed, lignanamides are also reported from *Mitrephora thorelii* (Annonaceae) and *Corydalis saxicola* (Papaveraceae), and, recently, five pairs of enantiomeric lignanamides were obtained from *Solanum nigrum*, and melongenamides A–D were isolated from the roots of *Solanum melongena* L. [14–16]. Furthermore, new lignanamides and neolignanamides were isolated from *Lycium chinense* [17,18], highlighting that the diversity of these compounds is so far to be really known. Moreover, (±)-sativamides A and B, two pairs of nor-lignanamide enantiomers featuring a unique benzo-angular triquinane skeleton, were isolated from the fruits of *Cannabis sativa*, and were observed to be able to reduce endoplasmic reticulum (ER) stress-induced cytotoxicity in neuroblastoma cells [19].

The ability of lignanamides to display interesting and diverse biological activities, including feeding deterrent activity, insecticidal effects, anti-inflammatory and neuroprotective activity [20,21], addresses the research in new analytical challenges for their ready exploitation from hempseed meal. In this context, starting on the great and renewed interest in hemp seeds and their by-products, as source of essential nutrients, in sufficient amount and ratio to satisfy the dietary human demand, commercial hempseeds underwent ultrasound assisted extraction first with *n*-hexane, and after an oil-like mixture recovery, the obtained by-product was investigated for its polyphenol content. A fraction rich in lignanamides (LnHS) was achieved and chemically profiled through HR-MS/MS tools operating in negative ion detection. Furthermore, taking into account recent literature, which highlights the protective effects of pure lignanamides on central nervous system cell lines [15,21], LnHS was investigated for its anti-cancer properties versus U-87 malignant glioblastoma (GMB) neuroepithelial cells. Indeed, U-87 cells are known to move very fast and to aggregate as clusters, showing rapid migration, and highest invasion ability [22]. As malignant gliomas are the most common primary brain tumors, among which GBM is the most malignant and highly aggressive [23], the aim of this study was to understand the effects of low doses of constituted LnHS fraction on U-87 viability cell line as well-established model of malignant glioblastoma cell line in terms of apoptosis and autophagic cell death. In addition, the effects of LnHS on U-87 colony formation efficacy and cell migration was analyzed. Finally, the ability of LnHS of inducing oxidative and inflammation processes was investigated through evaluation of the expression of Sirt1 and Sirt2, as well as of some pro-inflammatory cytokines. All tests were performed in comparison to non-tumorigenic human fibroblasts (thereafter indicated as HF).

2. Results and Discussion

The interest in the polyphenols contained in hemp seed was for a long time obscured by its high content in essential polyunsaturated fatty acids, mainly linoleic and α-linolenic acids, though different flavonoid glycosides were identified in cold-pressing hemp seed oil [24,25]. Indeed, after oil recovery, hemp seed meal yet represents an important food polyphenol source. Several phenylamides and lignanamides were previously isolated and structurally identified mainly by means of nuclear magnetic resonance (NMR) spectroscopic tools [20,26]. Herein, in order to get insight into the chemical composition of commercial hempseeds, ultrasound assisted maceration was applied on seeds first ground with a knife mill and crushed into liquid N_2 to better preserve the integrity of the fruit's constituents.

The cryo-crushed matrix, made as a friable powder, preliminarily underwent solid-liquid extraction process using *n*-hexane, obtaining an oil-like extract and the defatted-matrix. This latter was then ultrasound-assisted macerated using ethanol (Figure S1). The alcoholic fraction was further fractionated (please see Section 3) obtaining, among the others, a fraction, hereafter referred to as LnHS, able to strongly modify the morphology of SH-SY5Y cells (Figure S2). Thus, in the consciousness that extraction/fractionation steps are fundamental into defining the chemistry of a phytochemical extract, in order to get insight into this fraction chemical composition, negative HR-MS and HR-MS/MS spectra, as well as spectra by ultraviolet diode array detection (UV-DAD), were acquired. Phenylamides and lignanamides were found to be the main compounds (Tables 1 and 2), whereas flavonol glycosides were the minor constituents (Table 3). Thus, the fraction underwent an extensive cytotoxic screening on U-87 cells, exhibiting a promising behavior in fighting migration and invasion features of these glioblastoma cells.

2.1. HR-MS Analysis: Phenylamides

Compounds **1**, **2**, **5**, **8**, **11**, **17** and **33** were recognized as phenylamides, conjugates of aliphatic polyamines or arylmonoamines and hydroxycinnamic acids, suggested as defensive plant-specific molecules. Compound **1** was tentatively identified as *N*-caffeoyloctopamine (Figure 1), previously isolated from hempseed cakes in a screening aimed to identify novel arginase inhibitors [27], and further identified among hempseed constituents with potential anti-neuroinflammatory activity [25].

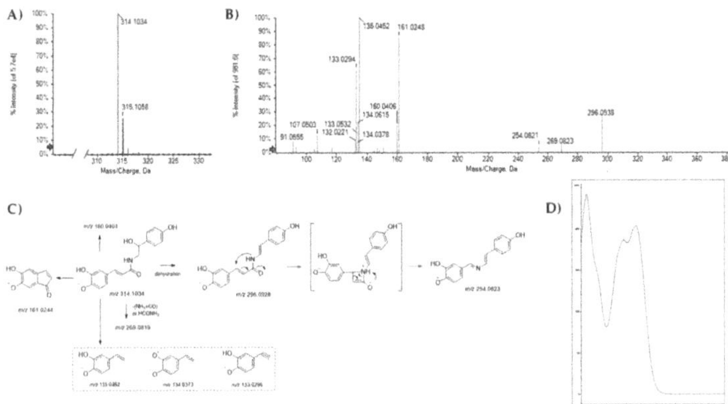

Figure 1. (**A**) and (**B**) TOF-MS and TOF-MS/MS spectra for compound **1**; (**C**) proposed fragmentation pathway of the [M – H]⁻ ion; (**D**) UV-DAD spectrum. In C panel, the theoretical *m/z* value is reported below each structure.

The [M – H]⁻ ion at *m/z* 314.1039 underwent dehydration (likely through octopamine moiety) providing the TOF-MS² fragment ion at *m/z* 296.0910, or rearranged to give, through 45 Da neutral loss, the fragment ion at *m/z* 269.0818.

Table 1. TOF-MS and TOF-MS/MS of tentatively identified phenylamides in the investigated hempseed fraction. RT = Retention Time; RDB = Ring Double Bond equivalent value. Compounds were numbered based on their RT in the whole LnHS chromatogram.

	RT (min)	Tentative Assignment	Formula	$[M - H]^-$ Calc. (m/z)	$[M - H]^-$ Found (m/z)	Error (ppm)	RDB	MS/MS Fragment Ions (m/z) and Relative Intensity (%)
1	4.190	*N*-caffeoyloctopamine	$C_{17}H_{17}NO_5$	314.1034	314.1039	1.6	10	296.0910(28.2); 269.0818(5.5); 254.0813(4.2); 161.0244(86.7); 160.0396(15.8); 135.0455(100); 134.0609(19.5); 134.0373(22.5); 133.0302(47.1); 132.0217(8.3); 123.0449(1.1); 107.0505(13.9); 93.0350(7.3)
2	5.141	*N*-p-coumaroyloctopamine	$C_{17}H_{17}NO_4$	298.1085	298.1086	0.4	10	280.0981(11.7); 160.0408(12.3); 145.0293(97.5); 134.0607(15.9); 133.0531(18.0); 119.0502(100); 117.0347(84.2); 93.0342(7.5)
5	6.426	*N*-caffeoyltyramine	$C_{17}H_{17}NO_4$	298.1085	298.1090 334.0871 $[M + Cl]^-$	1.7	10	298.1102(19.6); 256.0974(2.8); 190.0518(16.2); 178.0520(20.6); 161.0251(11.3); 148.0536(18.5); 147.0456(10.5); 136.0775(6.0); 135.0459(100); 134.0381(18.6); 133.0301(8.1); 107.0508(7.8)
8	7.151	*N*-feruloyltyramine	$C_{18}H_{19}NO_4$	312.1241	312.1243	0.5	10	312.1247(13.0); 297.1015(7.6); 190.0510(61.1); 178.0513(63.9); 176.0354(13.9); 148.0530(100); 147.0448(24.9); 135.0452(35.2); 134.0373(8.7)
11	7.755	*N*-p-coumaroyltyramine	$C_{17}H_{17}NO_4$	282.1136	282.1140	1.5	10	282.1140(4.9); 174.0559(7.8); 162.0558(9.4); 145.0294(4.3); 132.0580(7.4); 119.0503(100); 117.0341(9.4)
17	8.387	*N*-feruloyltyramine	$C_{18}H_{19}NO_4$	312.1238	312.1241	−1.1	10	312.1254(6.8); 297.1014(6.1); 190.0513(65.5); 178.0513(57.4); 176.0354(14.5); 148.0535(100); 147.0456(30.2); 135.0456(37.2); 134.0376(11.8)
33	12.209	tri-p-coumaroylspermidine	$C_{34}H_{37}N_3O_6$	582.2610	582.2640 618.2418 $[M + Cl]^-$ 628.2695 $[M + FA]^-$	5.2	18	582.2662(21.2); 462.2064(55.7); 436.2266(7.3); 342.1472(100); 316.1675(27.1); 299.1406(9.9); 145.0300(20.7); 119.0508(58.5)

Table 2. TOF-MS and TOF-MS/MS of tentatively identified lignanamides in the investigated hempseed fraction. RT = Retention Time; RDB = Ring Double Bond equivalent value. Compounds were numbered based on their RT in the whole LnHS chromatogram.

	RT (min)	Tentative Assignment	Formula	[M − H]⁻ calc. (m/z)	[M − H]⁻ Found (m/z)	Error (ppm)	RDB	MS/MS Fragment Ions (m/z) and Relative Intensity (%)
4	6.204	*N*-caffeoyltyramine/ *N*-caffeoyloctopamine dimer 1	$C_{34}H_{30}N_2O_9$	609.1879	609.1886	1.2	21	609.1909(63.7); 591.1788(7.0); 472.1058(14.2); 456.1103(100); 454.0944(8.2); 428.1152(14.4)
6	6.766	*N*-caffeoyltyramine/ *N*-caffeoyloctopamine dimer 2	$C_{34}H_{30}N_2O_9$	609.1879	609.1887	1.4	21	609.1897(5.5); 591.1799(18.4); 456.1102(100)
13	8.031	*N*-caffeoyltyramine dimer hydroxy derivative	$C_{34}H_{34}N_2O_9$	613.2192	613.2192	0.1	19	613.2192(56.0); 595.2080(82.8); 503.1798(9.9); 485.1700(35.8); 475.1864(44.5); 450.1552(5.1); 432.1146(72.5); 322.1071(84.9); 312.1227(100); 298.1065(5.2); 269.0809(65.2); 229.0500(30.0); 159.0448(10.1); 137.0248(85.3)
15	8.150	Cannabisin B	$C_{34}H_{32}N_2O_8$	595.2086	595.2095	1.5	20	595.2081(28.2); 485.1712(29.0); 432.1447(56.7); 430.1280 (7.5); 322.1076(60.3); 269.0815(100); 214.0507(7.4); 202.0513(3.9); 159.0450(7.9); 109.0299(7.5)
16	8.268	Cannabisin A	$C_{34}H_{30}N_2O_8$	593.1929	593.1941	2.0	25	593.1958(92.5); 575.1850(4.5); 473.1373(4.4); 456.1110(100); 430.1313(59.9); 428.1151(5.3); 349.0595(4.8); 336.0520(3.9); 322.0723(10.2); 293.0457(11.5); 263.0349(6.1)
19	8.840	Cannabisin H isomer 1	$C_{28}H_{31}NO_8$	508.1977	508.1968	1.8	14	460.1787(17.9); 312.1244(100); 311.1161(5.9); 297.1015(22.1); 195.0653 (6.2); 190.0506(8.1); 178.0511(23); 165.0557(25.4); 150.0327(19.9); 148.0531(22.8); 135.0457(6.2); 122.0363(5.1)
20	9.019	Cannabisin H isomer 2	$C_{28}H_{31}NO_8$	508.1977	508.1990	2.6	14	460.1781(6.5); 312.1248(100); 297.1009(26.3); 195.0669 (11.5); 190.0661(11.8); 178.0512(33.7); 165.0560(56.8); 150.0327(43.5); 148.0533(29.8); 135.0450(6.9)
21	9.118	Cannabisin I	$C_{26}H_{19}NO_7$	456.1089	456.1095	1.4	18	456.115 (100); 428.1167(6.0); 414.0996(16.1); 348.0529(25.2); 336.0527 (7.0); 320.0593 (7.0); 306.0416 (7.0); 293.0464 (13.1)
22	9.213	Cannabisin B isomer	$C_{34}H_{32}N_2O_8$	595.2086	595.2095	1.5	20	595.2089(33.6); 485.1708(22.8); 432.1449(53.7); 338.1032(1.2); 322.1077 (68.5); 295.0592(10.6); 269.0314(100); 254.0576(11.3); 214.0498(7.8); 202.0506(6.5); 159.0456(9.7); 147.0455(6.3); 109.0300(8.4)
23	9.622	Cannabisin C	$C_{35}H_{34}N_2O_8$	609.2242	609.2249	1.1	20	609.2267(83.6); 499.1891(41.8); 446.1622(100); 444.1453(18.2); 430.1304(30.9); 336.1241(57.4); 322.1030(12.7); 283.0977(20.3); 267.0661(14.5); 214.0507(21.7); 109.0296(15.9)

Table 2. *Cont.*

	RT (min)	Tentative Assignment	Formula	[M − H]⁻ calc. (*m/z*)	[M − H]⁻ Found (*m/z*)	Error (ppm)	RDB	MS/MS Fragment Ions (*m/z*) and Relative Intensity (%)
24	9.961	Cannabisin C isomer	$C_{35}H_{34}N_2O_8$	609.2242	609.2254	2.1	20	609.2290(36.5); 485.1757(35.7); 446.1642(100); 322.1097(67.8); 309.1002(6.1); 283.0976(12.8); 214.0517(5.6)
25	10.210	*N*-caffeoyltyramine/ *N*-feruloyltyramine dimer1	$C_{35}H_{34}N_2O_8$	609.2242	609.2255	1.9	20	609.2288(35.9); 446.1634(100); 431.1395(6.6); 322.1094(24.1); 310.1085(5.1); 283.0982(74.1); 268.0748(31.2); 267.0670(8.0)
26	10.407	*N*-caffeoyltyramine/ *N*-caffeoyloctopamine dimer2	$C_{34}H_{32}N_2O_9$	611.2035	611.2047	2	20	611.2067(1.5); 593.1973(1.2); 567.2161(1.3); 430.1310(3.0); 402.1355(1.1); 314.1044(31.8); 312.0881(6.2); 298.1088(100); 296.0931(58.5); 161.0241(8.4); 135.0450(13.7)
27	10.844	Cannabisin D	$C_{36}H_{36}N_2O_8$	623.2399	623.2414	2.4	20	623.2458(40.0); 460.1799(100); 458.1648(23.3); 445.1569(25.0); 444.1490(51.6); 443.1409(9.4); 350.1411(5.6); 336.1255(23.4); 322.1097(8.7); 283.0983(34.5); 282.0904(11.6); 267.0669(11.6); 214.0517(5.7)
28	11.023	3,3′-didemethylgrossamide	$C_{34}H_{32}N_2O_8$	595.2086	595.2110	4.0	20	595.2102(11.3); 458.1262(17.2); 432.1468(48.4); 338.1037(7.5); 295.0616(10.6); 269.0828(100); 147.0462(8.9); 121.0306(10.5)
29	11.319	Cannabisin D isomer	$C_{36}H_{36}N_2O_8$	623.2399	623.2408	1.5	20	623.2459(19.8); 499.1914(6.4); 487.1915(5.3); 460.1803(100); 445.1561(15.8); 339.1126(5.4); 336.1255(28.5); 322.1094(22.7); 283.0983(9.7); 216.0673(5.8); 123.0456(5.2)
30	11.676	Grossamide K	$C_{28}H_{29}NO_7$	490.1871	490.1875	0.8	15	472.1769(38.8); 460.1769(24.7); 457.1541(69.2); 445.1529(22.2); 442.1300(100); 440.1497(7.7); 430.1300(27.2); 414.1348(19.9); 338.1027(13.5); 323.0799(5.4); 308.1041(7.6); 297.1125(18.9); 293.0809(7.7); 283.0967(7.7); 276.0795(5.7); 267.0661(9.5)
31	11.715	Cannabisin E	$C_{36}H_{38}N_2O_9$	641.2505	641.2516	1.5	19	641.2540(10.4); 623.2435(16.7); 591.2164(6.6); 489.2053(35.3); 471.1576(6.3); 460.1779(27.4); 432.1824(18.4); 431.1986(43.3); 428.1508(7.4); 369.1455(8.5); 337.1191(12.9); 328.1191(33.8); 312.1241(93.4); 311.1396(21.4); 297.1003(10.2); 254.1183(5.8); 191.0349(6.7); 178.0508(8.0); 165.0555(14.8); 151.0404(100); 136.0170(34.8)
32	11.958	3,3′-demethyl-heliotropamide	$C_{34}H_{32}N_2O_8$	595.2086	595.2098	2.0	20	298.1083(100); 178.0503(2.8); 135.0444(5.7)

Table 2. *Cont.*

	RT (min)	Tentative Assignment	Formula	$[M-H]^-$ calc. (*m/z*)	$[M-H]^-$ Found (*m/z*)	Error (ppm)	RDB	MS/MS Fragment Ions (*m/z*) and Relative Intensity (%)
34	12.073	Cannabisin E isomer	$C_{36}H_{38}N_2O_9$	641.2505	641.2516	1.8	19	641.2555(12.3); 623.2452(10.6); 489.2064(42.0); 460.1795(45.6); 432.1839(12.1); 431.2002(37.9); 428.1527(6.0); 369.1470(9.8); 337.1203(15.8); 328.1202(13.5); 312.1253(85.8); 311.1405(20.9); 297.1015(8.4); 254.1193(5.8); 178.0510(7.4); 151.0406(100); 136.0170(36.8)
35	12.305	Demethylgrossamide	$C_{35}H_{34}N_2O_8$	609.2242	609.2252	1.6	20	609.2282(46.9); 472.1428(8.3); 446.1633(96.8); 431.1391(26.4); 283.0979(100); 282.0901(5.9); 268.0743(29.0)
36	12.721	*N*-caffeoyltyramine dimer (e.g., Cannabisin M)	$C_{34}H_{32}N_2O_8$	595.2086	595.2106	3.4	20	298.1088(100); 178.0505(2.6); 135.0457(5.4)
37	12.997	Cannabisin F	$C_{36}H_{36}N_2O_8$	623.2399	623.2416	2.7	20	623.2463(21.7); 486.1595(7.4); 460.1799(82.6); 445.1561(33.9); 430.1324(29.9); 297.1144(100); 296.1064(9.7); 282.0905(29.6); 267.0669(15.2)
38	13.251	*N*-caffeoyltyramine dimer (e.g., Cannabisin Q)	$C_{34}H_{32}N_2O_8$	595.2086	595.2094	1.4	20	298.1088(100); 178.0506(2.2); 135.0457(3.8)
39	14.178	Grossamide	$C_{36}H_{36}N_2O_8$	623.2399	623.2415 659.2180 $[M+Cl]^-$	2.6	20	623.2432(19.8); 486.1579(7.4); 460.1778(7.2); 445.1545(43.5); 430.1306(32.8); 297.1129(100); 296.1049(11.1); 282.0892(35.7); 267.0656(16.6)

Table 3. TOF-MS and TOF-MS/MS of tentatively identified flavonol glycosides in the investigated hempseed fraction. RT = Retention Time; RDB = Ring Double Bond equivalent value. Compounds were numbered based on their RT in the whole LnHS chromatogram.

	RT (min)	Tentative Assignment	Formula	$[M - H]^-$ calc. (m/z)	$[M - H]^-$ found (m/z)	Error (ppm)	RDB	MS/MS Fragment Ions (m/z) and Relative Intensity (%)
3	6.204	Quercetin pentoside	$C_{20}H_{18}O_{11}$	433.0776	433.0777	0.1	12	433.0823(9.8); 301.0362(29.4); 300.0273(100); 271.0250(69.4); 255.0304(31.7); 243.0304(12.2); 227.0358(5.3); 151.0036(10.3)
7	6.829	Quercetin-O-deoxyhexoside	$C_{21}H_{20}O_{11}$	447.0933	447.0937	0.9	12	447.0947(8.1); 301.0357(59.7); 300.0276(100); 271.0244(56.9); 255.0294(36.2); 243.0291(12.1); 227.0344(7.2); 178.9988(6.6); 151.0036(14.1)
9	7.157	Kaempferol pentoside 1	$C_{20}H_{18}O_{10}$	417.0827	417.0828	0.2	12	417.0837(16.3); 285.0406(14.1); 284.0328(80.4); 256.0378(8.1); 255.0299(100); 227.0352(77.8); 211.0398(5.4)
10	7.596	Kaempferol pentoside 2	$C_{20}H_{18}O_{10}$	417.0827	417.0832	1.1	12	417.0848(7.1); 285.0408(34.4); 284.0329 (78.6); 256.0373(7.7); 255.0303 (100); 229.0503(8.9); 227.0351(71.4); 183.0452(5.3)
12	8.011	Kaempferol-O-deoxyhexoside 1	$C_{21}H_{20}O_{10}$	431.0984	431.0984	0.1	12	431.1010(10.5); 285.0413(100); 284.0333(80.2); 257.0463(7.1); 256.0388(6.8); 255.0308(94.1); 229.0508(13.8); 227.0355(70.0); 211.0406(5.9); 183.0455(5.6)
14	8.032	Kaempferol-O-deoxyhexoside 2	$C_{12}H_{22}O_6$	431.0984	431.0990	1.5	12	431.0988(12.6); 285.0400(82.9); 284.0321(92.7); 272.9218(3.3); 257.0443(5.6); 256.0360(5.9); 255.0294(100); 239.0351(3.2); 229.0501(11.3); 227.0346(61.7); 211.0399(5.4); 197.0590(3.4); 187.0385(5.3); 183.0452(3.4); 169.0649(1.7); 163.0035(2.4); 159.0452(1.6)
18	8.561	Quercetin-7-O-acetyldeoxyhexose	$C_{23}H_{22}O_{12}$	489.1039	489.1040	0.3	13	489.1062(9.2); 429.0826(2.1); 409.1575(1.6); 397.1557(2.0); 315.0667(2.4); 301.0361(14.3); 300.0278(100); 299.0194(1.8); 271.0256(42.8); 269.0716(9.8); 255.0301(21.6); 243.0303(6.5); 227.0367(14.0); 178.9995(11.1); 151.0048(9.3)

This could be due to (NH_3 + CO) neutral loss eliminated in one step (as $HCONH_2$), or, more likely, eliminated in rapid sequential steps [28]. The N-CO α-cleavage, a characteristic fragmentation observed as a common pattern in all natural and synthetic amides [29], could drive the genesis of the ion at *m/z* 161.0244, whereas ions at *m/z* 135.0455 (base peak), 134.0373, and 133.0302, and 132.0217 were attributable to dihydroxycinnamoyl residue (Table 1). UV-DAD spectra of compound **1** showed absorption bands at 324, 296, and 216 nm, plus a shoulder at 242 nm. This latter, together with the broad band at 324 nm was attributable to aromatic moieties $\pi\rightarrow\pi^*$ transitions, whereas the $n\rightarrow\pi^*$ and $\pi\rightarrow\pi^*$ electronic transitions referred to the amidic group were responsible for the absorption at 296 and 216 nm. Compound **2** showed the [M − H]$^-$ ion at *m/z* 298.1086 according to the molecular formula $C_{17}H_{17}NO_4$. TOF-MS^2 spectrum suggested the occurrence of *N-p*-coumaroyloctopamine, before reported as an inducible phenolic amide in potato tuber tissue [30]. The deprotonated molecular ion gave rise to the ion [M − H − H_2O]$^-$ at *m/z* 280.0981, which in turn provided the ions at *m/z* 134.0607 and 119.0502. The ion at *m/z* 145.0293 confirmed coumaroyl moiety, as well as the fragment ion at *m/z* 160.0408, which could be from the cleavage at the N-Cα bond and the following 1,4 nucleophilic addition on the β'-carbon of the phenylpropanoid side chain. The presence of the coumaroyl moiety was further revealed through the UV-DAD spectrum of the compound, which is was similar to that previously reported for this hydroxycinnamoyl amide (HAA) [26] and it was consistent with the loss of catechol group as highlighted by the blue shift of the absorption band detectable at 324 nm in compound **1** (Figure S3).

A constitutional isomer of the previous compound was *N*-caffeoyltyramine, which was found to inhibit macrophage-mediated inflammatory responses through the suppression of the production of NO and pro-inflammatory cytokines [31]. This compound was tentatively identified under peak **5**. In this case, the [M − H]$^-$ ion at *m/z* 298.1090 dissociated providing the TOF-MS^2 fragment ion at *m/z* 135.0459 as base peak, likely corresponding to 2-hydroxy-4-vinylphenolate. Following the cleavage at the N-Cα bond, the fragment ion at *m/z* 178.0520 was formed, whereas the ion at *m/z* 190.0518 could be from Cα-Cβ bond breakdown [32]. The CH_2CO loss likely consisted in the fragment ion at *m/z* 256.0974 (Figure 2). In order to corroborate this latter hypothesis, hydrogen–deuterium (H/D) exchange reaction was carried out on pure *N*-caffeoyltyramine. The TOF-MS^2 spectra of the 3*d*-derived underwent rearrangement to give the ion at *m/z* 180.0619, whereas the loss of an ethen-1-one-2-d provided the fragment ion at *m/z* 258.1098. The electronic absorption spectrum of compound **5** (Figure 2C), similarly to that of compound **1**, with which shared the common caffeoyl component, evidenced the bathochromic and hyperchromic effect of -OH functional auxochromic groups of the double absorption band, which was at 317 and 294 nm, whereas the other band was at 237 nm. The compound was a constituent of other inestimable food sources such as hot pepper (*Capsicum annuum*), a spice used worldwide [33], Goji berry, the fruit of *Lycium barbarum* [34] and seeds of *Annona crassiflora* Mart., a fruitful tree native to the Brazilian Cerrado biome [35].

Compounds **8** and **17**, which showed the [M − H]$^-$ ion at *m/z* 312.1243 and 312.1241, respectively, were tentatively identified as *N*-feruloyltyramine geometrical isomers (Figure S4). The deprotonated molecular ion underwent in both TOF-MS^2 spectra methyl radical loss to achieve the fragment ion at *m/z* 297.1015(4), which gave rise to fragment ions at *m/z* 190.0510(3) and 178.0513, or more favorably led to ion at *m/z* 148.0530(5) (base peak) through CO-Cα' bond cleavage. The $CH_3^\bullet$ loss generated the radical ion at *m/z* 134.0373(6), as well as the ion at *m/z* 135.0452(6).

The [M − H]$^-$ ion at *m/z* 282.1140 for compound **11** was in accordance with the molecular formula $C_{17}H_{17}NO_4$. UV-DAD and TOF-MS/MS spectra were according to *N-p*-coumaroyltyramine, an antioxidant HAA compound with inhibiting effect on acetylcholinesterase, cell proliferation, platelet aggregation [36]. In particular, in TOF-MS/MS spectrum, beyond the fragment ion at *m/z* 162.0558, which was consistent with coumaramide, the abundance of 4-vinylphenolate further confirmed the acylic moiety identity (Figure S5). Finally, compound **33** was tentatively identified as tri-*p*-coumaroylspermidine (Figure S6). This latter, until now never reported among hemp seed constituents, was firstly reported to reduce the mycelial growth of the oat leaf stripe pathogen

m/z 595.2080 and 432.1446 could undergo phenyldihydronaphthalene moiety cleavage providing the ions at *m/z* 475.1864 and 312.1227, respectively. All the other detected fragment ions could be from 110, 163 or 120 Da losses. Furthermore, a dimer of *N*-caffeoyloctopamine and *N*-caffeoyltyramine was hypothesized to be compound **26**, whose deprotonated molecular ion at *m/z* 611.2047 gave rise the abundant ions at *m/z* 314.1044 and 298.1088, attributable to caffeoyloctopamine, and caffeoyltyramine, respectively (Figure S13).

H_2O and CH_2O losses were detectable in TOF-MS/MS spectra of peaks related to compounds **19** and **20**, which showed the deprotonated molecular ions at *m/z* 508.1968 and 508.1990, respectively, likely corresponding *erythro* and *threo*-diastereoisomers of cannabisin H [42]. The fragment ions at *m/z* 312.1244 and 312.1248, could be formed following the cleavage of β-aryl ether moiety, whereas the further methyl radical loss yielded the ions at *m/z* 297.1015 and 297.1009. Diagnostic ions at *m/z* 195.0653(69), likely α-hydroxyconiferyl alcohols, and the deformylation products at *m/z* 165.0557(60), appeared to support our hypothesis (Figure 4). The *erythro* diastereomer, together with grossamide K, was isolated for the first time as a phenolic constituent of the bark of the kenaf (*Hibiscus cannabinus* var. *Salvador*) [42].

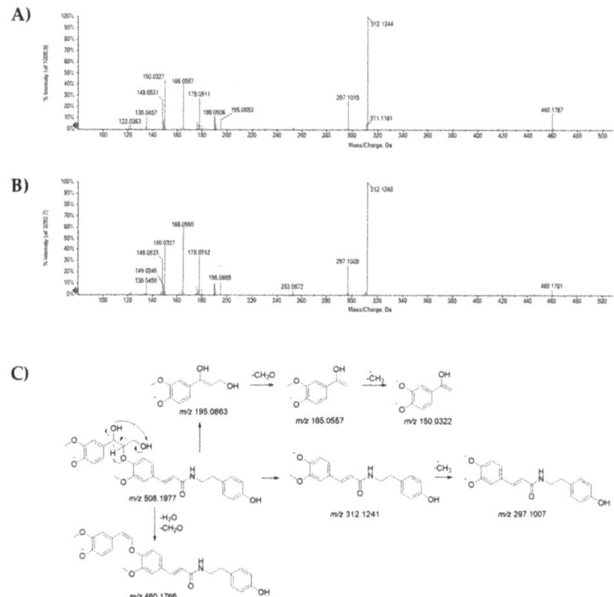

Figure 4. TOF-MS/MS spectra of compounds (**A**) **19** and (**B**) **20**, tentatively identified as cannabisin H isomers. The proposed fragmentation pathway of their [M − H]⁻ ion was reported (**C**); the theoretical *m/z* value is reported below each structure.

The deprotonated molecular ions of compounds **23–25** and **35** were in accordance with the molecular formula $C_{35}H_{34}N_2O_8$ and the occurrence of lignandiamides in which the hydroxycinnamoyl moieties were represented by caffeoyl and coniferyl alcohols, whereas tyramine constituted the amine part. In TOF-MS/MS spectra of compounds **23–25**, the base peak ion was at *m/z* 446.16, allowing us to confirm that the concurrent loss of isocyanic acid and *p*-hydroxystyrene could be advantageously observed as informative of tyramine presence. Moreover, other diagnostic fragment ions could be observed and differentiate the major bonding types encountered in hemp fruit lignandiamides. The aryldihydronaphtalene-type core likely characterized both compounds **23** and **24** which were distinguishable through fragment ions at *m/z* 499.1891 and 485.1757, respectively, which were in accordance with their relative catechol or guaiacol loss. This finding was in line monolignol cross-coupling with a caffeoyl-end-unit for compound **23** and a guaiacol-end unit for compound **24**. Based on previous observation, TOF-MS/MS spectrum of compound **35**, which appeared to fragment via a pathway similar to that of compound **28**, allowed us to hypothesize demethylgrossamide occurrence. In fact, also in this case, the [M − H]$^-$ ion underwent tyramine neutral loss with the genesis of the ion at *m/z* 472.1428, whereas the loss of 163.06 Da gave the most abundant ion at *m/z* 446.1633, which in turn lost a methyl radical, providing the radical anion at *m/z* 431.1391. Furthermore, as already observed in compound **28**, the loss of two 163.06 Da units could occur giving the ion at *m/z* 283.0979. This latter, which represented the base peak, also furnished the radical ion at *m/z* 268.0743 through methyl radical loss (Figure 5). UV-DAD spectra were in accordance with tentatively assigned lignandiamide skeleton, and represented a useful tool to unravel the lignan nucleus of compound **25**, which was supposed to belong to aryldihydronaphtalene class [26].

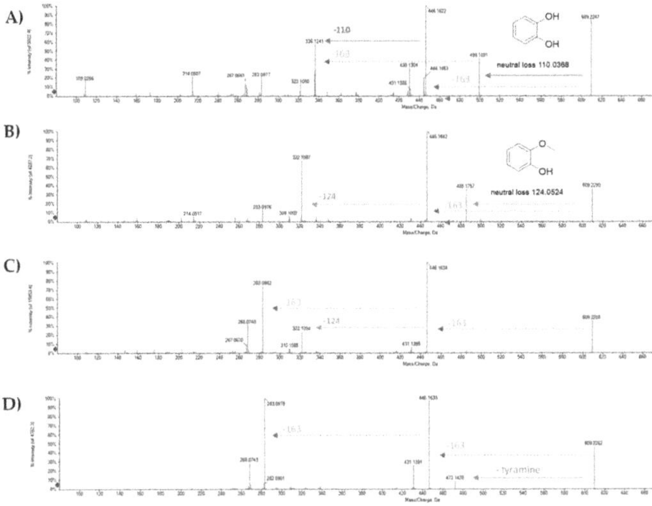

Figure 5. TOF-MS/MS spectra of compounds (**A**) **23**, (**B**) **24**, (**C**) **25**, and (**D**) **35**.

Based on previous MS observations, and UV-DAD spectra, compounds **27**, and **29**, whose pseudomolecular ions were in accordance with the $C_{36}H_{36}N_2O_8$ molecular formula, and showing the base peak at *m/z* 460.1799(1803), and its demethylated radical ion at *m/z* 445.1569(1), were suggested to be aryldihydronaphtalene-type lignanamide isomers, whereas compound **37** was tentatively identified as cannabisin F, and compound **39** was putatively assigned as grossamide (Figure S14). This latter compound, whose MS/MS fragmentation pattern in positive ion mode was previously reported [43], was found to exert anti-neuroinflammatory action, being able to inhibit the secretion of pro-inflammatory mediators (e.g., IL-6, and TNF-α), reducing LPS-mediated IL-6 and TNF-α mRNA levels [44]. Furthermore, neuroprotection by cannabisin F was ascertained against LPS-induced inflammatory response and oxidative stress in BV2 microglia cells [21].

Grossamide K, a phenylcoumaran-type lignanamide with previously reported antimelanogenic activity [45], was tentatively identified based on TOF-MS and TOF-MS/MS spectra related to peak **30**. The deprotonated molecular ion at *m/z* 490.1875 provided the fragment ions at *m/z* 472.1769 and 460.1769, following H_2O and formaldehyde neutral losses. Charge-driven CH_2O loss could be initiated when phenoxide ion abstracted the proton from aliphatic OH function. Both the ions underwent methyl radical loss to achieve ions at *m/z* 457.1541 and 445.1529, respectively, which, in turn, gave diradical anions at *m/z* 442.1300 (base peak) and 430.1300. This latter could provide the fragment ion at *m/z* 297.1125 through CO-Cα cleavage, or the ion at *m/z* 338.1027 by dehydrogenation and hydroxystirene loss. Methyl radical and CO losses were further detected (Figure S15).

Finally, compounds **31** and **34** were tentatively assigned as cannabisin E isomers based on their deprotonated molecular ion at *m/z* 641.2516, which provided the ion at *m/z* 489.2053(64) through 4-hydroxy-3-methoxy benzaldehyde loss. The detection of the ion at *m/z* 151.0404(6) as base peak, attributable to 4-formyl-2-methoxyphenolate, seemed to confirm the hypothesis (Figure S16).

2.3. HRMS Analysis: Flavonol Glycosides

Compounds **3**, **7**, **9**, **10**, **12**, **14** and **18** were putatively flavonol glycosides (Figures S17–S20, Table 3), whose presence was previously identified as bioactive components of hemp seeds, and its cold-pressed oil derived product [24]. In particular, collision-induced dissociation of the [M − H]⁻ ion at *m/z* 433.0777 for compound **3**, gave radical aglycone ion [aglycone − H]⁻ at *m/z* 300.0273 and its aglycone ion at *m/z* 301.0362, which were consistent with quercetin and the loss of a pentose moiety (−132.0415 Da). The presence of the pentose moiety was further suggested for compounds **9** and **10**, whose aglycone moiety was kaempferol, as suggested by the [aglycone − H]⁻ and aglycone radical ion at *m/z* 285.0406(08) and 284.0328(29), as well as the ions at *m/z* 255.0299(303) and 227.0352(1), which could come from the [aglycone − H]•⁻ ion. The loss of deoxyhexose moiety was shared by compounds **7**, **12** and **14**. Quercetin was identified as aglycone residue in compound **7**, whereas compounds **12** and **14** were kaempferol deoxyhexoside isomers. Finally, the detected loss of acetic acid moiety in the TOF-MS/MS spectrum of compound **18** as well as the occurrence of [aglycone–H]⁻, [aglycone − H]•⁻ and [aglycone − 2H]•⁻ at *m/z* 301.0361; 300.0278 and 299.0194, were in accordance with quercetin-7-O-acetyldeoxyhexose [46,47]. Quantifying flavonoid content as peak area percentage, it was found that the compounds constituted only the 2.8%, whereas lignanamides were the most representative compounds with the 79.0%. Compounds deriving from the coupling of two phenylamides, both preserving the amine moiety (e.g., metabolites showing the [M − H]⁻ at *m/z* 593, 595, 609, 623, 641), were the most abundant and they were present in comparable amount (Figure S21).

2.4. LnHS Inhibits Cell Survival of Glioblastoma U-87 Cell Line but not of HF

As some of compounds in LnHF were reported to exert neuroprotective and anti-neuroinflammatory effects [20,21], or also to be able to induce dramatic morphological changes and autophagic cell death [11], the effects of the treatment of LnHS hempseed mixture on primary glioblastoma cell lines was investigated. Indeed, a prior analysis of the effects of LnHS was carried out on undifferentiated SH-SY5Y cells. It was evidenced that LnHS at dose levels higher than 25 µg/mL, induced clear cell morphological changes as cells shrunk and loss cell adhesion (Figure S2). Thus, in order to better understand and define a preliminary LnHS-cell death inducing mechanism, the effects on cytotoxicity was further assessed on U-87 glioblastoma cells, which are characterized by a rapid migration, and a highest invasion capacity [48]. Human Fibroblast (HF) cells were used in order to evaluate LnHS effects on non-tumorigenic cells. MTT assay was performed to assess the effects of LnHS on cell viability, at seven dose levels (0.5, 1.25, 2.5, 5, 10, 25, 50 µg/mL), and at three exposure times (24 h, 48 h and 72 h). Data acquired showed that LnHS exerted toxic effects on U-87 cells. Indeed, U-87 cell viability was strongly compromised at the highest LnHS tested dose, at each exposure time considered (Figure 6). The dose level 25 µg/mL significantly affects the U-87 viability after 48 and 72 h of incubation, $p < 0.05$ (Figure 6A). LnHS did not show relevant toxic effects on HF cells, and only the highest dose (50 µg/mL) appeared to induce, after the longer 72 h exposure time, cell viability decrease (Figure 6B). In addition, the quantification of lactate dehydrogenase levels, considered an indicator of cell damage for necrotic cell death [49], evidenced that LnHS 25 µg/mL and 50 µg/mL dose levels exerted a time-dependent percentage increase of LDH release (Figure 6C,D). Furthermore, the colony formation assay [50] showed that LnHS-treated U-87 cells did not retain the capacity to produce colonies (Figure 6E), when LnHS dose exceeded only 5 µg/mL.

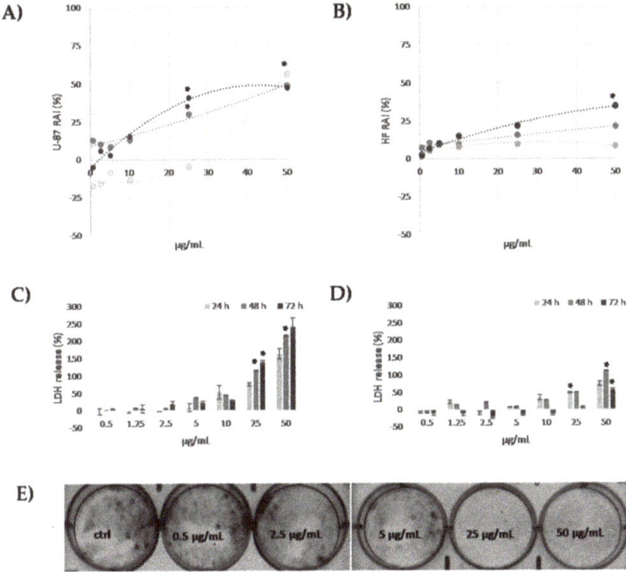

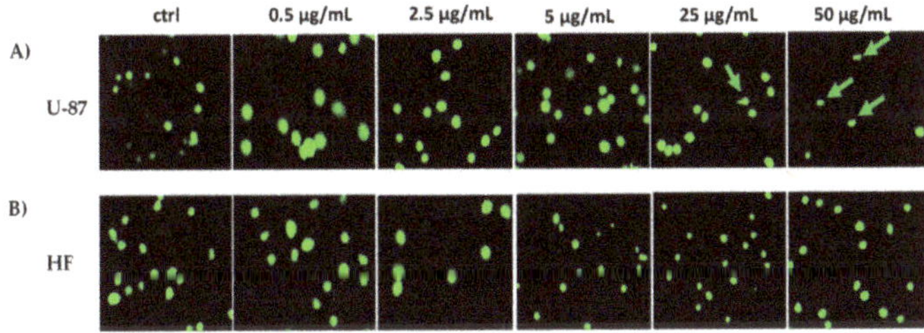

Figure 6. Cell viability of U-87 and human HF cell (**A** and **B**, respectively) was assessed by MTT assay after 24, 48 and 72 h of exposure. Data from LDH release assay at 24, 48 and 72 h exposure times were in panels **C** (U-87 cells) and **D** (HF cells). Values are the mean ± SE of two independent experiments performed in triplicate. $^{*}p < 0.05$ vs. untreated cells. (**E**) Representative images from colony forming efficiency of U-87 cells grown in presence of LnHS or vehicle control for ten days; the experiment was performed in duplicate.

2.5. LnHS Induced Genomic DNA Damage in U-87 Cell Line but not in HF

In order to verify if the observed cytotoxicity was the result of LnHS ability to induce a genotoxic damage, comet assay was performed on U-87 and HF cells, after 24 h of exposure to the hempseed mixture at 0.5, 2.5, 5, 25 and 50 µg/mL concentration levels (Figure 7). U-87 cells showed DNA damage when treated with LnHS 25 and 50 µg/mL (Figure 7A), whereas no genotoxic effects was detectable in human fibroblast cells.

Figure 7. Representative images of U-87 and HF cells treated with different LnHS doses and subjected to the comet assay (panels **A** and **B** respectively). ctrl: untreated cells; green arrows indicate comets with a tail. Experiments were performed in duplicate.

2.6. LnHS Inhibited Cell Migration of U-87 Cell Line

As invasiveness is one of the pathophysiological features of human malignant gliomas [51], the effects of LnHS on the U-87 cells migration, in comparison to HF, were tested through wound healing assay. Both the cell types were subjected to scraped wounds: U-87 cells were treated with five doses (0.5, 2.5, 5, 25 and 50 µg/mL), whereas HF cells with three LnHS doses (5, 25 and 50 µg/mL). Although LnHS significantly inhibited the migration of both cell types at the highest dose, U-87 appeared more sensitive to LnHS (Figure 8). In fact, already at a dose of 5 µg/mL, the hempseed fraction inhibited the migration of the U-87 cells; the maximum effect was evident at 50 µg/mL, where the cell toxicity is also evident.

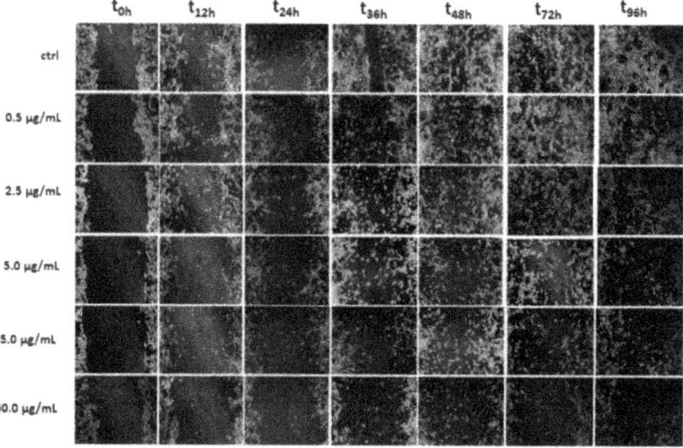

Figure 8. U-87 cells underwent a scraped wound and were then treated with different LnHS doses (0.5, 2.5, 5, 25 and 50 µg/mL). Cells were photographed immediately following the scratch (0 h), after 12, 24, 36, 48, 72 and 96 h. Untreated cells (ctrl), were used as a control. Representative figures are shown from one of two independent experiments.

2.7. LnHS did not Modify Sirtuins and Cytokines Expression in both U-87 and HF Cells

To verify LnHS ability to display cytotoxicity, interfering oxidative and inflammation processes, the expression of Sirt1 and Sirt2, as well as of some pro-inflammatory cytokines involved, was evaluated. In this context, cannabisin F was recently found to act as a modulator of SIRT1, whose activation is considered beneficial in attenuating neuro-inflammation and damage due to oxidative stress [21]. Thus, both U-87 and HF cells were treated with three LnHS doses (2.5, 25 and 50 µg/mL) for 24 and 48 h, and the expression level of Sirt1, Sirt2, IL-6 and IL-10 was assessed at mRNA level. Indeed, no effects were found on these markers in both the cell types (Figure S22).

2.8. LnHS Blocks Autophagy While Inducing Apoptosis in U-87 Cells

Autophagic cell death was hypothesized to be the mechanism through which cannabisin B exerts antiproliferative activity in human hepatocarcinoma HepG2 cells [11]. Autophagy is a cell catabolic program, and it is triggered following nutrient starvation, and requires for its initiation ULK kinases [52]. In particular, ULK-1 phosphorylates Beclin-1 is a multi-domain protein, which exerts a dual role in autophagy and apoptosis cellular processes [53]. LnHS treatment at dose levels equal to 5, 25 and 50 µg/mL, at 24 and 48 h exposure times, promoted the down-regulation of Beclin 1 and ULK-1 in U-87 cells, while in fibroblasts, both the enzymes were down-regulated only by 50 µg/mL dose (Figure 9). Simultaneously, it was found that Bcl-2 was also inhibited at the same incubation times and LnHS doses (Figure 9). Altogether, these results indicated a strong down-regulation of autophagy process in U-87 cells. These two processes are accompanied by reduction in AKT phosphorylation. E-cadherin expression was affected at 50 mL at 24 and 48 h in both cell types. This finding is in line with a potential double-edge behavior of LnHS, which, according to autophagy occurrence, at low doses appears to maintain E-cadherin expression, while, on the contrary, at the highest dose, LnHS may determine a cell-cell adhesion injury by the E-cadherin loss (Figure 9).

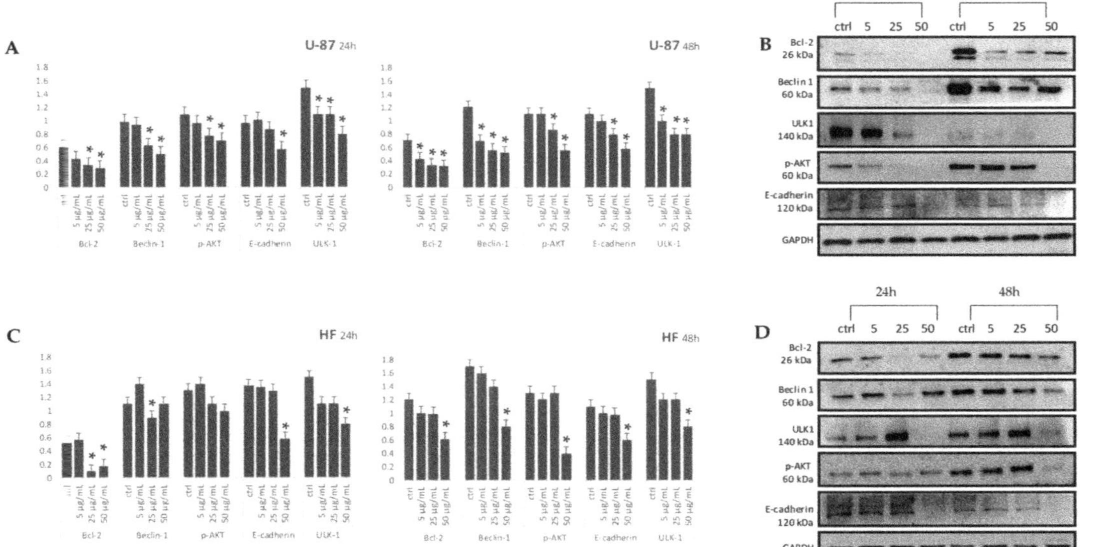

Figure 9. Levels of Bcl-2, Beclin-1, p-AKT, E-cadherin and ULK-1 in U-87 and HF cell types were detected using western blotting with respective antibodies. Graphical representation of pixel quantization of Bcl-2, Beclin-1, p-AKT, E-cadherin, and ULK-1 normalized to GAPDH (panels **A**, **C**). The intensities of signals were expressed as arbitrary units. *$p < 0.05$ vs. untreated cells (ctrl). Representative western blotting image of Bcl-2, Beclin-1, p-AKT, E-cadherin, ULK-1 and GAPDH are in panels **B** (U-87 cell type), and **D** (HF cell type).

3. Materials and Methods

3.1. Materials

All the solvents used for extraction and fractionation purposes, acetonitrile (LC–MS grade), formic acid (98%, for mass spectrometry) were purchased from Sigma-Aldrich (Buchs, Switzerland). Cell culture media and reagents for cytotoxicity testing were purchased from Invitrogen (Paisley, Scotland, UK), except MTT [3-(4,5-dimethyl-2-thiazolyl)-2,5-diphenyl-2H-tetrazolium bromide] which was from Sigma-Aldrich Chemie GmbH (Schnelldorf, Germany).

3.2. Plant Extraction and Fractionation

One hundred g of commercial hemp seeds (*Hanf & Natur* product, by Bioland, a certified company for the cultivation of ecological and sustainable hemp located in Lindlar, Germany), after mechanical reduction, underwent ultrasound assisted maceration (UAM) using sequentially *n*-hexane and ethanol as extracting solvents. The drug/solvent ratio used was 1:5. Ultrasound (mechanic waves able to propagate through an elastic medium) maceration cycles were in total 4 and performed using a Bransonic^{TM} M3800-E device (Branson Ultrasonics^{TM}, Danbury, CT, USA) operating in sweep-frequency mode at 40 kHz. The duration of each ultrasound cycle was 30 min. The alcoholic fraction (4.4 g) was then chromatographed through SiO_2 column chromatography (h 16 cm, Ø 2.0 cm), eluting with chloroform, ethyl acetate and methanol providing three subfractions. The methanol extract, named LnHS, underwent chemical investigation by means of UHPLC-HR-MS/MS and HPLC-UV-DAD analyses and cytotoxicity assessment towards U-87 human glioblastoma cells. The extraction/fractionation scheme is depicted in Figure S1.

3.3. UHPLC-ESI-QqTOF-MS/MS and HPLC-UV-DAD Analyses

LnHS was chemically characterized by ultrahigh-performance liquid chromatography coupled with high-resolution mass spectrometry (UHPLC-HRMS) techniques. A NEXERA UHPLC system (Shimadzu, Tokyo, Japan) was used with a Luna$^®$ Omega Polar C-18 columns (1.6 μm particle size, 50 × 2.1 mm, Phenomenex, Torrance, CA, USA). Separation was achieved with a linear gradient of water (A) and acetonitrile (B), both with 0.1% formic acid: 0–3 min, 2→12.5% B; 3–12.5 min, 12.5→30% B; 12.5–17.5 min, 30→45% B; 17.5–20 min, 45→75% B; held at 75% B for other 2 min; 22–23 min, 75→98% B. The mobile phase composition was maintained at 98% B for another 1 min, then returned to the starting conditions and allowed to re-equilibrate for 1 min. The total analysis time was 26 min, the flow rate was 0.5 mL/min, and the injection volume was 2.0 μL.

MS analysis was performed using a hybrid Q-TOF MS instrument, the AB SCIEX Triple TOF$^®$ 4600 (AB Sciex, Concord, ON, Canada), equipped with a DuoSpray^{TM} ion source (consisting of both electrospray ionization (ESI) and atmospheric pressure chemical ionization (APCI) probes), which was operated in the negative ESI mode. The APCI probe was used for automated mass calibration using the Calibrant Delivery System (CDS). The CDS injects a calibration solution matching the polarity of ionization and calibrates the mass axis of the TripleTOF$^®$ system in all scan functions used (MS and/or MS/MS). The Q-TOF HRMS method, which combines TOF-MS and MS/MS with Information Dependent Acquisition (IDA) for identifying non-targeted and unexpected compounds, consisted of a full scan TOF survey (dwell time 100 ms, 100–1000 Da) and a maximum number of eight IDA MS/MS scans (dwell time 50 ms, 80–850 Da). The MS parameters were as follows: curtain gas (CUR) 35 psi, nebulizer gas (GS 1) 60 psi, heated gas (GS 2) 60 psi, ion spray voltage (ISVF) 4.5 kV, interface heater temperature (TEM) 600 °C, declustering potential (DP) −80 V. Collision Energy (CE) applied was −45 V with a collision energy spread (CES) of 15 V. The instrument was controlled by Analyst$^®$ TF 1.7 software, while data processing was carried out using PeakView$^®$ software version 2.2. Hydrogen/deuterium exchange experiments were carried out on pure *N*-caffeoyltyramine as previously described [54]. Furthermore, in order to achieve UV-DAD information of each peak, separation was also performed by using a 1260 Infinity II LC System (Agilent, Santa Clara, CA, USA) equipped with an Agilent G711A quaternary pump and a WR G7115A diode array detector.

3.4. Cell Culture and Cytotoxicity Assessment

The human glioblastoma U-87 cell line and human fibroblasts (HF) were kindly provided by the Bank of Human and Animal Continuous Cell Lines-CEINGE Biotecnologie Avanzate – Napoli - Italy. The U-87 cell line was cultured in Dulbecco's medium (DMEM) supplemented with 10% of heat-inactivated Fetal Bovine Serum (FBS) and 1% L-glutamine (Sigma-Aldrich, St. Louis. MO, USA); HF were cultured in cultured in Dulbecco's medium (DMEM) supplemented with 20% of heat-inactivated Fetal Bovine Serum (FBS) and 1% L-glutamine (Sigma-Aldrich). Both cells were grown in a 5% CO_2 humidified incubator, at 37 °C. For treatments, the cells were incubated with LnHS at different concentrations in serum-free fresh medium for different incubation times.

3.4.1. MTT Cell Viability Assay

U-87 and HF cells were seeded in 10% FBS-containing medium in a 96-well plate at the density of 3×10^3 cells/well. The day after, the cells were treated as above described. Cell viability was evaluated by the 3-(4,5-dimethylthiazol-2-yl)-2,5-diphenyltetrazolium bromide (MTT; Sigma-Aldrich) assay. The next day, cells were treated as above described and, after incubation times, the assay was performed as previously described [55].

3.4.2. LDH Release Assay

The lactate dehydrogenase (LDH) leakage assay is a colorimetric test useful for quantifying cell death and lysis through the measurement of LDH released from the cytosol of damaged cells, which was into supernatant. The test was carried out as previously reported [56].

3.4.3. Colony Forming Assay

For colony forming assays, U-87 cells (1×10^3) were seeded per well of a 6-well plate and maintained for 10 days with medium changed every other day. Colonies were stained using crystal violet (Sigma-Aldrich) at room temperature for 20 min and washed repeatedly in water. Colonies were counted manually using a light microscope as previously described [57].

3.4.4. Comet Assay

DNA breakage was evaluated using a comet assay kit (Trevigen, Gaithersburg, MD, USA). Three independently reproduced experiments were performed. To determine DNA strand breakage, U-87 cells and HF were plated at 1×10^5 cells/mL in a 24-well plate and left to attach for 24 h. The next day, cells were treated with LnHS (0.2, 2.5, 5, 25, 50 µg/mL) for 24 h. Comet slides were stained with diluted SYBR green and examined under an automated robotic epifluorescence microscope with an excitation filter (510 to 550 nm). One hundred cells were analyzed per slide. Experiments were performed in duplicate. Untreated cells served as negative controls.

3.4.5. Wound Healing Assay

U-87 cells and HF were seeded at a density of 3×10^5 cells in a 6-well plate in complete culture media and grown to confluence. The day after, cells were treated with 4 µg/mL of mitomycin (Sigma-Aldrich) for 2 h to inhibit cell proliferation and then a wound was inflicted using a tip. After washing with phosphate-buffered saline (PBS), cells were incubated as above mentioned in comparison to untreated cells. The scratch wound was observed and photographed at different time points using an inverted-phase-contrast microscope (TS100 fluorescence microscope and video camera, Nikon, Tokyo, Japan). Three measurements per scratch were performed (two replicates/condition, experiments performed in duplicate).

3.4.6. RNA Extraction and Real Time Quantitative PCR

U-87 and HF cells, after 12 h starvation, were treated in 0% FBS medium as above described. After incubation, total RNA was isolated from using TRIzol (Invitrogen, Carlsbad, CA, USA) and Real-time PCR was performed as previously described using GAPDH as housekeeping gene [58]. The primers for Sirt1, Sirt2, IL-6, IL-10 and GAPDH are available on request. The experiments were performed two times in triplicate.

3.4.7. Preparation of Cell Extracts and Western Blotting Analysis

U-87 cells and HF cells, after 12-h starvation, were treated in 0% FBS medium as above described. After incubation, total proteins, extracted in RIPA buffer, were quantified by Bradford's method (Bio-Rad, Hercules, CA, USA). Total proteins were subjected to electrophoresis and transferred to PVDF membranes; the membranes were incubated with the following antibodies according to the manufacturer's instructions: E-cadherin and GAPDH (Santa Cruz Biotechnology, Dallas, TX, USA), p-AKT, ULK-1, Bcl-2, Beclin 1 (Cell Signaling Technology, ZA Leiden, Netherlands). The blots were developed by ECL (Amersham Biosciences, Piscataway, NJ, USA) and analyzed by densitometry as previously described [58]. Each sample was tested three times in duplicate.

3.5. Statistical Analysis

Experiments were performed three times with replicate samples, except where otherwise indicated. Data are expressed as mean ± SD (standard deviation). The means were compared using analysis of variance (ANOVA) plus Bonferroni's t-test. A p-value of < 0.05 was considered to indicate a statistically significant result.

4. Conclusions

The recovery of bioactive compounds (pure or in their mixture form) from hempseed meal could be the driving force for developing new hemp seed-based goods, in which the processing of hemp fruits waste is valuable for fully exploiting the innumerable advantages of this crop [59,60]. In this context, high resolution negative tandem mass spectrometric techniques could be a useful tool in the structure elucidation of hemp seeds compounds such as phenylamides and lignanamides.

The main phenylamides in hemp seeds are hydroxycinnamoyl amides in which the amine moiety was octopamine or tyramine. They are readily differentiable based on their TOF-MS^2 spectra as the octopamine conjugate promptly lost a water molecule and the base peak corresponds to a hydrocinnamoyl moiety. Caffeoyl and feruloyl-derived phenyldihydronaphthalene lignanamides are distinguishable by a characteristic loss of 110 and 124 Da, whereas phenylcoumaran lignandiamide were characterized by a first direct tyramine moiety. Moreover, isocyanic acid (HNCO) + p-hydroxystyrene represent a common neutral loss for all the identified lignanamides. The presence of an α,β-unsaturated function favors a facile CO-Cα cleavage, and low fragment ions give additional structural information on the investigated lignanamides. Cytotoxicity assessments of LnHS on the U-87 glioblastoma cell line and on human fibroblasts provided new insight into the molecular effects of this lignanamide extract. Indeed, even if further studies are necessary, our data strongly suggest that LnSH negatively and specifically regulates U-87 cell line survival and migration then reinforcing the need to fully analyze its biochemical behavior, and lignanamides purified therefrom. Indeed, the compounds' purification will be addressed in order to deeply investigate their quantitation in hemp seeds products from the different cultivars available on the market, as well as to achieve a clearer picture of their already promising anti-cancer activity. In this context, the ability of LnHS (and pure compounds therefrom) to cross the blood brain barrier (BBB) is aimed to be promptly pursued. This is in line with several recent evidences which highlight the ability of dietary polyphenols, and their known physiologically relevant metabolites, to enter the brain endothelium, cross the BBB and to impact brain health and cognition [61–63]. The BBB-crossing feature in phenylamides and lignanamides could be improved by the intrinsic amide function occurrence. In fact, recent studies showed that the introduction of an amide function was a strategy to enhance BBB transport of antineoplastic drug [64], whereas the development of *N*-acetylcysteine amide (NACA) preserved *N*-acetylcysteine antioxidant ability improving its permeability through cell membranes [65]. Furthermore, considering the identified compounds mainly as polyamine derivatives, it was recently shown that tyramine analogue in *Gingko biloba* extract were identified among compounds able to cross BBB [66].

Supplementary Materials: The following are available online: Figure S1. Simplified extraction and fractionation scheme of cryo-crushed hempseeds. Figure S2. Morphological changes in SH-SY5Y cells treated with LnHS fraction in respect to untreated cells. Representative images were acquired by Inverted Phase Contrast Brightfield Zeiss Primo Vert Microscope. Figure S3. A) TOF-MS/MS spectrum; B) proposed fragmentation pathway of the [M − H]⁻ ion; C) UV-DAD spectrum for compound **2**. In B panel, the theoretical *m/z* value is reported below each structure. Figure S4. A) Extracted ion chromatogram (XIC) of the [M − H]⁻ ion at *m/z* 312.124 ±0.025; TOF-MS/MS spectra of B) compound **8**, and C) compound **17**. In the grey panel the tentatively proposed fragmentation pathway of the [M − H]⁻ ions. Representative UV-DAD spectrum, acquired under peak **17**, is reported in panel D. Figure S5. A) TOF-MS spectrum for compound **11**; B) TOF MS/MS spectrum of the [M − H]⁻ ion; C) UV-DAD spectrum. Figure S6. A) TOF-MS/MS spectrum and B) proposed fragmentation pathway of the [M − H]⁻ ion for compound **33**; theoretical *m/z* values are reported below each structure. Figure S7. Compound 16 A) UV-DAD spectrum; B) TOF-MS/MS spectrum (in grey panel TOF-MS spectrum is reported. C) Proposed fragmentation pathway of the [M − H]⁻ ion (theoretical *m/z* values are reported below each structure). Figure S8. Proposed fragmentation pathway of the [M − H]⁻ ion relative to cannabisin B isomers (theoretical *m/z* values are reported below each structure). Figure S9. A) Proposed fragmentation pathway of the [M − H]⁻ ion of the compound **28** (theoretical *m/z* values are reported below each structure); B) UV-DAD, and C) TOF-MS/MS spectra. Figure S10. TOF-MS/MS spectra of compounds **32**, **36** and **38** (A, B, and C, respectively). Proposed fragmentation pathway of the [M − H]⁻ ion of the compound **32** is reported in grey panel, whereas that of **36** and **38** is in light green panel (theoretical *m/z* values are reported below each structure). Figure S11. TOF-MS/MS spectra of compounds **4** and **6** (A, and B, respectively). Proposed fragmentation pathway of the [M − H]⁻ ion of both the compounds are reported in grey panels (theoretical *m/z* values are reported below each structure). Figure S12. TOF-MS/MS spectrum (A) and proposed fragmentation pathway of the [M − H]⁻ ion of compound **13** (theoretical *m/z* values are reported below each structure). Figure S13. TOF-MS/MS and proposed fragmentation pathway of the [M − H]⁻ ion for compound **26**. The theoretical *m/z* value is reported below each structure. Figure S14. TOF-MS/MS spectra of compounds A) **27**, B) **29**, C) **37**, and D) **39**. Figure S15. A) TOF-MS/MS spectrum of compound **30**, and B) proposed fragmentation pathway of its [M − H]⁻ ion (theoretical *m/z* values are reported below each structure). Figure S16. Proposed fragmentation pathway of the [M − H]⁻ ion for compounds **31** and **34**. Figure S17. Flavanol glycosides in LnHS hempseed fraction. Figure S18. TOF-MS/MS spectra of quercetin derivatives A) **3**, B) **7** and C) **18**. Figure S19. A) Extracted ion chromatogram (XIC) of the [M − H]⁻ ion at *m/z* 417.083 ±0.025; TOF-MS/MS spectra of B) compound **9**, and C) compound **10**. Figure S20. TOF-MS/MS spectra of kaempferol derivatives A) **12**, and B) **14**. Figure S21. A) Relative content of each class of the tentatively identified compounds: HAAs – hydroxycinnamoyl amides; LnAs – lignanamides; Fls – Flavonols; B) relative content of lignanamides sharing a common [M − H]⁻ ion in LnHS fraction. Figure S22. A) mRNA expression of Sirt1 and Sirt2 by real-time; B) mRNA expression of IL6 and IL10 by real-time in U-87 and HF cells after 24 and 48 h exposure times.

Author Contributions: Conceptualization, S.P. (Severina Pacifico); methodology, E.N., S.P. (Simona Piccolella), M.F., M.T.P. and G.C.; formal analysis, S.P. (Simona Piccolella), M.M., M.F., M.T.P., and G.C.; data curation, E.N., M.F., S.P. (Severina Pacifico); writing—original draft preparation, S.P. (Severina Pacifico) and E.N.; writing—review and editing, M.F, A.D., S.P. (Severina Pacifico); supervision, A.D and S.P. (Severina Pacifico). All authors have read and agreed to the published version of the manuscript.

Funding: This research received no external funding.

Acknowledgments: This research was supported by grant L.R. n. 5/2017. DRD n. 92 del 21.06.2018 ("Progetto per la Rivalutazione Olistica della canapa oltre il PIL", PROHEMPIL PROJECT) from the Campania Region (Italy).

Conflicts of Interest: The authors declare no conflict of interest.

References

1. Adefegha, A. Functional Foods and Nutraceuticals as Dietary Intervention in Chronic Diseases; Novel Perspectives for Health Promotion and Disease Prevention. *J. Diet. Suppl.* **2017**, *15*, 977–1009. [CrossRef] [PubMed]

2. Piccolella, S.; Crescente, G.; Candela, L.; Pacifico, S. Nutraceutical polyphenols: New analytical challenges and opportunities. *J. Pharm. Biomed. Anal.* **2019**, *175*, 112774. [CrossRef]

3. Piccolella, S.; Pacifico, S. Plant-Derived Polyphenols: A Chemopreventive and Chemoprotectant Worth-Exploring Resource in Toxicology. In *Advances in Molecular Toxicology*; Fishbein, J.C., Heilman, J.M., Eds.; Elsevier: Cambridge, MA, USA, 2015; Volume 9, pp. 161–241.

4. García, C.R.; Sánchez-Quesada, C.; Toledo, E.; Rodríguez-Delgado, M.; Gaforio, J.J. Naturally Lignan-Rich Foods: A Dietary Tool for Health Promotion? *Molecules* **2019**, *24*, 917. [CrossRef] [PubMed]

5. Gaya, P.; Medina, M.; Sanchez, A.; Landete, J.M. Phytoestrogen Metabolism by Adult Human Gut Microbiota. *Molecules* **2016**, *21*, 1034. [CrossRef]

6. Zálešák, F.; Bon, D.; Pospisil, J. Lignans and Neolignans: Plant secondary metabolites as a reservoir of biologically active substances. *Pharmacol. Res.* **2019**, *146*, 104284. [CrossRef] [PubMed]

7. Kajla, P.; Sharma, A.; Sood, D.R. Flaxseed—a potential functional food source. *J. Food Sci. Technol.* **2014**, *52*, 1857–1871. [CrossRef] [PubMed]

8. Sicilia, T.; Niemeyer, H.B.; Honig, D.M.; Metzler, M. Identification and Stereochemical Characterization of Lignans in Flaxseed and Pumpkin Seeds. *J. Agric. Food Chem.* **2003**, *51*, 1181–1188. [CrossRef]

9. André, C.M.; Hausman, J.-F.; Guerriero, G. Cannabis sativa: The Plant of the Thousand and One Molecules. *Front. Plant Sci.* **2016**, *7*, 1. [CrossRef]

10. Crescente, G.; Piccolella, S.; Esposito, A.; Scognamiglio, M.; Fiorentino, A.; Pacifico, S. Chemical composition and nutraceutical properties of hempseed: an ancient food with actual functional value. *Phytochem. Rev.* **2018**, *17*, 733–749. [CrossRef]

11. Chen, T.; Hao, J.; He, J.; Zhang, J.; Li, Y.; Liu, R.; Li, L. Cannabisin B induces autophagic cell death by inhibiting the AKT/mTOR pathway and S phase cell cycle arrest in HepG2 cells. *Food Chem.* **2013**, *138*, 1034–1041. [CrossRef]

12. Moccia, S.; Siano, F.; Russo, G.L.; Volpe, M.G.; La Cara, F.; Pacifico, S.; Piccolella, S.; Picariello, G. Antiproliferative and antioxidant effect of polar hemp extracts (Cannabis sativa L., Fedora cv.) in human colorectal cell lines. *Int. J. Food Sci. Nutr.* **2019**, 1–14. [CrossRef] [PubMed]

13. Gallagher, R.S.; Ananth, R.; Granger, K.; Bradley, B.; Anderson, J.; Fuerst, E.P. Phenolic and Short-Chained Aliphatic Organic Acid Constituents of Wild Oat (*Avena fatua* L.) Seeds. *J. Agric. Food Chem.* **2010**, *58*, 218–225. [CrossRef]

14. Zhang, B.; Huang, R.; Hua, J.; Liang, H.; Pan, Y.; Dai, L.; Liang, D.; Wang, H.-S. Antitumor lignanamides from the aerial parts of Corydalis saxicola. *Phytomedicine* **2016**, *23*, 1599–1609. [CrossRef] [PubMed]

15. Li, C.-X.; Song, X.-Y.; Zhao, W.-Y.; Yao, G.-D.; Lin, B.; Huang, X.-X.; Li, L.-Z.; Song, S.-J. Characterization of enantiomeric lignanamides from *Solanum nigrum* L. and their neuroprotective effects against MPP+-induced SH-SY5Y cells injury. *Phytochemistry* **2019**, *161*, 163–171. [CrossRef] [PubMed]

16. Sun, J.; Gu, Y.-F.; Su, X.-Q.; Li, M.-M.; Huo, H.-X.; Zhang, J.; Zeng, K.-W.; Zhang, Q.; Zhao, Y.-F.; Li, J.; et al. Anti-inflammatory lignanamides from the roots of Solanum melongena L. *Fitoterapia* **2014**, *98*, 110–116. [CrossRef]

17. Zhang, J.-X.; Guan, S.-H.; Feng, R.-H.; Wang, Y.; Wu, Z.-Y.; Zhang, Y.-B.; Chen, X.-H.; Bi, K.-S.; Guo, D.-A. Neolignanamides, Lignanamides, and Other Phenolic Compounds from the Root Bark of *Lycium chinense*. *J. Nat. Prod.* **2013**, *76*, 51–58. [CrossRef]

18. Zhang, J.; Guan, S.; Sun, J.; Liu, T.; Chen, P.; Feng, R.; Chen, X.; Wu, W.-Y.; Yang, M.; Guo, D.-A. Characterization and profiling of phenolic amides from Cortex Lycii by ultra-high performance liquid chromatography coupled with LTQ-Orbitrap mass spectrometry. *Anal. Bioanal. Chem.* **2014**, *407*, 581–595. [CrossRef]

19. Zhu, G.-Y.; Yang, J.; Yao, X.; Yang, X.; Fu, J.; Liu, X.; Bai, L.-P.; Liu, L.; Jiang, Z.-H. (±)-Sativamides A and B, Two Pairs of Racemic Nor-Lignanamide Enantiomers from the Fruits of Cannabis sativa. *J. Org. Chem.* **2018**, *83*, 2376–2381. [CrossRef]

20. Yan, X.; Tang, J.; Passos, C.D.S.; Nurisso, A.; Simoes-Pires, C.; Ji, M.; Lou, H.; Fan, P. Characterization of Lignanamides from Hemp (*Cannabis sativa* L.) Seed and Their Antioxidant and Acetylcholinesterase Inhibitory Activities. *J. Agric. Food Chem.* **2015**, *63*, 10611–10619. [CrossRef]

21. Wang, S.; Luo, Q.; Fan, P. Cannabisin F from Hemp (*Cannabis sativa*) Seed Suppresses Lipopolysaccharide-Induced Inflammatory Responses in BV2 Microglia as SIRT1 Modulator. *Int. J. Mol. Sci.* **2019**, *20*, 507. [CrossRef]

22. Diao, W.; Tong, X.; Yang, C.; Zhang, F.; Bao, C.; Chen, H.; Liu, L.; Li, M.; Ye, F.; Fan, Q.; et al. Behaviors of Glioblastoma Cells in in Vitro Microenvironments. *Sci. Rep.* **2019**, *9*, 85. [CrossRef] [PubMed]

23. Louis, D.N.; Perry, A.; Reifenberger, G.; Deimling, A.; Figarella-Branger, M.; Cavenee, W.K.; Ohgaki, H.; Wiestler, O.D.; Kleihues, P.; Ellison, D.W. The 2016 World Health Organization Classification of Tumors of the Central Nervous System: a summary. *Acta Neuropathol.* **2016**, *131*, 803–820. [CrossRef] [PubMed]

24. Faugno, S.; Piccolella, S.; Sannino, M.; Principio, L.; Crescente, G.; Baldi, G.M.; Fiorentino, N.; Pacifico, S. Can agronomic practices and cold-pressing extraction parameters affect phenols and polyphenols content in hempseed oils? *Ind. Crop. Prod.* **2019**, *130*, 511–519. [CrossRef]

25. Smeriglio, A.; Galati, E.M.; Monforte, M.T.; Lanuzza, F.; D'Angelo, V.; Circosta, C. Polyphenolic Compounds and Antioxidant Activity of Cold-Pressed Seed Oil from Finola Cultivar of *Cannabis sativa* L. *Phytotherapy Res.* **2016**, *30*, 1298–1307. [CrossRef] [PubMed]

26. Zhou, Y.; Wang, S.; Lou, H.; Fan, P. Chemical constituents of hemp (*Cannabis sativa* L.) seed with potential anti-neuroinflammatory activity. *Phytochem. Lett.* **2018**, *23*, 57–61. [CrossRef]

27. Bourjot, M.; Zedet, A.; Demange, B.; Pudlo, M.; Girard-Thernier, C. In Vitro Mammalian Arginase Inhibitory and Antioxidant Effects of Amide Derivatives Isolated from the Hempseed Cakes (*Cannabis sativa*). *Planta Medica Int. Open* **2017**, *3*, 64–67. [CrossRef]

28. Simón-Manso, Y.; Neta, P.; Yang, X.; Stein, S.E. Loss of 45 Da from a2 Ions and Preferential Loss of 48 Da from a2 Ions Containing Methionine in Peptide Ion Tandem Mass Spectra. *J. Am. Soc. Mass Spectrom.* **2011**, *22*, 280–289. [CrossRef]

29. Fokoue, H.H.; Marques, J.V.; Correia, M.V.; Yamaguchi, L.F.; Qu, X.; Aires-De-Sousa, J.; Scotti, M.T.; Lopes, N.P.; Kato, M. Fragmentation pattern of amides by EI and HRESI: study of protonation sites using DFT-3LYP data. *RSC Adv.* **2018**, *8*, 21407–21413. [CrossRef]

30. Matsuda, F.; Miyagawa, H.; Ueno, T. Beta-1,3-glucooligosaccharide induced activation of four enzymes responsible for N-p-coumaroyloctopamine biosynthesis in potato (*Solanum tuberosum* cv.) tuber tissue. *Z. für Nat. C* **2000**, *55*, 373–382. [CrossRef]

31. Ko, H.-J.; Ahn, E.-K.; Oh, J.S. N-trans-*p*-caffeoyl tyramine isolated from *Tribulus terrestris* exerts anti-inflammatory effects in lipopolysaccharide-stimulated RAW 264.7 cells. *Int. J. Mol. Med.* **2015**, *36*, 1042–1048. [CrossRef]

32. Takayama, M. N-Cα bond cleavage of the peptide backbone via hydrogen abstraction. *J. Am. Soc. Mass Spectrom.* **2001**, *12*, 1044–1049. [CrossRef]

33. Chen, C.-Y.; Yeh, Y.-T.; Yang, W.-L. Amides from the stem of *Capsicum annuum*. *Nat. Prod. Commun.* **2011**, *6*, 227–229. [CrossRef] [PubMed]

34. Wang, S.; Suh, J.H.; Hung, W.-L.; Zheng, X.; Wang, Y.; Ho, C.-T. Use of UHPLC-TripleQ with synthetic standards to profile anti-inflammatory hydroxycinnamic acid amides in root barks and leaves of *Lycium barbarum*. *J. Food Drug Anal.* **2018**, *26*, 572–582. [CrossRef] [PubMed]

35. Santos, L.; Boaventura, M.; De Oliveira, A.; Cassady, J. Grossamide and *N-trans*-caffeoyltyramine from *Annona crassiflora* Seeds. *Planta Medica* **1996**, *62*, 76. [CrossRef] [PubMed]

36. Neelam, S.; Gokara, M.; Sudhamalla, B.; Amooru, D.G.; Subramanyam, R. Interaction studies of coumaroyltyramine with human serum albumin and its biological importance. *J. Phys. Chem. B* **2010**, *114*, 3005–3012. [CrossRef] [PubMed]

37. Walters, D.; Meurer–Grimes, B.; Rovira, I. Antifungal activity of three spermidine conjugates. *FEMS Microbiol. Lett.* **2001**, *201*, 255–258. [CrossRef]

38. Gericke, S.; Lübken, T.; Wolf, D.; Kaiser, M.; Hannig, C.; Speer, K. Identification of new compounds from sage flowers (*Salvia officinalis* L.) as markers for quality control and the influence of the manufacturing technology on the chemical composition and antibacterial activity of sage flower extracts. *J. Agric. Food Chem.* **2018**, *66*, 1843–1853. [CrossRef]

39. Sakakibara, I.; Katsuhara, T.; Ikeya, Y.; Hayashi, K.; Mitsuhashi, H. Cannabisin A, an arylnaphthalene lignanamide from fruits of *Cannabis sativa*. *Phytochemistry* **1991**, *30*, 3013–3016. [CrossRef]

40. Hao, G.; Wang, D.; Gu, J.; Shen, Q.; Gross, S.S.; Wang, Y. Neutral loss of isocyanic acid in peptide CID spectra: A novel diagnostic marker for mass spectrometric identification of protein citrullination. *J. Am. Soc. Mass Spectrom.* **2009**, *20*, 723–727. [CrossRef]

41. Morreel, K.; Dima, O.; Kim, H.; Lu, F.; Niculaes, C.; Vanholme, R.; Dauwe, R.; Goeminne, G.; Inze, D.; Messens, E.; et al. Mass spectrometry-based sequencing of lignin oligomers. *Plant Physiol.* **2010**, *153*, 1464–1478. [CrossRef]

42. Seca, A.M.L.; Silva, A.M.S.; Silvestre, A.J.D.; Cavaleiro, J.A.S.; Domingues, F.M.; Neto, C. Lignanamides and other phenolic constituents from the bark of kenaf (*Hibiscus cannabinus*). *Phytochemistry* **2001**, *58*, 1219–1223. [CrossRef]

43. Bolleddula, J.; Fitch, W.; Vareed, S.K.; Nair, M. Identification of metabolites in *Withania sominfera* fruits by liquid chromatography and high-resolution mass spectrometry. *Rapid Commun. Mass Spectrom.* **2012**, *26*, 1277–1290. [CrossRef] [PubMed]

44. Luo, Q.; Yan, X.; Bobrovskaya, L.; Ji, M.; Yuan, H.; Lou, H.; Fan, P. Anti-neuroinflammatory effects of grossamide from hemp seed via suppression of TLR-4-mediated NF-κB signaling pathways in lipopolysaccharide-stimulated BV2 microglia cells. *Mol. Cell. Biochem.* **2017**, *428*, 129–137. [CrossRef] [PubMed]

45. Kim, K.H.; Moon, E.; Kim, S.Y.; Lee, K.R. Lignans from the tuber-barks of *Colocasia antiquorum* var. *esculenta* and their antimelanogenic activity. *J. Agric. Food Chem.* **2010**, *58*, 4779–4785.

46. Davis, B.D.; Brodbelt, J. An investigation of the homolytic saccharide cleavage of deprotonated flavonol 3-O-glycosides in a quadrupole ion trap mass spectrometer. *J. Mass Spectrom.* **2008**, *43*, 1045–1052. [CrossRef]

47. Pacifico, S.; Piccolella, S.; Nocera, P.; Tranquillo, E.; Poggetto, G.D.; Catauro, M. New insights into phenol and polyphenol composition of *Stevia rebaudiana* leaves. *J. Pharm. Biomed. Anal.* **2019**, *163*, 45–57. [CrossRef]

48. Jalili-Nik, M.; Sabri, H.; Zamiri, E.; Soukhtanloo, M.; Roshan, M.K.; Hosseini, A.; Mollazadeh, H.; Vahedi, M.M.; Afshari, A.R.; Mousavi, S.H. Cytotoxic effects of *Ferula latisecta* on human glioma U87 Cells. *Drug Res.* **2019**, *69*, 665–670. [CrossRef]

49. Chan, F.K.; Moriwaki, K.; De Rosa, M.J. Detection of necrosis by release of lactate dehydrogenase activity. *Breast Cancer* **2013**, *979*, 65–70.

50. Franken, N.A.; Rodermond, H.M.; Stap, J.; Haveman, J.; Van Bree, C. Clonogenic assay of cells in vitro. *Nat. Protoc.* **2006**, *1*, 2315–2319. [CrossRef]

51. Gu, J.-J.; Gao, G.-Z.; Zhang, S.-M. miR-218 inhibits the migration and invasion of glioma U87 cells through the Slit2-Robo1 pathway. *Oncol. Lett.* **2015**, *9*, 1561–1566. [CrossRef]

52. Russell, R.C.; Tian, Y.; Yuan, H.; Park, H.W.; Chang, Y.-Y.; Kim, J.; Kim, H.; Neufeld, T.P.; Dillin, A.; Guan, K.-L. ULK1 induces autophagy by phosphorylating Beclin-1 and activating VPS34 lipid kinase. *Nature* **2013**, *15*, 741–750. [CrossRef] [PubMed]

53. Menon, M.B.; Dhamija, S. Beclin 1 Phosphorylation - at the Center of Autophagy Regulation. *Front. Cell Dev. Boil.* **2018**, *6*, 137. [CrossRef] [PubMed]

54. Ricci, A.; Fiorentino, A.; Piccolella, S.; Golino, A.; Pepi, F.; D'Abrosca, B.; Letizia, M.; Monaco, P. Furofuranic glycosylated lignans: a gas-phase ion chemistry investigation by tandem mass spectrometry. *Rapid Commun. Mass Spectrom.* **2008**, *22*, 3382–3392. [CrossRef] [PubMed]

55. Nigro, E.; Colavita, I.; Sarnataro, D.; Scudiero, O.; Zambrano, G.; Granata, V.; Daniele, A.; Carotenuto, A.; Galdiero, S.; Folliero, V.; et al. An ancestral host defence peptide within human β-defensin 3 recapitulates the antibacterial and antiviral activity of the full-length molecule. *Sci. Rep.* **2015**, *5*, 18450. [CrossRef] [PubMed]

56. Pacifico, S.; Gallicchio, M.; Lorenz, P.; Duckstein, S.M.; Potenza, N.; Galasso, S.; Marciano, S.; Fiorentino, A.; Stintzing, F.C.; Monaco, P. Neuroprotective potential of *Laurus nobilis* antioxidant polyphenol-enriched leaf extracts. *Chem. Res. Toxicol.* **2014**, *27*, 611–626. [CrossRef]

57. Hynds, R.E.; Gowers, K.H.C.; Nigro, E.; Butler, C.R.; Bonfanti, P.; Giangreco, A.; Prele, C.M.; Janes, S.M. Cross-talk between human airway epithelial cells and 3T3-J2 feeder cells involves partial activation of human MET by murine HGF. *PLoS ONE* **2018**, *13*, e0197129. [CrossRef]

58. Nigro, E.; Schettino, P.; Polito, R.; Scudiero, O.; Monaco, M.L.; De Palma, G.D.; Daniele, A. Adiponectin and colon cancer: evidence for inhibitory effects on viability and migration of human colorectal cell lines. *Mol. Cell. Biochem.* **2018**, *448*, 125–135. [CrossRef]

59. Benelli, G.; Pavela, R.; Petrelli, R.; Cappellacci, L.; Santini, G.; Fiorini, D.; Sut, S.; Zengin, G.; Canale, A.; Maggi, F. The essential oil from industrial hemp (*Cannabis sativa* L.) by-products as an effective tool for insect pest management in organic crops. *Ind. Crop. Prod.* **2018**, *122*, 308–315. [CrossRef]

60. Fiorini, D.; Molle, A.; Nabissi, M.; Santini, G.; Benelli, G.; Maggi, F. Valorizing industrial hemp (*Cannabis sativa* L.) by-products: cannabidiol enrichment in the inflorescence essential oil optimizing sample pre-treatment prior to distillation. *Ind. Crop. Prod.* **2019**, *128*, 581–589. [CrossRef]

61. Figueira, I.; Garcia, G.; Pimpão, R.C.; Terrasso, A.; Costa, I.; Almeida, A.F.; Tavares, L.; Pais, T.F.; Pinto, P.; Ventura, M.R.; et al. Polyphenols journey through blood-brain barrier towards neuronal protection. *Sci. Rep.* **2017**, *7*, 11456. [CrossRef]

62. Youdim, K.A.; Dobbie, M.S.; Kuhnle, G.; Proteggente, A.R.; Abbott, N.J.; Rice-Evans, C. Interaction between flavonoids and the blood-brain barrier: in vitro studies. *J. Neurochem.* **2003**, *85*, 180–192. [CrossRef] [PubMed]

63. Pacifico, S.; Piccolella, S.; Marciano, S.; Galasso, S.; Nocera, P.; Piscopo, V.; Fiorentino, A.; Monaco, P. LC-MS/MS profiling of a mastic leaf phenol enriched extract and its effects on H_2O_2 and Aβ (25–35) oxidative injury in SK-B-NE (C)-2 cells. *J. Agric. Food Chem.* **2014**, *62*, 11957–11966. [CrossRef] [PubMed]

64. Turunen, B.J.; Ge, H.; Oyetunji, J.; Desino, K.E.; Vasandani, V.; Güthe, S.; Himes, R.H.; Audus, K.L.; Seelig, A.; Georg, G.I. Paclitaxel succinate analogs: anionic and amide introduction as a strategy to impart blood–brain barrier permeability. *Bioorganic Med. Chem. Lett.* **2008**, *18*, 5971–5974. [CrossRef] [PubMed]

65. Price, T.O.; Uras, F.; Banks, W.A.; Ercal, N. A novel antioxidant *N*-acetylcysteine amide prevents gp120- and Tat-induced oxidative stress in brain endothelial cells. *Exp. Neurol.* **2006**, *201*, 193–202. [CrossRef]

66. Könczöl, Á.; Rendes, K.; Dékány, M.; Müller, J.; Riethmüller, E.; Balogh, G.T. Blood-brain barrier specific permeability assay reveals N -methylated tyramine derivatives in standardised leaf extracts and herbal products of *Ginkgo biloba*. *J. Pharm. Biomed. Anal.* **2016**, *131*, 167–174. [CrossRef]

Sample Availability: Samples of the compounds are available from the authors.

 *molecules*

Article

Simultaneous LC/MS Analysis of Carotenoids and Fat-Soluble Vitamins in Costa Rican Avocados (*Persea americana* Mill.)

Carolina Cortés-Herrera [1],*, Andrea Chacón [1], Graciela Artavia [1] and Fabio Granados-Chinchilla [2]

1   Centro Nacional de Ciencia y Tecnología de Alimentos, Universidad de Costa Rica, Ciudad Universitaria Rodrigo Facio, 11501-2060 San José, Costa Rica; andrea28.chacon@gmail.com (A.C.); graciela.artavia@ucr.ac.cr (G.A.)
2   Centro de Investigación en Nutrición Animal (CINA), Universidad de Costa Rica, Ciudad Universitaria Rodrigo Facio, 11501-2060 San José, Costa Rica; fabio.granados@ucr.ac.cr
*   Correspondence: carolina.cortesherrera@ucr.ac.cr; Tel.: +506-2511-7226

Academic Editor: Severina Pacifico
Received: 26 August 2019; Accepted: 5 October 2019; Published: 10 December 2019

Abstract: Avocado (a fruit that represents a billion-dollar industry) has become a relevant crop in global trade. The benefits of eating avocados have also been thoroughly described as they contain important nutrients needed to ensure biological functions. For example, avocados contain considerable amounts of vitamins and other phytonutrients, such as carotenoids (e.g., β-carotene), which are fat-soluble. Hence, there is a need to assess accurately these types of compounds. Herein we describe a method that chromatographically separates commercial standard solutions containing both fat-soluble vitamins (vitamin A acetate and palmitate, Vitamin D_2 and D_3, vitamin K_1, α-, δ-, and γ-vitamin E isomers) and carotenoids (β-cryptoxanthin, zeaxanthin, lutein, β-carotene, and lycopene) effectively (i.e., analytical recoveries ranging from 80.43% to 117.02%, for vitamins, and from 43.80% to 108.63%). We optimized saponification conditions and settled at 80 °C using 1 mmol KOH L^{-1} ethanol during 1 h. We used a non-aqueous gradient that included methanol and methyl *tert*-butyl ether (starting at an 80:20 ratio) and a C_{30} chromatographic column to achieve analyte separation (in less than 40 min) and applied this method to avocado, a fruit that characteristically contains both types of compounds. We obtained a method with good linearity at the mid to low range of the mg L^{-1} (determination coefficients 0.9006–0.9964). To determine both types of compounds in avocado, we developed and validated for the simultaneous analysis of carotenoids and fat-soluble vitamins based on liquid chromatography and single quadrupole mass detection (LC/MS). From actual avocado samples, we found relevant concentrations for cholecalciferol (ranging from 103.5 to 119.5), δ-tocopherol (ranging from 6.16 to 42.48), and lutein (ranging from 6.41 to 15.13 mg/100 g dry weight basis). Simmonds cultivar demonstrated the higher values for all analytes (ranging from 0.03 (zeaxanthin) to 119.5 (cholecalciferol) mg/100 g dry weight basis).

Keywords: avocado; LC/MS; fat-soluble vitamins; carotenoids

---

1. Introduction

Avocado represents a billion-dollar industry, the projection of the apparent *per capita* consumption of avocado sets the top six world importers of avocado to be US, Netherlands, France, United Kingdom, Spain, and Canada with 3.64, 1.62, 2.10, 2.21, 2.30, and 2.55 kg in a given year, respectively [1]. Avocado exports have made some countries like Mexico, Dominican Republic, Peru, Chile, Colombia, and Costa Rica increase their cultivated area [1].

The topmost exporter countries have directed their offer to destinations such as the US, Europe, and Asia, especially China [2], where import growth is in the order of 250%, from 154 tons in 2012, to 25,000 tons in 2016 [3,4]. For example, Costa Rican avocado harvested area was estimated at 1 888 ha in 2014 and increased to 3092 ha in 2017, which represents 845 tons of Costa Rican avocado (or 448900 USD) [5].

As a complex matrix, performing chemical analysis on the avocado flesh presents an additional difficulty. The ripe avocado fruit has a firm, oily, and yellow to light green colored mesocarp that contains both fat-soluble vitamins and carotenoids associated with other lipids [6]. Usually, both types of analysis have to be performed separately using different chromatographic conditions altogether, which represents an additional expense, both economic and in terms of labor. Analyzing both types of fat-soluble compounds is relevant, especially in fruits such as avocado in which there is evidence that carotenoid absorption might be improved by the addition of avocado and avocado oil [7–9].

Interestingly, the Association of Official Analytical Chemists (AOAC) Official Methods of Analysis (OMA^{SM}) does not have any method established for neither fat-soluble vitamins nor carotenoids in fruits. Several papers have already annotated the relevance of including avocado in the diet as it has been related to health benefits [10–19]. Previously, another report analyzed carotenoid content in avocado using spectrophotometry [20]. A later report has used a similar approach to measure total carotenoid content [21]. On another hand, a research group has analyzed avocado pigments in oil [22] and tissue [22,23] using HPLC coupled with a photodiode array detector using a triphasic organic solvent gradient. Carotenoids in avocado seed have also been described [24,25].

The application of chromatography and mass spectrometry to carotenoid analysis in fruits is not new [23,26,27]. However, few papers have been dedicated solely to the study of both fat-soluble and carotenoid content in avocado fruits. For example, research assayed both types of compounds (i.e., carotenoids and tocopherol) using two independent chromatographic techniques [28].

Among the papers dedicated to assessing specifically carotenoids or vitamins in avocado, the most relevant include advantages such as that most researchers use acetone (a versatile and low boiling point solvent) during primary extraction [22,23,28,29] and ethyl ether or hexane after saponification [22,23,29]. For saponification at room temperature, the use of 2,6-di-*tert*-butyl-4-methylphenol, and nitrogen flushing seems to be a norm [28,29], thus protecting the target analytes. Diversity of carotenoids studies is ample (including epoxides [28], isomers of carotene [20], chlorophylls [22], phytoene [23]). Mobile phases are usually simple and environmentally friendly [22,23,29], which include methanol, water, methyl *tert*-butyl ether, and ethyl acetate. Yano and coworkers were able to apply their method to 75 and 15 different fresh and processed fruits, respectively [23]. Solid-phase extraction has been used to reduce interferences [22]. Mass spectrometry has been used to recognize unidentified compounds [20,28].

Disadvantages of these methods include the presence of water in the mobile phase (which increases mobile phase polarity), several approaches exhibit some issues with chromatographic resolution for some of the signals [28,29] and in some cases saponification time [29], base concentration [29], and chromatographic run [23] are excessive or the identification of compounds is based on light absorption [20,22,29]. Finally, vitamin analysis of the fruit is scarce at best [11,28].

Lastly, very little information has been gathered regarding varieties of Costa Rican avocados, proximate analysis, mineral content, and some vitamins have been explored [30]. Some papers have also focused on standing out differences between avocado varieties [21,28], including varieties of Guatemalan race (e.g., Hass [11] and Nabal [20]), as avocadoes originated from New Zealand [22], California [28], and Mexico [29] and those commercially available from Israel [20] and Japan [23].

Herein, we report a liquid chromatography and single quadrupole mass detection (LC/MS) based method using a C$_{30}$ column and methanol and methyl *tert*-butyl ether to assay and quantitatively separate both fat-soluble vitamins (vitamin A acetate and palmitate, Vitamin D$_2$ and D$_3$, vitamin K$_1$, α-, δ-, and γ-vitamin E isomers) and carotenoids (β-cryptoxanthin, zeaxanthin, lutein, β-carotene, and lycopene) simultaneously in avocado fruit.

2. Results and Discussion

2.1. Stationary Phase Selection and Green Chemistry

Where other alkyl modified stationary phases failed (Table 1), the C_{30} allowed an excellent separation of fat-soluble compounds even between structurally related molecules using a MeOH/*tert*-butyl methyl ether (MTBE)-based mobile phase (Table 2). The solvent selection not only ensured good compound solubility and chromatographic separation, but it also helped improve column life span as no water was involved and the column could be safely stored under 80% MeOH.

Table 1. Performance of other stationary phases and conditions tested to try to separate calciferol and tocopherol isomers.

Stationary Phase	C_8		C_{18}		C_{30}	
Solvent System	95:5 MeOH:H_2O		95:5 MeOH:H_2O		90:7:3 MeOH:CH_3CN:2-propanol	
Flow, mL min^{-1}	0.75		1.00		0.50	
Temperature, °C	50		50		35	
Compound	t_R, min	R_s	t_R, min	R_s	t_R, min	R_s
Retinyl acetate	3.50		4.75		5.42	
Ergocalciferol	4.74	0	8.41		8.96	0.78
Cholecalciferol	4.74	0	8.81	1.12	9.27	0.78
δ-tocopherol	4.67	1.47	8.05	1.20	7.42	2.96
γ-tocopherol	5.17	1.47	9.52	2.07	8.28	1.85
α-tocopherol	7.05		15.14		12.19	
Phylloquinone	7.84		19.87		13.17	
Retinyl palmitate	13.09		48.33		35.22	

Table 2. Optimized Mass Spectrometry (MS) parameters for the assayed compounds, in order of *m/z*.

Detector Set Time, min	Compound	t_R, min	Selected SIM Ion, *m/z*	Fragmentor, V	Dwell Time, ms
From 0 to 5	Retinyl acetate	2.99	269.3 $[C_{20}H_{29}]^{\bullet +}$/325.2 $[C_{20}H_{29}OH + K]^+$	100	95
	Ergocalciferol	3.59	398.3 $[M + H]^+$	220	
	Cholecalciferol	4.02	385.3 $[M + H]^+$	160	
From 5 to 8	δ-tocopherol	5.26	402.5 [M+]	220	71
	γ-tocopherol	5.78	416.4 [M+]	140	
	α-tocopherol	6.61	430.4 [M+]	80	
From 8 to 13	Phylloquinone	8.07	451.4 $[M + H]^+$	140	56
	Astaxanthin	9.07	597.4 $[M + H]^+$	160	
	Lutein	9.97	569.4 $[M + H]^+$	140	
	Zeaxanthin	11.49	568.4 [M+]	140	
After 13	Retinyl palmitate	11.19	269.3 $[C_{20}H_{29}]^{\bullet +}$/563.4 $[M + K]^+$	100	95
	β-cryptoxanthin	16.58	552.6 [M+]	120	
	β-carotene	22.51	536.4 [M+]	120	
	Lycopene	37.39	536.1 [M+]	160	

Some methods have selected chlorinated solvents as an effective way to extract [28] or separate carotenoids [31]. However, we chose MTBE as a greener alternative to chlorinated solvents [32]. Additionally, our chromatographic separation was mostly based on MeOH. The high degree of shape recognition of the C_{30} was validated early [33] and was, once again, here, demonstrated. It is recommended for analysis of retinoid as well as carotenoid molecules [34]. However, herein, we exploited the versatility of the C_{30} column further. Furthermore, as avocado lacks the presence of lycopene [20,28], this compound when introduced into the separation (especially in its deuterated form), can be used as an internal standard (IS). Though lycopene is considered a relative liable carotenoid, it has been used successfully as an IS [35]. Other more stable molecules have been used as well and could be considered (e.g., canthaxanthin [35], sudan I [36]; 8′-apo-8′-β-carotenal [37], echinenone [38]).

2.2. Singular Ion Monitoring Parameter Selection

As expected, using reverse phase chromatography, fat-soluble vitamins eluted during the first 8 min (except for retinyl palmitate, an esterified compound with an extra C_{15} alkyl chain) (Table 2). Both retinoids assessed were prone to retain monovalent cations such as K^+ ($[M + K]^+$). Retinyl acetate and retinyl palmitate share a similar fragmentation pattern, which includes ion 269 *m/z* as a base peak (Table 2). The retinoid alkyl chain did not seem to affect ionization voltage, indicating that the $C_{20}H_{29}OH$ base was governing their behavior. Ion 296 *m/z* found for retinoids corresponded to the protonation and elimination of water and acetate during positive ion electrospray [34].

All tocopherol isomers, as well as zeaxanthin, β-cryptoxanthin, β-carotene, and lycopene (the heavier analogs), exhibited deprotonated species as molecular ions ($[M+]$). Electrospray ionization (ESI^+) ions are usually preformed in solution by acid/base reactions (i.e., $[M + nH]^{n+}$), some ions are probably formed by a field desorption mechanism at the surface. Hence, the production of abundant cations, with little fragmentation [34]. Meanwhile, phylloquinone, the acidic carotenoid astaxanthin [34], and lutein all ionized through protonated species ($[M + H]^+$) (Table 2). Interestingly, the highly related compounds α-, γ-, and β-tocopherols exhibited very different ionization energies (i.e., 80, 140, and 220 V, respectively) (Table 2).

2.3. Method Performance Data

For the vitamin group, the higher sensitivity was exhibited by α-, γ-, and retinyl palmitate (Table 2). On the other hand, carotenoids with a lower limit of detection were astaxanthin, lutein, and zeaxanthin (Table 3). Electrospray analysis of carotenoids, since some years ago, has demonstrated high sensitivity (i.e., in the pmol range) [39]. These compounds, as expressed in the matrix of interest, could be detected as low as 1 μg/100 g dry matter (Table 3). The resolution obtained between the compounds analyzed ranged from 2.18–71.90; the lowest value recommended is 2 (Table 2, Figure 1A) [40]. Several compounds were very close to the theoretical value of 1 (i.e., a perfect peak with a Gaussian distribution and completely symmetrical) though ergocalciferol (1.707), and lutein (1.348) exhibited some tailing. Fronting (leading peak) can also be observed for retinyl acetate (0.709) (Table 3). Column efficiency was very high for compounds eluted above 8 min, such as phylloquinone, zeaxanthin, retinyl palmitate, β-cryptoxanthin, and β-carotene (Table 3). A good baseline definition was observed for all compounds analyzed (i.e., α_s ranging from 1.07 to 1.76) (Table 3).

Table 3. Method performance parameters obtained during validation.

Compound	LoD, μg L^{-1}	LoQ, μg L^{-1}	LoD, μg/100 g fat	LoQ, μg/100 g fat	LoD, μg/100 g dry matter	LoQ, μg/100 g dry matter
			Sensitivity			
Retinyl acetate	3.00×10^2	9.20×10^2	1.00×10^2	3.07×10^2	1.50×10^1	4.60×10^1
Ergocalciferol	1.00×10^2	2.90×10^2	3.30×10^1	9.70×10^1	0.50×10^1	1.50×10^1
Cholecalciferol	2.70×10^2	8.20×10^2	9.00×10^1	2.73×10^2	1.40×10^1	4.10×10^1
δ-tocopherol	1.70×10^2	5.10×10^2	5.70×10^1	1.70×10^2	0.90×10^1	2.60×10^1
γ-tocopherol	1.30×10^1	3.80×10^1	0.40×10^1	1.30×10^1	0.10×10^1	0.20×10^1
α-tocopherol	0.70×10^1	2.40×10^1	0.20×10^1	0.80×10^1	0.10×10^1	0.10×10^1
Phylloquinone	4.30×10^2	1.29×10^3	1.43×10^2	4.30×10^2	2.20×10^1	6.50×10^1
Astaxanthin	2.20×10^1	1.25×10^2	0.70×10^1	4.20×10^1	0.10×10^1	0.60×10^1
Lutein	1.00×10^1	2.90×10^1	0.30×10^1	1.00×10^1	0.10×10^1	0.10×10^1
Zeaxanthin	0.90×10^1	2.80×10^1	0.30×10^1	0.90×10^1	0.10×10^1	0.10×10^1

Table 3. *Cont.*

Retinyl palmitate	2.40×10^1	7.40×10^2	0.80×10^1	2.47×10^2	0.10×10^1	3.70×10^1
β-cryptoxanthin	3.30×10^2	9.90×10^2	1.10×10^2	3.30×10^2	1.70×10^1	5.00×10^1
β-carotene	8.80×10^2	2.67×10^3	2.93×10^2	8.90×10^2	4.40×10^1	1.34×10^2
Lycopene	1.56×10^2	4.71×10^2	5.20×10^1	1.57×10^2	0.80×10^1	2.40×10^1

			Sensitivity			
Compound	LoD, µg L^{-1}	LoQ, µg L^{-1}	LoD, µg/100 g fat	LoQ, µg/100 g fat	LoD, µg/100 g dry matter	LoQ, µg/100 g dry matter

Chromatographic Parameters

Compound	t_R, min	[[], mg L^{-1}	Area, ×10^5	Height, ×10^4	Peak Width	Symmetry	R$_s$	k	α	N, ×10^4
Retinyl acetate	3.00	5.00	8.98	12.96	0.12	0.71	3.67	0.35	1.61	1.08
Ergocalciferol	3.48	0.50	5.34	5.12	0.15	1.71	3.54	0.57	1.39	0.92
Cholecalciferol	3.97	1.00	1.76	1.78	0.13	0.91	7.96	0.79	1.71	1.45
δ-tocopherol	5.21	1.00	0.67	0.63	0.18	0.95	2.85	1.35	1.16	1.34
γ-tocopherol	5.71	1.00	9.07	7.65	0.17	0.77	3.52	1.57	1.22	1.87
α-tocopherol	6.49	1.00	21.99	13.27	0.28	0.99	7.20	1.92	1.33	0.88
Phylloquinone	7.89	1.00	0.29	0.32	0.11	0.80	4.75	2.56	1.18	7.77
Astaxanthin	8.94	0.05	0.78	0.40	0.33	0.88	2.18	3.03	1.11	1.19
Lutein	9.68	0.05	4.19	1.95	0.36	1.35	4.59	3.37	1.19	1.17
Zeaxanthin	11.09	1.00	0.66	0.44	0.25	1.06	2.87	3.40	1.07	3.09
Retinyl palmitate	11.75	0.20	5.68	4.52	0.21	1.08	26.91	4.30	1.49	5.03
β-cryptoxanthin	16.40	0.20	0.03	0.04	0.14	0.80	27.10	6.40	1.40	23.14
β-carotene	22.14	1.00	0.38	0.22	0.29	0.76	71.90	8.98	1.76	9.54
Lycopene	37.39	1.00	0.28	0.34	0.14	1.04	-	-	-	118.31

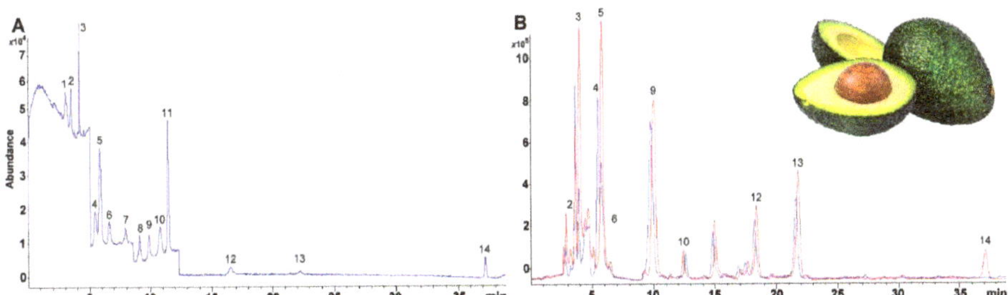

Figure 1. Selected ion monitoring (SIM) chromatogram for (**A**). (**A**) A standard mixture of 14 analytes at 1 µg mL^{-1} for liposoluble vitamins and 0.05 µg mL^{-1} for carotenoids. 1. Retinyl acetate. 2. Ergocalciferol. 3. Cholecalciferol. 4. δ-Tocopherol. 5. γ-Tocopherol. 6. α-Tocopherol. 7. Phylloquinone. 8. Astaxanthin. 9. Lutein. 10. Zeaxanthin. 11. Retinyl palmitate. 12. β-Cryptoxanthin. 13. β-Carotene. 14. Lycopene. (**B**) The saponified avocado sample analyzed using the proposed chromatographic method (blue line), and a sample spiked with 5 µmol for each vitamin and 1 µmol for each carotenoid (red line).

The method was successfully applied to avocado fruits (Figure 2B). The lack of available certified reference materials for matrices such as fruits obliges the use of spike solutions to demonstrate both matrix effects (if any) and extraction efficiency of vitamins and carotenoids in avocado specifically.

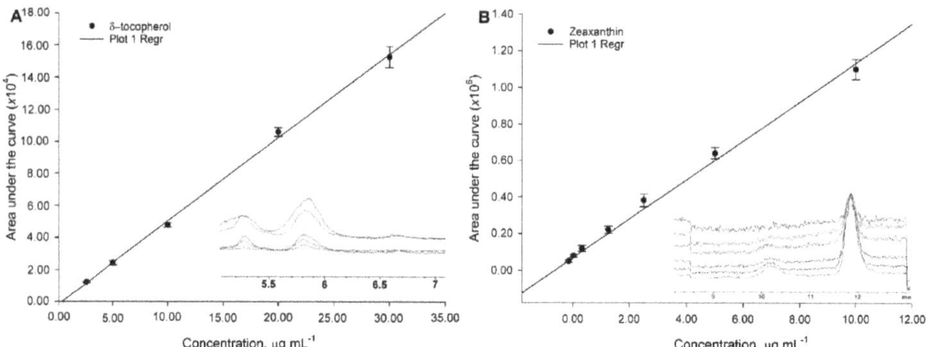

Figure 2. Average calibration curves and error bars depicting variability obtained for two of the target analytes (**A**) δ-tocopherol and (**B**) zeaxanthin. Mean and standard deviations (used as error) calculated from $n = 3$ independently constructed calibration curves injected on different days.

2.4. Performance during Saponification

Saponification-wise, using the same starting mass and sample, we proceeded to optimize the conditions needed (i.e., base concentration and time) to improve analyte recovery. The differential analysis showed that, overall, samples treated at 80 °C for one hour and using 1 mol KOH L⁻¹ exhibited the best results in the case of the analytes of interest, for avocado (Table 4). Variables tested showed a profound and significant effect over analyte recovery ($p < 0.05$ for all cases) (Table 4). Reaction parameters must be optimized to ensure proper hydrolysis within the complex, considerably oily (ranging from 35.3 to 39.1 g fat/100 g dry weight basis), food matrix that is found in the ripe avocado fruit [41]. Carotenoids are increasingly sensitive to heat. Hence, hot saponification was carried out using an organic solvent with a relatively low boiling point [6] (i.e., 78.37 °C at Standard Pressure and Temperature [STP] for ethanol) and pyrogallol was used as a radical sink to protect the compounds of interest. However, the improvement in recovery was achieved by increasing the temperature to 80 °C. A procedure that might be justified as (i) the thermal effect must be sufficient to break the cell walls from the avocado fruit and provide a rapid molecular diffusion to promote reaction; (ii) higher temperatures increase solubility of lipophilic compounds and enhance kinetics of saponification (i.e., favors localized "hot spots" which deliver sufficient energy for the molecules to react) [42]. We chose hot saponification to diminish reaction time; a similar approach has been reported elsewhere [43].

Table 4. Optimization of saponification conditions.

Temperature, °C		60		80		95	
Base Concentration, mol KOH L⁻¹		1	2	1	2	1	2
Compound	t_R, min	mg/100 g [a]					
Ergocalciferol	3.59	18.89 (−0.14)	10.98 (−0.50)	22.06	8.38 (−0.62)	1.74 (−0.92)	ND (−1.00)
δ-tocopherol	5.26	9.65 (−0.89)	88.22 (−0.02)	90.06	67.32 (−0.25)	0.24 (−1.00)	88.17 (−0.02)
γ-tocopherol	5.78	0.45 (−0.93)	0.50 (−0.92)	6.14	2.48 (−0.60)	0.11 (−0.98)	9.62 (0.57)
α-tocopherol	6.61	180.96 (−0.32)	102.35 (−0.62)	267.77	90.77 (−0.66)	74.00 (−0.72)	70.63 (−0.74)
Astaxanthin	9.07	18.69 (−0.26)	16.82 (−0.33)	25.14	17.12 (−0.32)	7.64 (−0.70)	14.16 (−0.44)
Lutein	9.97	0.08 (0.00)	0.07 (−0.13)	0.08	0.12 (0.50)	0.02 (−0.75)	0.12 (0.50)
Zeaxanthin	11.49	26.60 (−0.13)	37.70 (0.23)	30.73	47.44 (0.54)	7.85 (−0.74)	20.38 (−0.34)
β-cryptoxanthin	16.58	24.66 (−0.22)	12.13 (−0.61)	31.50	27.02 (−0.14)	5.54 (−0.82)	16.93 (−0.46)
β-carotene	22.51	18.89 (−0.14)	10.98 (−0.50)	22.06	8.38 (−0.62)	1.74 (−0.92)	ND (−1.00)
p values		0.044	0.040	-	0.037	0.011	0.028

[a] Brackets indicate bias, expressed as a fraction, with respect to the saponification treatment of 1 h at 80 °C using a 1 mol L⁻¹ base concentration. ND: not detected.

2.5. Quantification, Linearity, and Calibration Curve Construction

Five-point calibration curves were constructed to quantitate the analytes. Slopes (m_x) ranging from 4.05×10^3 (cholecalciferol) to 5.51×10^4 (ergocalciferol) and 5.45×10^4 (β-cryptoxanthin) to 6.49×10^6 (zeaxanthin) for vitamins and carotenoids, respectively, were obtained. Determination coefficients spoke toward excellent linearity ranging from 0.9906 to 0.9964 (Figure 2A,B). For vitamins, standard calibration curves were constructed as follows: from 1.00 to 10.00 mg L^{-1} for retinyl acetate, ergocalciferol, γ-, and α-tocopherol and retinyl palmitate; from 2.50 to 30.00 mg L^{-1} for cholecalciferol, δ-tocopherol, and phylloquinone. Finally, considering lower concentrations found in the target matrix and carotenoid sensitivity, standard calibration curves were constructed as follows: 0.05 to 0.80 mg L^{-1} for astaxanthin, β-cryptoxanthin, and β-carotene and from 0.1 to 2 mg L^{-1} for zeaxanthin and lutein (Figure 2A,B). Concentrated stock solutions were prepared by dissolving the solid standard in 2-propanol for vitamins and chloroform for carotenoids; dilutions performed after that were matched with the mobile phase (i.e., the gradient at the start of the chromatographic run).

2.6. Analyte Recovery and Method Accuracy

Overall, vitamin recovery, in spiked avocadoes, exhibited better performance than for carotenoids (except for β-carotene with a recovery of 108.63%). Structurally, the lower recoveries (ranging from 43.80% to 63.68%) were found for those carotenoids that were oxygenated (Table 5). Noteworthy, when zeaxanthin and β-cryptoxanthin were extracted using a mixture of ethyl ether, and hexane (50:50) obtained/experimental mass increased between 12.35% and 15.56%. Furthermore, when nitrogen flushing was performed, in conjunction, an additional increment between 13.35% to 19.66%, was observed (data not shown). Carotenoids, among several mechanisms of transformation [44,45], are susceptible to thermal degradation [46,47], β-carotene may suffer from symmetrical oxidative cleavage (which generates, the apocarotenoid, retinol) [48], and β-cryptoxanthin can undergo light-induced oxidation and isomerization [49].

Table 5. Spiked avocado samples and recovery for representatives for fat-soluble vitamins and carotenoids.

Compound	t_R, min	Concentration, µmol		Recovery, % [a]
		Theoretical/Added	Experimental/Obtained	
Ergocalciferol	3.59	5.00×10^1	4.00×10^1	81.21 (70–110)
α-tocopherol	6.61	4.60×10^1	3.70×10^1	80.43 (70–110)
Phylloquinone	8.07	3.50×10^1	4.20×10^1	117.02 (70–110)
Astaxanthin	9.07	1.70×10^1	1.00×10^1	62.27 (60–120)
Lutein	9.97	0.35×10^1	0.23×10^1	63.68 (60–120)
Zeaxanthin	11.49	0.23×10^1	0.10×10^1	43.80 (60–120)
β-cryptoxanthin	16.58	1.09×10^2	6.50×10^1	59.51 (70–110)
β-carotene	22.51	1.23×10^1	1.33×10^1	108.63 (60–120)

[a] Brackets represent recovery values recommended by AOAC [50].

2.7. Mass Spectra Analysis

Retinyl acetate and β-cryptoxanthin showed, under our conditions, a higher degree of fragmentation compared to tocopherol and an oxygenated carotenoid. Additional to the signals analyzed above for retinoids, ion 369.3 represented the retinyl acetate molecular ion plus a potassium ion, $[M + K]^+$ (Figure 3A). On another hand, tocopherols usually are characterized to present few fragmentation processes [51] (Figure 3B). Major ions are formed by cleavage through the non-aromatic portion of the chromanol ring, both with and without hydrogen transfer and by a loss of the isoprenoid side chain [51]. In contrast, we found none of the usual fragments reported elsewhere for the fragmentation of γ-tocopherol (e.g., $C_9H_{11}O_2^+$ *m/z* 151 amu; [51]) probably because the total ion chromatogram was obtained using lower energy than required. Also, there may have been some instrument limitations as a single quadrupole was used throughout the experiments, and the fragmentation ions were the result of molecule degradation in the ion source not in the collision chamber used during tandem mass spectrometry. Hence, the lack of tandem mass detection justified the absence, in some cases, of multiple fragmentation patterns [52]. However, the molecular ion [M+] for γ-tocopherol was unmistakable and could be used to differentiate nuances among tocopherols (e.g., the difference between γ-tocopherol and α-tocopherol is a methyl group in the aromatic ring (Δ15 amu)). As an example of calciferol identification, cholecalciferol total ion chromatogram (TIC)-mass spectra showed relevant fragments at 365.1 (loss of H_2O), 337.1 (subsequent loss of both $-CH_3$ from the isopropyl moiety), 301.2, and 227.0 (loss of $-CH_3$ and $=CH_2$ or $C_{17}H_{23}^{3\bullet}$) m/z (data not shown).

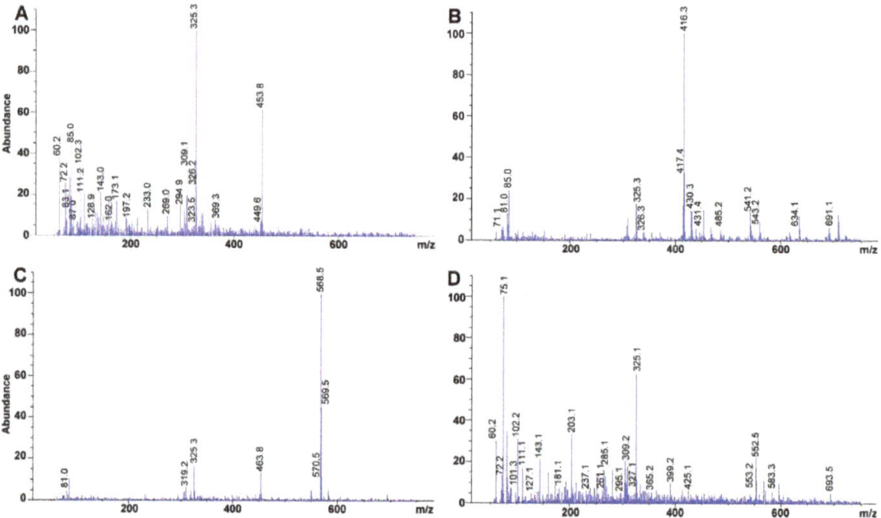

Figure 3. Experimental mass spectra obtained for (**A**) retinyl acetate, (**B**) γ-tocopherol, (**C**) Lutein, (**D**) β-cryptoxanthin. Total ion chromatograms at 100 mg L^{-1} each.

On another hand, Atmospheric-pressure chemical ionization (APCI), Fast atom bombardment (FAB), electron ionization (EI), and ESI all have mass-based techniques used to assess carotenoids [53]. While APCI is a popular approach for the ionization of lipophilic compounds, we selected ESI^+. In this scenario, some modifications were introduced into the mobile phase to enhance ionization. In this specific case, formic acid was selected as it has been described to decrease signal suppression [54]. For the case of lutein the signal 551 m/z (i.e., [M + H − 18] which corresponds to the loss of H_2O) and 463 m/z ([M + H − 106], loss of two water molecules and xylene from the polyene chain) were observable ([53,55]; Figure 3C). Additionally, as zeaxanthin and lutein differed from each other by a position of a double bond, in a ε-ring, they rendered similar spectra [53]. Further fragmentation for both compounds responded to the loss of said rings. Finally, for β-cryptoxanthin, signal 552.5 ([M+]) and 553.2 m/z ([M + H]$^+$) were evident, fragments 425.1 [$C_{32}H_{40}$]$^+$ and 399 [M + H − 153]$^+$ m/z responded to the partial breakup of both β-rings and the complete loss of the oxygenated β-ring plus a methyl group (Figure 3D).

2.8. Chromatographic Separation of Tocopherol Isomers

Under our chromatographic conditions, tocopherol isomers were easily segregated (Figure 4A). Using the tocopherol mixture containing α-, β-, δ-, and γ-tocopherols, we were able to assess further that no additional [M+] signal was necessary for the detection of β-tocopherol, as β- and γ-tocopherol share the same molecular mass (Figure 4B,C). The signal for β-tocopherol was observed at a t_R of 5.387 min (Figure 4A). Hence, the four tocopherol isomers eluted as follows: δ-, β-, γ-, and α-tocopherol (Table 1, Figure 4A).

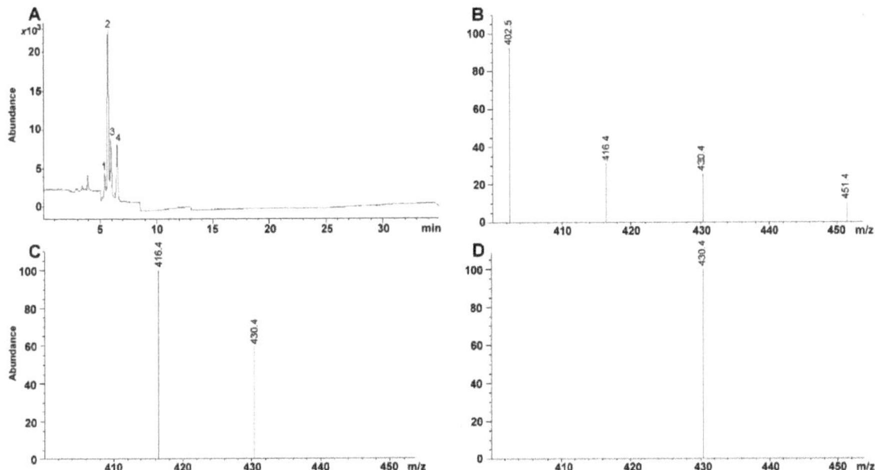

Figure 4. (**A**) Chromatogram of a mixture of tocopherols (W530066 70, 13, 523, 105 mg g^{-1} for α- (peak 4), β- (peak 2), γ- (peak 3), and δ-tocopherol (peak 1), respectively). Selected [M + H]$^+$ ion for (**B**) δ-tocopherol (402.1 *m/z*), (**C**) β/γ-tocopherol (416.4 *m/z*), and (**D**) α-tocopherol (430.4 *m/z*). Selected ion monitoring (SIM) at 700, 130, 5230, 1050 µg mL^{-1}.

2.9. Method Application in Real Samples

Simmonds variety characterizes itself for having an oblong oval to pear-shape large-sized fruit of light green-colored smooth skin, and a seed of medium size, usually tight. Meanwhile, Guatemala has a medium to large size and nearly round shape, smooth, thick, granular, and green skin with a small seed. Finally, Hass fruit possesses an ovoid to pear shape, is of medium size, with tough, leathery, pebbled, thin, dark purple to black (when ripe) skin, with a small seed [30,56].

We found in avocado both non-provitamin A (lutein and zeaxanthin) and provitamin A (β-carotene and β-cryptoxanthin) carotenoids [57] (Table 6). All carotenoids encountered have been previously identified for the fruit (see, for example, [11,20,21,23,28,29]). Our data indicated that, overall, all three varieties of avocado exhibited concentrations, of carotenoids and vitamins, in line (except for provitamin A carotenoids which levels were considerably higher for Costa Rican avocadoes) with those reported elsewhere [28]. Considering the above, Costa Rican avocado would supply to diet retinol equivalents [58] ranging from 1085.21 to 425.01 µg on a dry weight basis. Though three different varieties of avocado have been examined, it should be clear that any number of factors can affect the concentrations of such molecules, including edaphoclimatic conditions, genotypic differences, and nutritional status of the plants [59]. Hass and Guatemala varietals showed a somewhat similar profile for both fat-soluble vitamins as carotenoids. Meanwhile, the Simmonds variety showed higher levels for almost all analytes tested (except for lutein) (Table 6). Interestingly, the δ/α ratio of isomers is close to 1 for all varieties (i.e., 0.75–1.22).

Table 6. Fat-soluble vitamins and carotenoids obtained from Costa Rican avocadoes.

Compound	Simmonds	Hass	Guatemala
	Concentration, mg/100 g dry Weight Basis		
Ergocalciferol	1.84 ± 0.18	0.31 ± 0.16	0.53 ± 0.33
Cholecalciferol	119.50 ± 1.20	103.50 ± 15.50	108.50 ± 33.40
δ-tocopherol	42.48 ±7.96	8.31 ±1.63	6.16 ± 2.88
γ-tocopherol	4.81 ± 0.48	2.42 ± 0.37	0.92 ± 0.16
α-tocopherol	34.80 ± 14.50	8.16 ± 0.97	8.20 ± 3.67
Astaxanthin	2.23 ± 0.14	0.64 ± 0.23	0.98 ± 0.55
Lutein	6.41 ± 2.03	15.13 ± 8.66	10.79 ± 2.77
Zeaxanthin	0.03 ± 0.01	0.02 ± 0.01	0.02 ± 0.01
β-cryptoxanthin	6.10 ± 1.39	3.37 ± 0.20	1.66 ± 0.85
β-carotene	3.43 ± 2.09	2.28 ± 1.56	1.71 ± 0.21

2.10. Method Application in Other Samples

As stated before, no standard reference material is available for avocado. However, as a quality control material, we subjected our method to infant formula to further assess method accuracy. We obtained values according to the provider, and no appreciable matrix effects were observed for this food either (Figure 5A). Also, our method showed promise to extrapolate to other matrices, especially fruits. Unambiguous signals of several vitamins of interest can be seen when a chloroform extract obtained from green tomatoes was injected (Figure 5B).

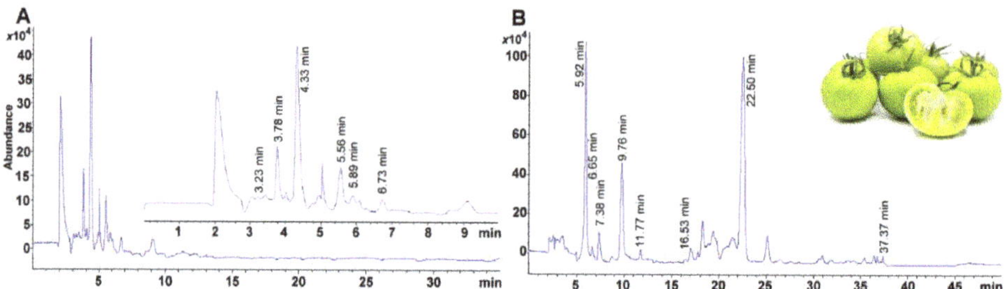

Figure 5. (**A**) Infant formula with certified values for fat-soluble vitamins. The box shows the amplification of the region from 1 to 10 min of the said chromatogram. FAPAS®® reference material TYG009RM. Vitamin A, vitamin D$_3$, and vitamin E at 508 ± 12, 7.16 ± 0.33, and 10,900 ± 400 µg/100 g. (**B**) Non-saponified sample of green tomatoes extracted after mechanical shearing and chloroform extraction and analyzed using the proposed method.

3. Materials and Methods

3.1. Reagents

tert-Butyl methyl ether (99%, catalog 34875, MTBE, chromatographic grade), methanol (≥99.9%, catalog 646377, MeOH, chromatographic grade) and potassium hydroxide (ACS reagent, catalog 1050210250) were acquired from Merck Millipore (Burlington, MA, USA). Retinyl acetate (catalog 46958), retinyl palmitate (catalog 46959-U), cholecalciferol (catalog C9774), ergocalciferol (catalog 47768), 3-phytylmenadione catalog (95271), α-tocopherol (catalog 47783), δ-tocopherol (catalog 47784), γ-tocopherol (catalog T1782), tocopherols (mixed, W530066), lutein (catalog 071068), β-carotene (catalog C4582), β-cryptoxanthin (≥97%, catalog C6368), zeaxanthin (catalog 14681), all-*trans*-astaxanthin (catalog 41659), and lycopene (catalog 75051) from Sigma-Aldrich (unless stated otherwise, all standards were of analytical grade, St. Louis, MO, USA). Chloroform (ACS reagent, catalog 366919), ethanol (200 proof, ACS reagent, ≥99.5%, 459844), 2-propanol (HPLC Plus, 650447), pyrogallol (ACS reagent, ≥99%, catalog 16040), and sodium sulfate (anhydrous, granular, free-flowing, Redi-Dri™, ACS reagent, ≥99%, catalog 798592) were also acquired from Sigma-Aldrich.

3.2. Sample Treatment and Preparation

Three different varieties of avocado collected from Costa Rica farms, two varietals from low and one highland, i.e., Simmonds (from 0 to 1000 m amsl), Guatemala (from 600 to 1500 m amsl), and Hass (from 1000 to 2000 m amsl), respectively [30,56]. Three batches of each variety were analyzed, each with eight days of maturation. A previously homogenized subsample (2 g) of freeze-dried material was weighed, and chloroform (20 mL, ACS reagent, Sigma-Aldrich 366919) was added. After that, the mixture was stirred continuously (30 min) using an Ultraturrax® (T25, at 7500 rpm, IKA Works Staufen, Germany). The remnant suspension was centrifuged, and the supernatant liquid recollected. The extraction procedure was repeated twice. The solvent was evaporated using a rotary evaporator (Multivapor™ P-6, Büchi, Flawil, Switzerland) until the lipid fraction was attained, which was then processed immediately for saponification.

3.3. Optimization of Saponification Conditions

An additional experiment was performed to enhance the recovery of the analytes of interest during the saponification reaction. The experimental design consisted of maintaining constant reaction time (1 h), the concentration of radical protection agent (0.1 g/100 mL), and the extraction solvent (hexane). Base concentration was contrasted (1 vs. 2 mol KOH L ethanol^{-1} [43] for each temperature), and the temperature was progressively increased (60, 80, and 95 °C). The conditions that rendered the most recoveries were selected to process the samples.

3.4. Sample Saponification

A small portion of the fat fraction (0.3 g) was weighed and quantitatively transferred to a conical centrifuge tube (50 mL, CLS430829, polypropylene, Corning®, New York, USA). Afterward, KOH in ethanol (1 mmol mL^{-1}) and containing pyrogallol (0.1 g/100 g) was added (10 mL). The mixture was let to saponify (during 1 h at 80 °C) in a heat bath (1229U55, Boekel Scientific, Feasterville, PA, USA). The resulting mixture was let to cool to room temperature and transferred to a Squibb separatory funnel (PYREX®, 250 mL, Corning® 6402). The above ethanol/aqueous layer was subjected to a liquid-liquid extraction using hexane. The extraction procedure was repeated twice. The totality of the organic solvent layer was then collected and filtered through sodium sulfate (used as a desiccant). The resulting solution was evaporated to dryness under a nitrogen flow (Ultra-High Pure Nitrogen was purchased from Praxair Technology Inc., Danbury, Connecticut, USA) and then reconstituted with MTBE (1 mL) and 2-propanol (1 mL) used at the start of the chromatographic separation and transferred to an HPLC vial (Agilent technologies, Santa Clara, CA, USA).

3.5. Stationary Phase and Selection of Chromatographic Conditions

The major obstacle in vitamin separation is the segregation of isomers. With this in mind, as a starter setup, we used a mobile phase based on MeOH and water using an eight carbon-based alkyl stationary phase (0.75 mL min^{-1}, Eclipse Plus C$_8$, 4.6 mm ID × 150 mm, 3 µm, Agilent Technologies). As the resolution was insufficient, a C$_{18}$ column was selected, and only flow was modified (1 mL min^{-1}, Eclipse Plus C$_{18}$, 4.6 mm ID × 150 mm, 3 µm, Agilent Technologies). Then we substituted the column for a C$_{30}$, removed water, and used a less polar solvent in acetonitrile and 2-propanol, reducing yet again the solvent flow (0.5 mL min^{-1}) (Table 1). Finally, retaining a similar proportion of MeOH, we substituted acetonitrile and isopropanol for MTBE. The C$_{30}$ column was kept as it already had excellent capabilities reported for highly lipophilic compounds (e.g., carotenoids) [39].

3.6. Chromatographic Conditions

All assays performed using an Agilent Technologies LC/MS system equipped with 1260 infinity quaternary pump (61311C), column compartment (G1316A), automatic liquid sampler modules (ALS, G7129A) and a 6120-single quadrupole mass spectrometer with electrospray ionization ion source (Agilent Technologies, Santa Clara, CA, USA). Gradient elution was used to separate all the compounds. The solvent gradient was optimized using MeOH (solvent A) and MTBE (solvent B), both acidified with formic acid (0.1 mL/100 mL). Solvent proportions were set as follows: at 0 min 80% A, at 5 min 80% A, at 7 min 73% A, at 15 min 62.5% A, at 20 min 62.5% A, at 30 min 45% A, at 35 min 10% A, at 40 min 10% A, at 45 min 80% A and 50 min 80% A. Flow rate was kept constant at 0.6 mL min^{-1}. Injection volume was held at 10 µL. The column compartment was held at a temperature of 10.0 ± 0.8 °C. Considering the need for the separation of structurally similar compounds, a 30-carbon alkyl chain based chromatographic column was used to achieve the analytical separation (YMC Carotenoid, 4.6 mm ID × 150 mm, S-3 µm, YMC Co., Ltd., Kyoto, Japan).

3.7. MS Detection System Conditions

The fragmentor was initially cycled to assess the voltage (from 20 to 300 V) that rendered the highest sensitivity for the compounds; omitting column interaction (Figure 6A). Afterward, total ion chromatographs (TIC) allowed us to obtain the MS spectra for each of the compounds (scan mode using a mass range and detector gain set to 50–750 *m/z*, and 10.00, respectively) (Figure 6B,C). Each TIC was used to identify the molecular ion signal. Drying gas, nebulizer pressure, drying gas temperature, and capillary voltage was set, respectively, to 12.0 L min^{-1}, 50 psi, 350 °C, 4000 V for positive ion mode electrospray ionization (ESI$^+$). Selected ion monitoring was used to corroborate each compound identity, remove interferences and improve sensitivity (SIM mode with peak width and cycle time set to 0.05 min, and 0.30 s cycle^{-1}, respectively) (Table 2, Figure 6D).

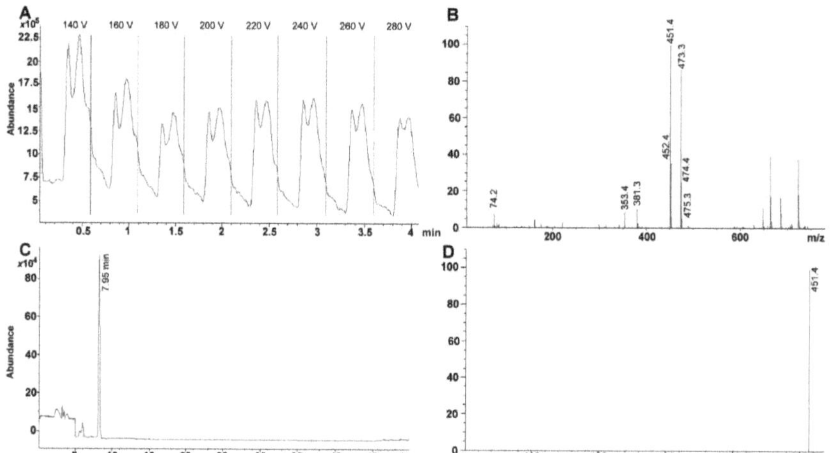

Figure 6. Example of the voltage cycling for phylloquinone to obtain the most sensitivity, data obtained at 100 µg mL^{-1}. (**A**) A parameter was selected where the signal delivered the most area under the curve. (**B**) Mass spectra obtained from a total ion chromatogram for phylloquinone at 100 µg mL^{-1}. Fragmentation as follows (molar mass 450.7 g mol^{-1}): 473.3 ([M + K]$^+$), 451.4 ([M + H]$^+$), 381.3 ([C$_{26}$H$_{35}$O$_2$]$^\bullet$), 353.4 ([C$_{24}$H$_{30}$O$_2$]$^{2\bullet}$), 225 ([C$_{15}$H$_{13}$O$_2$]$^\bullet$ and partial alkyl isoprenoid chain [C$_{16}$H$_{33}$]$^\bullet$), and 186 (quinone ring, [C$_{12}$H$_9$O$_2$]$^\bullet$) *m/z* [60]. (**C**) Chromatogram for phylloquinone was obtained using the selected [M + H]$^+$ 451 *m/z* and (**D**) selected ion monitoring (SIM) for the target analyte at 20 µg mL^{-1}.

Sensitivity is greatly improved using a SIM targeted scan. For example, the same standard 34.4 mg phylloquinone L^{-1} in TIC throws 37870 vs. 457785 area under the curve in SIM. Furthermore, within curve sensitivity reaches only 24.1 mg L^{-1} for TIC while the signal for 4.31 mg L^{-1}, in SIM, is still appreciable (i.e., 80471 area under the curve, 0.54 mg L^{-1} within curve sensitivity) (Figure 7A,B). Absolute sensitivity to vitamin K increases almost 50 fold (24.1/0.54). For carotenoids, the change is more dramatic as 100 mg L^{-1} standard has to be prepared in TIC for a detectable signal while 0.136 mg L^{-1} is still noticeable (Figure 2B), which represents ca. 750-fold in increased sensitivity.

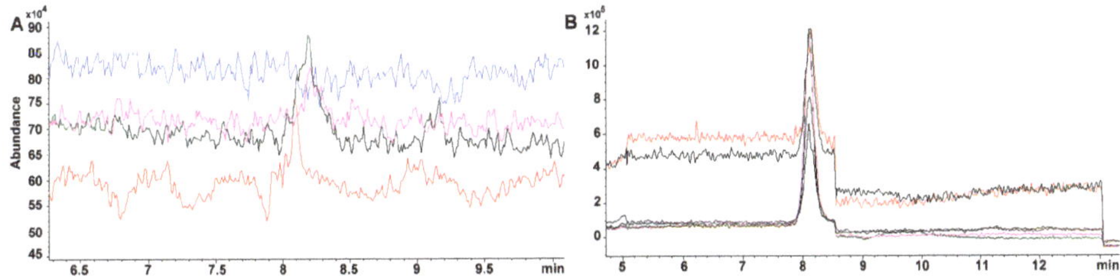

Figure 7. Calibration curve and sensitivity comparison using (**A**). TIC and SIM modes for vitamin K_1 both signals tested at 137.8, 68.9, 34.4, 17.2, 8.61, and 4.31 mg L^{-1} and using a fragmenter of 140 V. No signal is noticeable at 17.2 mg L^{-1} for TIC (blue line in panel (**A**)). Meanwhile, a calibration curve is easily constructed in SIM mode with the lowest point in 4.31 mg L^{-1} (purple line panel (**B**)).

Retention times and mass spectra were collected by the centroid of the chromatographic peak. Quantitation was carried out by comparing the peak areas found in the samples with those of standard solutions. The identification and quantification of targeted compounds analyzed by LC-ESI$^+$-MS were performed using OpenLab Chemstation C.01.07 (Agilent Technologies) for the processing of MS data sets. Confirmation of target analytes was based on the retention time (± 0.2 min as accepted time deviation), measurement of the molecular ion in a specific timeframe (Table 2).

3.8. Statistical Analysis

Calibration curves parameters (i.e., slopes and intercepts), coefficients of determination, limits of detection, and standard errors were computed as a linear fit model using SAS JMP 13 (Marlow, Buckinghamshire, England). An ANOVA with a post-hoc Dunnet test was used to assess differences among treatments during the optimization of the conditions during saponification. Concentrations obtained using the conditions 1 h and 1 mmol KOH mL ethanol^{-1} at 80 °C, were used as the control parameters; the test considered if the data was below the control, with $\alpha = 0.05$ significance level. The statistical analysis was performed using IBM SPSS®® Statistics 23 (Armonk, NY, USA).

4. Conclusions

The proposed method was regarded as a greener option by replacing chlorinated solvents and allowed two nutritionally relevant families of bioactive compounds (i.e., carotenoids and fat-soluble vitamins) to be analyzed together (which is not usually the case) and offered an adequate resolution in the case of tocopherol and calciferol isomers to improve their differential quantification. We obtained an accurate, sensitive, robust, and highly specific multi-analyte method that was successfully applied to avocadoes, a fruit of high economic value, of dietary interest, and a staple of Latin-American cuisine. The use of separation based entirely on organic solvents and the C_{30} column retention capability rendered a versatile method that can facilitate the incorporation of other pigments that have been reported present in avocado fruit (e.g., neoxanthin, trollichrome, chrysanthemaxanthin) [14,21] to further extend its chemical characterization. Saponification was paramount in the recovery of fat-soluble compounds. Therefore, optimized conditions should be assessed for each matrix to be tested. The method may be extended to evaluate fat-soluble vitamins and carotenoids in other matrices.

Mass spectrometry was a crucial tool in enabling the discrimination of structurally related compounds (e.g., all three tocopherol isomers could be easily accounted for in avocado).

Author Contributions: Conceptualization, F.G.-C., G.A., and C.C.-H.; methodology, F.G.-C. and C.C.-H.; software, F.G.-C. and C.C.-H.; validation, F.G.-C., G.A., A.C., and C.C.-H.; formal analysis, F.G.-C., G.A., A.C., and C.C.-H.; investigation, F.G-C.; resources, F.G.-C., G.A., and C.C.-H.; data curation, C.C.-H., F.G-C., and G.A; writing—original draft preparation, F.G.-C.; writing—review and editing, F.G.-C., G.A., A.C., C.C.-H.; visualization, F.G.-C.; supervision, F.G.-C. and C.C-H; project administration, F.G.-C. and C.C-H.; funding acquisition, C.C-H.

Funding: This research received no external funding except for the APC, which was funded by the Vice Provost Office for Research of the Universidad de Costa Rica.

Acknowledgments: Laura Arroyo is acknowledged for acquiring the avocado samples and preliminary integration of a few avocado samples and María Sabrina Sánchez for their suggestions, revising the manuscript and for language editing.

Conflicts of Interest: The authors declare no conflict of interest.

References

1. Arias, F.; Montoya, C.; Velásquez, O. Dinámica del Mercado mundial de aguacate. *Rev. Virtual Univ. Católica Del Norte* **2018**, *55*, 22–35. [CrossRef]
2. Xiong, B.; Song, Y. Big data and dietary trend: The case of avocado imports in China. *J. Int. Food Agribus. Mark.* **2018**, *30*, 343–354. [CrossRef]
3. Macías Macías, A. Mexico in the International Avocado Market. *Rev. De Cienc. Soc. (RCS)* **2011**, *17*, 517–532.
4. Díaz Vasquez, J.; Ardila Lopez, C.; Guerra Aranguren, M.A. Case Study on the Eligibility of Colombian Hass Avocado in the US Market: Opportunities in East Asia. *Online J. Mundo Asia Pac.* **2019**, *8*, 5–27.
5. PROCOMER [Promotora de Comercio Exterior de Costa Rica]. Anuario Estadístico 2017. PROCOMER: San José, Costa Rica, 2018. Available online: https://procomer.com/en/estudios/anuario_estadistico_2018 (accessed on 25 August 2019).
6. Butnariu, M. Methods of analysis (extraction, separation, identification, and quantification) of carotenoids from natural products. *J. Ecosys. Ecograph.* **2016**, *6*, 2. [CrossRef]
7. Unlu, N.Z.; Bohn, T.; Clinton, S.K.; Schwartz, S.J. Carotenoid Absorption from Salad and Salsa by Humans Is Enhanced by the Addition of Avocado or Avocado Oil. *J. Nutr.* **2005**, *135*, 431–436. [CrossRef]
8. Brown, M.J.; Ferruzzi, M.G.; Nguyen, M.L.; Cooper, D.A.; Schwartz, S.J.; White, W.S. Carotenoid bioavailability is higher from salads ingested with full-fat than with fat reduced salad dressings as measured with electrochemical detection. *Am. J. Clin. Nutr.* **2004**, *80*, 396–403. [CrossRef]
9. Desmarchelier, C.; Borel, P. Overview of carotenoid bioavailability determinants: From dietary factors to host genetic variations. *Trends Food Sci Technol.* **2017**, *69*, 270–280. [CrossRef]
10. Meyer, M.D.; Landahl, S.; Donetti, M.; Terry, L.A. Avocado. In *Health-Promoting Properties of Fruits and Vegetables*, 1st ed.; Terry, L.A., Ed.; CABI: Oxfordshire, UK, 2011; pp. 27–50.
11. Dreher, M.L.; Davenport, A.J. Hass avocado composition and potential health effect. *Crit. Rev. Food Sci. Nutr.* **2013**, *53*, 738–750. [CrossRef]
12. Fulgoni, V.L.; Dreher, M.L.; Davenport, A.J. Avocado consumption is associated with better diet quality and nutrient intake, and lower metabolic syndrome risk in US adults: Results from the National Health and Nutrition Examination Survey (NHANES) 2001–2008. *Nutr. J.* **2013**, *12*. [CrossRef]
13. Comerford, K.B.; Ayoob, K.T.; Murray, R.D.; Atkinson, S.A. The role of Avocados in Maternal Diets during the periconceptional Period, Pregnancy, and Lactation. *Nutrients* **2016**, *8*, 313. [CrossRef] [PubMed]
14. Duarte, P.F.; Chaves, M.A.; Borges, C.D.; Mendoca, C.R.B. Avocado: Characteristics, Health Benefits and Uses. *Ciência Rural* **2016**, *46*, 747–754. [CrossRef]
15. Noorul, H.; Nesar, A.; Zafar, K.; Khalid, M.; Zeeshan, A.; Vartika, S. Health benefits and pharmacology of *Persea americana mill.* (Avocado). *Int. J. Res. Pharmacol. Pharmacother.* **2016**, *5*, 132–141.
16. Scott, T.M.; Rasmussen, H.M.; Chen, O.; Johnson, E.J. Avocado Consumption Increases Macular Pigment Density in Older Adults: A Randomized, Controlled Trial. *Nutrients* **2016**, *9*, 919. [CrossRef]

17. Lu, Q.-Y.; Arteaga, J.R.; Zhang, Q.; Huerta, S.; Go, V.L.W.; Heber, D. Inhibition of prostate cancer cell growth by an avocado extract: Role of lipid-soluble bioactive substances. *J. Nutr. Biochem.* **2005**, *16*, 23–30. [CrossRef]

18. Gross, J.; Gabai, M.; Lifshitz, A. The carotenoid of the avocado pear. Persea americana, Nabal Variety. *J. Food Sci.* **1972**, *37*, 589–591. [CrossRef]

19. Moran, N.E.; Johnson, E.J. Closer to clarity on the effect of lipid consumption on fat-soluble vitamin and carotenoid absorption: Do we need to close in further? *Am. J. Clin. Nutr.* **2017**, *106*, 969–970. [CrossRef]

20. Grune, T.; Lietz, G.; Palou, A.; Catharine Ross, A.; Stahl, W.; Tang, G.; Thurnham, D.; Yin, S.; Bielaski, H.K. β-Carotene is an important vitamin A source for human. *J. Nutr.* **2010**, *140*, 2269S–2285S. [CrossRef]

21. Mardigan, L.P.; dos Santos, V.J.; da Silva, P.T.; Visentainer, J.V.; Gomes, S.T.M.; Matsuchita, M. Investigation of bioactive compounds from various avocado varieties (*Persea americana* Miller). *Food Sci. Technol.* **2018**, *39* (Suppl. 1), 15–21. [CrossRef]

22. Ashton, O.B.O.; Wong, M.; McGhie, T.K.; Vather, R.; Wang, Y.; Requejo-Jakcman, C.; Ramankutty, P.; Woolf, A.B. Pigments in Avocado Tissue and Oil. *J. Agric. Food Chem.* **2006**, *54*, 10151–10158. [CrossRef]

23. Yano, M.; Kato, M.; Ikoma, Y.; Kawasaki, A.; Fukazawa, Y.; Sugiura, M.; Matsumo, H.; Oohara, Y.; Nagao, A.; Ogawa, K. Quantitation of Carotenoids in Raw and Processed Fruits in Japan. *Food Sci. Technol. Res.* **2005**, *11*, 13–18. [CrossRef]

24. Ramos, A. *Phytochemical analysis of avocado seeds (Persea amerciana Mill., c.v. Hass)*, 1st ed.; Cuvillier Verlag: Göttingen, Germany, 2007; pp. 26–28.

25. Talabi, J.Y.; Osukoya, O.A.; Ajayi, O.O.; Adegoke, G.O. Nutritional and antinutritional compositions of processed Avocado (*Persea americana* Mill) seeds. *Asian J. Plant. Sci. Res.* **2016**, *6*, 6–12.

26. Khoo, H.-E.; Prasad, K.N.; Kong, K.-W.; Jiang, Y.; Ismail, A. Carotenoids and Their Isomers: Color Pigments in Fruits and Vegetables. *Molecules* **2011**, *16*, 1710–1738. [CrossRef]

27. Pérez-Gálvez, A.; Roca, M. Recent Developments in the Analysis of Carotenoids by Mass Spectrometry. In *Progress in Carotenoid Research*, 1st ed.; Zepka, L.Q., Ed.; IntechOpen Limited: London, UK, 2018; pp. 17–44.

28. Lu, Q.-Y.; Zhang, Y.; Wang, Y.; Wang, D.; Lee, R.-P.; Gao, K.; Byrns, R.; Heber, D. California Hass Avocado: Profiling of Carotenoids, Tocopherol, and Fat Content during Maturation and from Different Growing Areas. *J. Agric. Food Chem.* **2009**, *57*, 10408–10413.

29. Jacobo-Velázquez, D.A.; Hernández-Brenes, C. Stability of avocado paste carotenoids as affected by high hydrostatic pressure processing and storage. *Innov. Food Sci. Emerg. Technol.* **2012**, *16*, 121–128. [CrossRef]

30. Morera, J.A. *El Aguacate*, 1st ed.; Unidad de Recursos Fitogenéticos CATIE/GTZ: Turrialba, Costa Rica, 1983; pp. 10–19.

31. McGraw, K.J.; Toomey, M.B. Carotenoid accumulation in the tissues of Zebra Finches: Predictors of Intergumentary Pigmentation and Implication for Carotenoid Allocation strategies. *Physiol. Biochem. Zool.* **2010**, *83*, 97–109. [CrossRef]

32. Taygerly, J.P.; Miller, L.M.; Yee, A.; Peterson, E.A. A convenient guide to help select replacement solvents for dichloromethane in chromatography. *Green Chem.* **2012**, *14*, 3020–3025. [CrossRef]

33. Sander, L.C.; Sharpless, K.E.; Craft, N.E.; Wise, S.A. Development of Engineered Stationary Phase for the Separation of Carotenoid Isomers. *Anal. Chem.* **1994**, *66*, 1667–1674. [CrossRef]

34. Van Breemer, R.B.; Huang, C.-H. High-performance liquid chromatography-electrospray mass spectrometry of retinoids. *FASEB J.* **1996**, *10*, 1098–1101. [CrossRef]

35. Fraser, P.D.; Enfissi, E.M.A.; Goodfellow, M.; Eguchi, T.; Bramley, P.M. Metabolite profiling of plant carotenoids using the matrix-assisted laser desorption ionization time-of-flight mass spectrometry. *Plant J.* **2007**, *49*, 552–564. [CrossRef]

36. Xu, F.; Yuan, Q.P.; Dong, H.R. Determination of lycopene and β-carotene by high-performance liquid chromatography using sudan I as internal standard. *J. Chrom. B* **2006**, *838*, 44–49. [CrossRef] [PubMed]

37. Ding, J.; Hui, B. Quantification of all-*trans*-lycopene and β-carotene from tomato and its products by internal standard method on C30-HPLC. *Food Sci.* **2010**, *31*, 348–354.

38. Tan, J.; Leong Neo, J.G.; Setiawati, T.; Zhang, C. Determination of carotenoids in human serum and breast milk using high performance liquid chromatography coupled with a Diode Array Detector (HPLC-DAD). *Separations* **2017**, *4*, 19. [CrossRef]

39. Van Breemen, R.B. Liquid chromatography/mass spectrometry of carotenoids. *Pure Appl. Chem.* **1997**, *69*, 2061–2066.

40. Borman, P.; Elder, D. Q2 (R1) Validation of Analytical Procedures. In *ICH Quality Guidelines: An Implementation Guide*; John Wiley & Sons, Inc.: Hoboken, NJ, USA, 2017; pp. 127–166.

41. Kumar Saini, R.; Keum, Y.-S. Carotenoid extraction methods: A review of recent developments. *Food Chem.* **2018**, *240*, 90–103. [CrossRef]

42. Binnal, P.; Nirguna Babu, P. Production of high biodiesel through direct saponification of wet biomass of *Chlorella protothecoides* in a low cost microwave reactor: Kinetic and thermodynamic studies. *Korean J. Chem. Eng.* **2017**, *34*, 1027–1036. [CrossRef]

43. Li, T.; Xu, J.; Wu, H.; Wang, G.; Dai, S.; Fan, J.; He, H.; Xiang, W. A saponification method for chlorophyll removal from microalgae biomass as oil feedstock. *Mar. Drugs* **2016**, *14*, 162. [CrossRef]

44. Boon, C.S.; McClements, D.J.; Weiss, J.; Decker, E.A. Factors Influencing the Chemical Stability of Carotenoids. *Crit. Rev. Food Sci. Nutr.* **2010**, *50*, 515–532. [CrossRef]

45. Pénicaud, C.; Achir, N.; Dhuique-Mayer, C.; Dornier, M.; Bohuon, P. Degradation of β-carotene during fruit and vegetable processing or storage: Reaction mechanisms and kinetic aspects: A review. *Fruits* **2011**, *66*, 417–440. [CrossRef]

46. Hadjal, T.; Dhuique-Mayer, C.; Madani, K.; Dornier, M.; Achir, N. Thermal degrdation kinetics of xanthophylls from blood orange in model and real food systems. *Food Chem.* **2013**, *138*, 2442–2450. [CrossRef]

47. Aparicio-Ruiz, R.; Mínguez-Mosquera, M.; Gandul-Rojas, B. Thermal degradation kinetics of lutein, β-carotene and β-cryptoxanthin. *J. Food Comp. Anal.* **2011**, *24*, 811–820. [CrossRef]

48. Silva Fernades, A.; Casagrande do Nascimento, T.; Jacob-Lopes, E.; Vera de Roso, V.; Queiroz Zepka, L. Carotenoids: A Brief Overview on Its Structure, Biosynthesis, Synthesis, and Applications. In *Progress in Carotenoid Research*, 1st ed.; Queiroz Zepka, L., Jacob-Lopes, E., Vera de Roso, V., Eds.; IntechOpen: London, UK, 2018; pp. 1–15.

49. Li, D.; Xiao, Y.; Zhang, Z.; Liu, C. Light-induced oxidation and isomeration of all-*trans*-β-cryptoxanthin in a model system. *J. Photochem. Photobiol. B Biol.* **2015**, *142*, 51–58. [CrossRef] [PubMed]

50. AOAC [Association of Official Analytical Chemists]. AOAC Guidelines for Single Laboratory Validation of Chemical Methods for Dietary Supplements and Botanicals. 2002. Available online: https://www.aoac.org/aoac_prod_imis/AOAC_Docs/StandardsDevelopment/SLV_Guidelines_Dietary_Supplements.pdf (accessed on 25 August 2019).

51. Scheppele, S.E.; Mitchum, R.K.; Rudolph, C.J.; Kinneberg, K.F.; Odell, G.V. Mass Spectra of Tocopherols. *Lipids* **1971**, *7*, 297–304. [CrossRef]

52. Djoumbou-Feunang, Y.; Pon, A.; Karu, N.; Zheng, J.; Li, C.; Arndt, D.; Gautam, M.; Allen, F.; Wishart, D.S. CFM-ID 3.0: Significanlty improved ESI-MS/MS prediction and compound identification. *Metbolites* **2019**, *9*, 72. [CrossRef]

53. Rivera, S.M.; Christou, P.; Canela-Garyoa, R. Identification of carotenoids using mass spectrometry. *Mass Spectrom. Rev.* **2013**, *9999*, 1–20. [CrossRef]

54. Liigand, J.; Laaniste, A.; Kruve, A. pH effects on electrospray ionization efficiency. *J. Am. Soc. Mass Spectrom.* **2017**, *28*, 461–469. [CrossRef]

55. De Rosso, V.V.; Mercadante, A.Z. Identification and quantification of carotenoids, by HPLC-PDA_MS/MS, from Amazonian fruits. *J. Agric. Food Chem.* **2007**, *55*, 5062–5072. [CrossRef]

56. Morera, J.A. Caracterización agronómica de una colección de variedades de aguacate (*Persea americana* Miller) en la subestación fraijanes, Alajuela, Costa Rica. *Rev. Agr. Trop.* **2004**, *34*, 19–25.

57. Toti, E.; Oliver Chen, C.-Y.; Palmery, M.; Villañ Valencia, D.; Peluso, I. Non-Provitamin A and Provitamin A Carotenoids as Inmunomodulators: Recommneded Diatary Allowance, Therapeutic Index, or Personalized Nutrition? *Oxidative Med. Cell. Longev.* **2018**, *2018*. [CrossRef]

58. Aremu, S.O.; Nweze, C.C. Determination of vitamin content form selected Nigerian fruits using spectrophotometric method. *Bangladesh J. Sci. Ind. Res.* **2017**, *52*, 153–158. [CrossRef]

59. Anary, P.M.; Santos, A.C. Nutritional Value of the Pulp of Different Sugar Apple Cultivars (*Annona squamosa* L.). In *Nutritional composition of fruit cultivars*, 1st ed.; Simmonds, M.S.J., Preedy, V.R., Eds.; Academic Press: Cambridge, MA, USA, 2015; p. 208.

60. Marinova, M.; Lütjohann, D.; Westhofen, P.; Watzka, M.; Breuer, O.; Oldenburg, J. A validated HPLC Method for the determination of vitamin K in human serum–First Application in a Pharmocological Study. *Open Clin. Chem. J.* **2011**, *4*, 17–27. [CrossRef]

Sample Availability: Samples of the fat-soluble vitamins and carotenoids are available from the authors.

 *molecules*

Article

Exposure of Human Gastric Cells to Oxidized Lipids Stimulates Pathways of Amino Acid Biosynthesis on a Genomic and Metabolomic Level

Mathias Zaunschirm [1,†], Marc Pignitter [1,*,†], **Antonio Kopic [1], Claudia Keßler [1], Christina Hochkogler [1], Nicole Kretschy [2], Mark Manuel Somoza [2] and Veronika Somoza [1]**

[1] Department of Physiological Chemistry, Faculty of Chemistry, University of Vienna, 1090 Vienna, Austria; Mathias.Zaunschirm@gmx.at (M.Z.); kopic_5@hotmail.com (A.K.); kessler_claudia@web.de (C.K.); christina.hochkogler@gmx.at (C.H.); veronika.somoza@univie.ac.at (V.S.)

[2] Department of Inorganic Chemistry, Faculty of Chemistry, University of Vienna, 1090 Vienna, Austria; nicole.kretschy@univie.ac.at (N.K.); mark.somoza@univie.ac.at (M.M.S.)

* Correspondence: marc.pignitter@univie.ac.at; Tel.: +43-14277-70621

† These authors contributed equally to this work.

Academic Editor: Severina Pacifico

Received: 15 October 2019; Accepted: 12 November 2019; Published: 14 November 2019

Abstract: The Western diet is characterized by a high consumption of heat-treated fats and oils. During deep-frying processes, vegetable oils are subjected to high temperatures which result in the formation of lipid peroxidation products. Dietary intake of oxidized vegetable oils has been associated with various biological effects, whereas knowledge about the effects of structurally-characterized lipid peroxidation products and their possible absorption into the body is scarce. This study investigates the impact of linoleic acid, one of the most abundant polyunsaturated fatty acids in vegetable oils, and its primary and secondary peroxidation products, 13-HpODE and hexanal, on genomic and metabolomic pathways in human gastric cells (HGT-1) in culture. The genomic and metabolomic approach was preceded by an up-to-six-hour exposure study applying 100 µM of each test compound to the apical compartment in order to quantitate the compounds' recovery at the basolateral side. Exposure of HGT-1 cells to either 100 µM linoleic acid or 100 µM 13-HpODE resulted in the formation of approximately 1 µM of the corresponding hydroxy fatty acid, 13-HODE, in the basolateral compartment, whereas a mean concentration of 0.20 ± 0.13 µM hexanal was quantitated after an equivalent application of 100 µM hexanal. An integrated genomic and metabolomic pathway analysis revealed an impact of the linoleic acid peroxidation products, 13-HpODE and hexanal, primarily on pathways related to amino acid biosynthesis ($p < 0.05$), indicating that peroxidation of linoleic acid plays an important role in the regulation of intracellular amino acid biosynthesis.

Keywords: linoleic acid peroxidation products; hexane; gastric cells; metabolomics; cDNA microarray

---

1. Introduction

In industrialized countries, the habitual diet is characterized by a high intake in dietary fats, mainly originating in heat-treated foods [1]. Among them, the popularity of deep-fried products is based on convenience, their crispy texture, and pleasant mouth feel compared to non-fried foods [2]. During the deep-frying process, the frying oils are severely heated, resulting in multiple chemical reactions of the oils' constituents [3]. One of the major determinants of the quality of a frying oil are lipid peroxidation products. Here, primary and secondary lipid peroxidation products are formed through autoxidation or photo-oxidation reactions of unsaturated fatty acids, resulting in the generation of lipid hydroperoxides and further decomposition products, such as aldehydes, carboxylic acids, alcohols, or hydrocarbons [4]. Under thermal treatment, free fatty acids undergo faster oxidation processes than under non-heated conditions [5]. The free fatty acid, linoleic acid, is an essential unsaturated fatty acid and the predominant omega-6 polyunsaturated fatty acid in the Western diet. Linoleic acid can be found in vegetable oils commonly used for deep-frying, such as rice bran, safflower, sunflower, soybean, corn, and canola oil [6,7]. Other prominent sources of linoleic acid, and therefore prone to oxidation, are walnuts, pine nuts, and pecans [8].

In several animal feeding studies, highly oxidized fats with peroxide values >10 meq/kg oil have been administered [9–12]. On the one hand, oxidized fats and oils have been hypothesized to be harmful to health and associated with, e.g., the development of atherosclerosis [13]. On the other hand, animal feeding studies have also shown that dietary oxidized fats lead to a decrease in triacylglycerols and cholesterol in liver and plasma, and to the regulation of genes involved in lipid metabolism [14]. However, studies on the effects of structurally-characterized primary and secondary peroxidation products of linoleic acid are lacking. Moreover, it has been shown that the primary peroxidation product of linoleic acid, 13-hydroperoxy-9Z,11E-octadecadienoic acid (13-HpODE), is mainly decomposed in the stomach into secondary peroxidation products such as its corresponding alcohols, epoxyketones and aldehydes, which were demonstrated to be partially incorporated into the intestinal lumen and further absorbed by enterocytes [10,15].

Since the stomach is a very reactive environment for the chemically rather instable dietary lipid peroxidation products [16], the presented work addresses whether linoleic acid and its primary and secondary peroxidation products, 13-HpODE and hexanal, are absorbed by human gastric cells (HGT-1) and show pre-absorptive cellular effects on genomic and metabolic levels.

2. Results

2.1. Quantitative Recovery of Linoleic Acid and Its Peroxidation Products in the Basolateral Compartment of HGT-1 Cells after Apical Exposure

After six hours of incubation with 100 µM linoleic acid, 13-HODE but no linoleic acid could be quantitated in the apical and basolateral compartment of the HGT-1 cells, respectively (Table 1). Treatment with 100 µM 13-HpODE revealed a 71% higher concentration of 13-HODE in the basolateral compartment of the cells compared to the apical compartment ($p \leq 0.05$), whereas the concentration of 13-HpODE did not reach the limit of detection (LOD: 0.1 µg/mL). Moreover, a substantially higher amount of 13-HODE (+ 87%) was quantitated in the cells' basolateral compartment after exposure to 13-HpODE compared to the treatment with linoleic acid ($p \leq 0.05$). After a 30 min incubation period with 100 µM hexanal, hexanal could be quantitated in the apical as well as basolateral compartments. Neither 13-HODE nor hexanal were detected in the lysates of HGT-1 cells.

Table 1. Quantitation of 13-HODE and hexanal in different compartments (apical, lysate, basolateral) of HGT-1 cells after six-hours incubation with 100 µM linoleic acid and 100 µM 13-HpODE or 0.5 h incubation with 100 µM hexanal [a].

Incubation Substance	Linoleic Acid	13-HpODE	Hexanal
Quantitation of	13-HODE [µM]	13-HODE [µM]	Hexanal [µM]
apical	1.11 ± 0.05 [a]	1.22 ± 0.05 [a]	3.15 ± 0.62 [b]
lysate	n.d.	n.d.	n.d.
basolateral	1.12 ± 0.05 [a]	2.09 ± 0.53 [b,*]	0.20 ± 0.13 [c,*]

[a] Data are displayed as mean ± SD (n = 3–4, tr = 1–2). Statistically significant differences were analyzed using two-way ANOVA ($p \leq 0.01$), followed by the Holm–Sidak post hoc test ($p \leq 0.05$). [a,b,c] Different letters in a row indicate significant differences between the three treatments ($p \leq 0.05$). Asterisks (*) indicate significant differences within one treatment between apical and basolateral compartments ($p \leq 0.05$).

2.2. Genomic Analysis of RNA

The impact of linoleic acid, 13-HpODE, and hexanal at the genomic level of HGT-1 cells was analyzed using a customized cDNA microarray [17]. The scatterplots show log base 2 fluorescence intensity after a six-hour incubation period with 100 µM linoleic acid, 100 µM 13-HpODE or 100 µM hexanal in HGT-1 cells (Figure 1). Treatment with 100 μM linoleic acid resulted in 1303 regulated probes (2.78% of all probes). Exposure of the cells to 100 μM 13-HpODE revealed 420 regulated probes (0.90% of all probes), whereas cells treated with 100 μM hexanal showed 193 regulated probes (0.41% of all probes) compared to medium-only treated cells. The standard threshold for regulation, either a ≤0.8 or ≥1.2-fold change benchmark for an altered gene expression, was used [18]. The software DAVID (database for annotation, visualization and integrated discovery) was used for analysis and to generate individual functional annotation clusters for gene expression changes of genes with similar biological properties and to show different enrichment scores (ES). Resulting annotation clusters with an enrichment score ≥1.3 are considered to be of statistical relevance [19]. After six hours of incubation with 100 μM linoleic acid, 100 μM 13-HpODE or 100 μM hexanal, selected annotation clusters with enrichment scores ≥1.3 were found (supplementary data, Table S1) to be involved in gene regulation of "tRNA splicing endonuclease subunit 2 (TSEN2)" (in annotation cluster: "mRNA processing", ES: 1.33)), "general transcription factor IIH subunit 1 (GTF2H1)" (in annotation cluster: "transcription initiation from RNA polymerase II promoter processing", ES: 1.62)) and "trace amine associated receptor 1 (TAAR1)" (in annotation cluster: "topological domain: Cytoplasmic", ES: 3.28)), respectively.

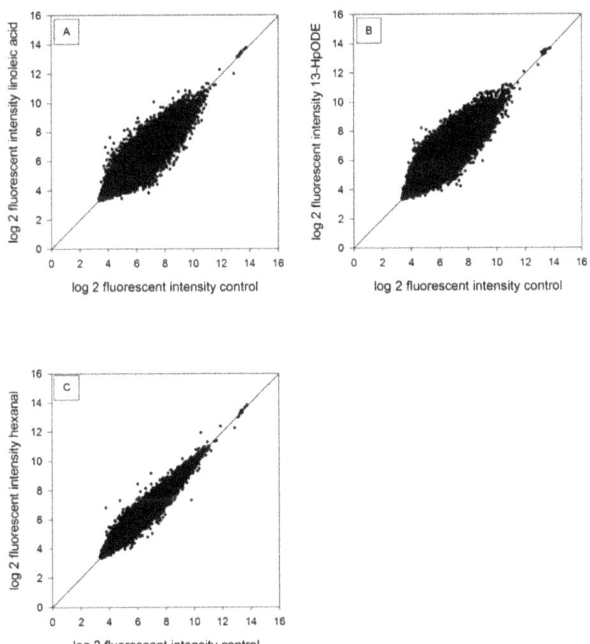

Figure 1. Scatterplots of log2 fluorescence intensity after six-hours incubation with (**A**) 100 μM linoleic acid, (**B**) 100 μM 13-HpODE or (**C**) 100 μM hexanal in HGT-1 cells. Diagonal represents equal regulation in untreated control and treated samples ($n = 3$).

2.3. Metabolomic Analysis

The impact of linoleic acid, 13-HpODE, and hexanal at the metabolic level of HGT-1 cells was examined by applying a pathway analysis at the metabolomic level. The scatterplots (Figure 2) show the metabolomic regulation [μM] after a six-hour incubation with 100 μM linoleic acid, 100 μM 13-HpODE, or 100 μM hexanal in HGT-1 cells. Treatment with 100 μM linoleic acid resulted in 49 regulated probes (30.1% of all probes), with 100μM 13-HpODE in 40 regulated probes (24.5% of all probes) and with 100 μM hexanal in 32 regulated probes (19.6% of all probes), which showed either ≤0.8 or ≥1.2-fold change used as a benchmark for potential changed metabolites [20]. With these probes, a metabolic pathway analysis was performed using MetaboAnalyst 3.6, revealing different pathways affected by changed metabolites, showing different *p*-values (*p*-values less than 0.05 show statistical relevance). Metabolic pathway analysis revealed that treatment with 100 μM linoleic acid altered metabolites affecting "Glycerophospholipid metabolism" ($p = 0.007$), "Aminoacyl-tRNA biosynthesis" ($p = 0.020$) and "Linoleic acid metabolism" ($p = 0.049$). Top pathways influenced after incubation with 10 0 μM 13-HpODE were shown to be "Aminoacyl-tRNA biosynthesis" ($p < 0.001$) and "Glycerophospholipid metabolism" ($p = 0.011$). An incubation with 100 μM hexanal resulted in a major impact of the metabolic pathways "Aminoacyl-tRNA biosynthesis" ($p < 0.001$), "D-Arginine and D-Ornithine metabolism" ($p < 0.001$) and "Valine, Leucine, Isoleucine" ($p = 0.008$) (supplementary data, Table S2). These results indicate an alteration of endogenous metabolites affecting especially the metabolic pathway of the aminoacyl-tRNA biosynthesis after a six-hour incubation period.

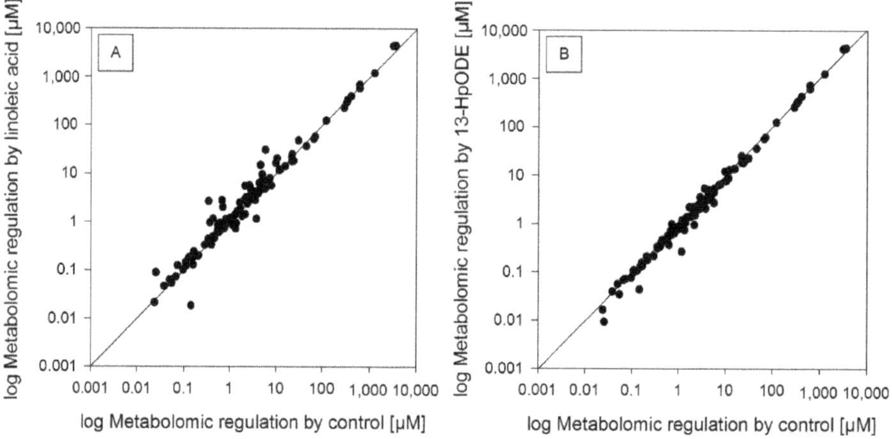

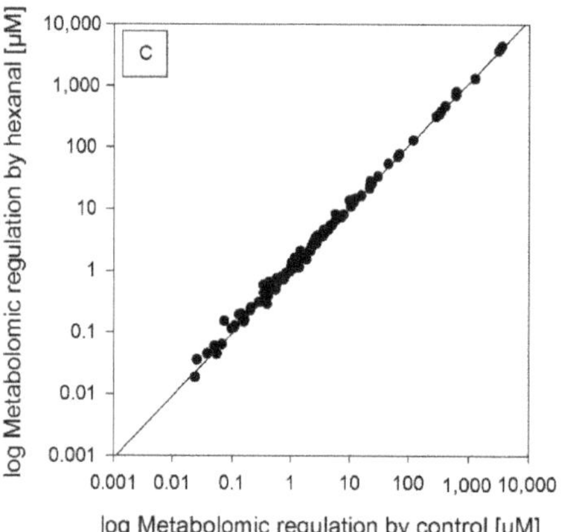

Figure 2. Scatterplots of metabolomic regulation [log μM] after six-hours incubation with (**A**) 100 μM linoleic acid, (**B**) 100 μM 13-HpODE, and (**C**) 100 μM hexanal in HGT-1 cells. Bisector represents equal regulation in untreated control and treated samples ($n = 3–4$).

2.4. Combined Genomic and Metabolomic Effects of Linoleic Acid and Its Lipid Peroxidation Products on HGT-1 Cells: An Integrated Pathway Analysis

Since a regulation of gene expression does not necessarily result in functional changes of metabolic pathways, an integrated pathway analysis of the combined genomic and metabolomic results was performed by means of the freely available MetaboAnalyst 3.6 software, a comprehensive tool for metabolomics analysis and interpretation [20]. After incubation with linoleic acid, 13-HpODE, or hexanal, genes and metabolites showing a fold change of either ≤0.8 or ≥1.2 were uploaded and an integrated pathway analysis was carried out. The data was mapped to KEGG (Kyoto encyclopedia of genes and genomes) metabolic pathways and enrichment and topology analyses were performed. Enrichment analysis evaluates the appearance of affected genes or metabolites in specific pathways. Topology analysis estimates the biological importance of affected genes or metabolites based on their position within a pathway. Table 2 shows selected enriched genes and metabolites, with a *p*-value less than 0.05, being involved in certain pathways after six hours of incubation with 100 µM 13-HpODE or 100 µM hexanal in HGT-1 cells. The top-ranked pathways being influenced by enriched genes and metabolites in combination after incubation with 100 µM 13-HpODE were "Aminoacyl-tRNA biosynthesis" ($p < 0.001$, 6 hits out of 87), "Linoleic acid metabolism" ($p = 0.03$, 3 hits out of 34) and "Arginine and proline metabolism" ($p < 0.05$, 5 hits out of 102). Incubation with 100 µM 13-HpODE increased levels of the amino acids glycine, methionine, proline leucine/isoleucine by 1.30, 1.26, 1.21, 1.28-fold, respectively, compared to non-treated controls (= 1). Incubation with 100 µM hexanal led to an enrichment of the pathways "Valine, leucine and isoleucine biosynthesis" ($p < 0.001$, 3 hits out of 13) and "Aminoacyl-tRNA biosynthesis" ($p < 0.001$, 5 hits out of 87). Moreover, treatment with 100 µM hexanal led to an upregulation of cellular levels of arginine (1.34-fold), glycine (1.20-fold), methionine (1.36-fold), valine (1.31-fold), leucine/isoleucine (1.33-fold) compared to untreated control cells (= 1). However, incubation with 100 µM linoleic acid did not result in a significant enrichment of any amino acid pathway. These outcomes indicate that predominantly pathways of the amino acid and protein biosynthesis were affected by the primary and secondary linoleic acid peroxidation products, 13-HpODE and hexanal, respectively, but not by the unoxidized linoleic acid (Figure 3).

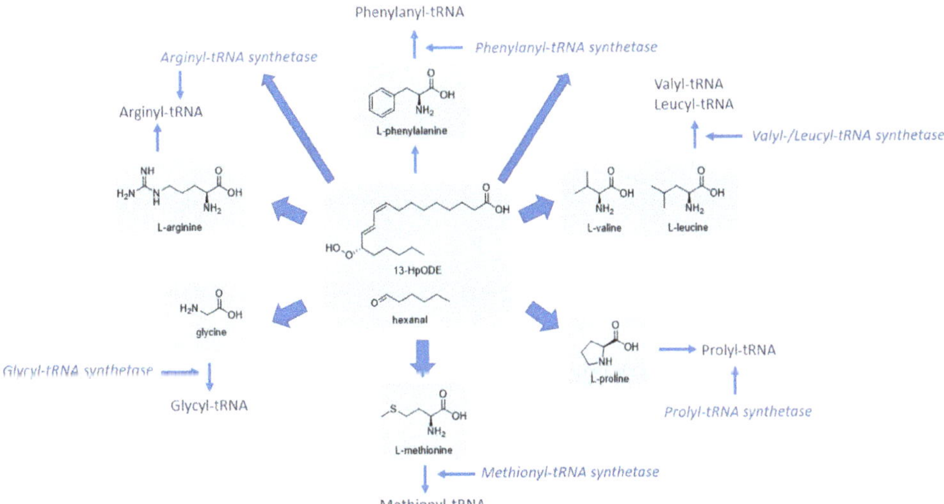

Figure 3. Integrated pathway analysis of genomic and metabolomic data revealed pathways related to amino acid biosynthesis being influenced by 13-HpODE and hexanal in gastric cells as indicated by thick arrows.

Table 2. Significantly enriched pathways ($p < 0.05$) based on the integrated methods pathway analysis after six-hours incubation with 100 µM linoleic acid, 100 µM 13-HpODE, or 100 µM hexanal in HGT-1 cells [a].

Compound	Pathway	Hits (Official Gene Symbol, KEGG Compound Entry)	p Value	Topology
Linoleic acid	Pyrimidine metabolism	UMPS, POLR1A, CANT1, NTSC3, TXNRD2, TYMP, TYMS	0.036	0.37
	Cyanoamino metabolism	5HMT1, C00037	0.040	0.75
13-HpODE	Aminoacyl-tRNA biosynthesis	C00123, C00037, RARS2, C00148, FARSB, C00073	< 0.001	0.12
	Arginine and proline metabolism	GLS2, ASS1, C00148, GOT2, ALDH3A2	0.004	0.17
	Linoleic acid metabolism	C00157, PLA2G6, CYP2C8	0.006	0.88
Hexanal	Valine, leucine and isoleucine biosynthesis	PDHA1, C00183, C00123	< 0.001	0.45
	Aminoacyl-tRNA biosynthesis	C00183, C00123, C00037, C00062, C00073	< 0.001	0.07
	Pantothenate and CoA biosynthesis	VNN2, C00183	0.030	0.13
	Arginine and proline metabolism	NOS2, C00062, C00077	0.044	0.27

[a] *p*-values less than 0.05 indicate a statistical significance compared to non-treated control cells.

3. Discussion

Consumption of thermally-treated and/or stored vegetable oils is accompanied by an intake of lipid peroxidation products generated during these processes [3,21]. Dietary intake of food-derived lipid peroxides is hypothesized to have adverse effects on human health, whereas studies on the effects of structurally-characterized lipid peroxidation products are lacking. There is evidence that ingested lipid peroxides undergo further reactions in the stomach during digestion, either forming further peroxidation products, or being decomposed, or partially incorporated by the gastric tissue [10,15,16]. Since linoleic acid is one of the quantitatively dominating fatty acids of vegetable oils, the exposure of human gastric cells in culture to linoleic acid and its primary and secondary peroxidation products, 13-HpODE and hexanal, was tested after six (linoleic acid and 13-HpODE) or 0.5 h (hexanal) of incubation, thereby simulating the range of retention times of fat during the gastric digestion in humans [22–24]. It has been shown in animal studies that the main degradation processes of linoleic acid hydroperoxides occur in the gastric environment [10,15]. Moreover, it has been demonstrated that 13-HpODE is not absorbed by intact Caco-2 intestinal cells. Instead they are likely to be metabolized by gastrointestinal glutathione peroxidase, which has been shown to be present in rats [25]. Therefore, the exposure study presented here was conducted in HGT-1 cells, based on the hypothesis that this cell model is more relevant than intestinal cells for investigating absorption mechanisms of linoleic acid and its peroxidation products. After six hours of incubation with 100 µM linoleic acid or 100 µM 13-HpODE, neither linoleic acid nor 13-HpODE were detectable in the apical compartment of HGT-1 cells, whereas 13-HODE could be quantitated. Also, HGT-1 cell exposure to 100 µM hexanal resulted in a mean detectable concentration of 0.20 ± 0.13 µM in the basolateral compartment, indicating the majority of hexanal might have been decomposed or evaporated after 0.5 h of incubation under the experimental conditions. It has to be noted that the degree of evaporation and/or decomposition was not investigated in the current study. It might also be conceivable that a lack of diffusion of hexanal in the system might be an explanation for the limited amount of hexanal detected on the basolateral side. Since linoleic acid and its peroxidation products are highly reactive compounds, their degradation during the six hours of incubation is conceivable, likely resulting in a conversion of linoleic acid and 13-HpODE to 13-HODE and many other decomposition products, which will need to be identified in future studies. Moreover, we cannot exclude metabolic or chemical modifications of the test compounds during the exposure study, e.g., formation of other, yet not identified peroxidation products or non-/covalent binding to proteins. An earlier study by Schieberle and Grosch [26] could show that, after three hours incubation at 38 °C in the presence of oxygen and di-tert-butyl peroxyoxalate, the incubated methyl 13-hydroperoxy-cis-9-trans-11-octadecadienoate was completely decomposed into a mixture consisting of oxo-, oxo-/epoxy-, hydroxy-, hydroxy-/epoxy-, oxo-/dihydroxy- compounds in a cell-free system. In our study, we demonstrated that 13-HODE and hexanal, both degradation products of 13-HpODE, were quantitated at the basolateral side of HGT-1 cells cultured in a transwell system and exposed to linoleic acid, 13-HpODE, or hexanal. This experiment revealed an 87% higher amount of 13-HODE at the basolateral side of the cells after incubation with 13-HpODE than with linoleic acid at the equimolar concentrations applied. An explanation for this result could be that 13-HpODE is a direct precursor of 13-HODE [5]. Linoleic acid, in contrast, would need to be first oxidized into 13-HpODE before its possible cleavage into 13-HODE, making 13-HpODE a better source of 13-HODE [5]. The incorporation of linoleic acid hydroxide into gastric tissue was also shown after intragastric administration of linoleic acid hydroperoxides to rats. Moreover, an application of linoleic acid hydroxide as single compound led to an 80% decrease during four hours in the gastric environment [10]. Ramsden et al. [27] could show that after a twelve-week intervention study lowering dietary linoleic acid intake, the circulating levels of 13-HODE significantly decreased in the plasma of the participants. Furthermore, our study showed a transfer of hexanal from the apical to the basolateral side of HGT-1, which is in accordance with findings from previous studies showing, first, an incorporation of hexanal by gastric rat tissue after administration of linoleic acid hydroperoxides during four hours, and, second, even an accumulation thereof in the rats' livers after intragastric administration of a radioactive labelled mixture of aldehydes 15 h post load [10,28]. Here we presented results indicating that the oxidation products of linoleic acid and their decomposition products might be absorbed by human gastric cells, thereby making 13-HODE available for further biological effects, although quantitative studies using stable isotope labelled lipid peroxidation products subjected to mimicked gastric digestion are needed to quantitate metabolic conversions and uptake kinetics. Furthermore, 13-HODE could function as a biomarker for the assessment of the oxidative status caused by dietary intake of oxidized lipids in vivo and in early detection of diseases, which was demonstrated in one study showing elevated levels of 9-/13-HODE in patients with nonalcoholic steatohepatitis compared to patients with steatosis [29,30]. Although hexanal has been shown to be a potential biomarker for some cancer types, such as lung or breast cancer [31,32], it also could be used as a marker of oxidized dietary fat intake.

In addition to our results indicating absorption of linoleic acid oxidation products by gastric cells, we could show that linoleic acid peroxidation products, 13-HpODE and hexanal, as opposed to unoxidized linoleic acid, influenced parameters related to amino acid and protein biosynthesis after a six-hour incubation in human gastric cells on genomic and metabolic levels. Genomic RNA regulation showed that a treatment with linoleic acid, 13-HpODE, and hexanal in tested concentrations affected genes, among others, which have been shown to be involved in transcription and translation [33]. However, a regulation on the genomic level does not necessarily result in an expression on the metabolic level. Moreover, metabolomic analysis showed that an incubation with linoleic acid had more pronounced effects on glycerophospholipid metabolism.

The integrated pathway analysis combining genomic and metabolomic results of linoleic acid and its peroxidation products, 13-HpODE and hexanal, demonstrated a significant impact of the linoleic acid peroxidation products on pathways related to amino acids in this pre-absorptive cell model. However, combining the effects of linoleic acid on genomic and metabolic levels eliminates the impact of linoleic acid on the regulation of amino acid metabolism. Instead, it could be shown that 13-HpODE and hexanal increased the concentrations of valine, leucine, and isoleucine, which are branched-chain amino acids. Branched-chain amino acids are associated with a broad spectrum of physiological effects in vivo, such as protein synthesis [34]. It is suggested that branched-chain amino acids, e.g., leucine, act as signaling molecules regulating protein synthesis by influencing mRNA translation, thus enhancing protein synthesis [35]. Zhang et al. [36] could also demonstrate that a thermally processed diet significantly affected amino acid profiles of fish. In the current study, the coincident rise of the branched chain amino acids after incubation of HGT-1 cells with 100 μM peroxidized lipids for six hours suggests a regulation of the branched-chain amino acid metabolism. Literature evidence has suggested elevated leucine levels to decrease lipogenic enzymes, fatty acid synthase and acetyl-coenzyme A carboxylase and corresponding upstream proteins [37]. Thus, leucine has been suggested to improve lipid metabolism and metabolic disorders [38], known to be associated with gastric cancer. Therefore, peroxidation of linoleic acid might be crucial in influencing cellular amino acid metabolism, thereby affecting metabolic health.

4. Materials and Methods

4.1. Chemicals

13-Hydroperoxy-9Z,11E-octadecadienoic acid (13-HpODE, purity: 95.1%) and ($^{13}C_{18}$)-13-hydroperoxy-9Z,11E-octadecadienoic acid ([$^{13}C_{18}$]-13-HpODE, purity: 92.0%) were synthesized as described in a previous publication of our group [39]. (±)-13-Hydroxy-9Z,11E-octadecadienoic acid (13-HODE, purity: ≥98%) was bought from Cayman Chemical (Cayman Europe, Tallinn, Estonia). Hexanal-d_{12} (98.5 atom% D, purity: 96%) was purchased from C/D/N Isotopes Inc. (Quebec, Canada). All other chemicals were purchased from Sigma-Aldrich (Vienna, Austria) unless otherwise stated.

4.2. Cell Culture

The human gastric adenocarcinoma (HGT-1) cell line was first characterized in 1982 [40]. Although HGT-1 cells do not secrete mucus, this parietal cell line does express all functional genes needed for proton secretion as key mechanism of gastric secretion. The applicability of the HGT-1 cell model to evaluate proton secretion was confirmed in an in vivo study [41]. Specifically, formation of intracellular cAMP by histamine is a common feature of HGT-1 cells and parietal cells in healthy gastric mucosa. We therefore consider the characteristics of HGT-1 cells being comparable with the physiology of parietal cells in the human stomach. HGT-1 cells were obtained as gift from Dr. C. Laboisse (Laboratory of Pathological Anatomy, Nantes, France) and cultured in DMEM with 4 g/L glucose supplemented with 10% fetal bovine serum, 2% L-glutamine and 1% penicillin/streptomycin under standardized conditions (37 °C, 5% CO_2 and 95% humidity) until reaching confluency. Effects of the tested substances on the cells' viability at a concentration of 100 μM revealed maximum exposure times of 30 min for hexanal and 6 h for linoleic acid and 13-HpODE not showing statistically different effects from medium-only treated cells by means of an MTT assay (data not shown). The MTT assay determines the reduction of 3-[4,5-dimethylthiazole-2-yl]-2,5-diphenyltetrazolium bromide (MTT) to the MTT-formazan which is catalyzed by mitochondrial succinate dehydrogenase.

4.3. Exposure Study

After reaching about 80% confluency, the HGT-1 cells were seeded onto Snapwell inserts with a polycarbonate membrane (tissue culture treated, 12 mm diameter, 0.4 μm pore size) (Costar, Corning Inc., New York, NY, USA) using DMEM supplemented with 20% fetal bovine serum (FBS) with a density of 1×10^5 cells/cm^2. On day one post-seeding, 250 μL of DMEM supplemented with 20% FBS were added to each apical compartment of the Snapwell inserts. On day three post seeding, a complete medium change (DMEM + 20% FBS) was performed in the apical and basolateral compartment, whereas the experiment was conducted on Day 5. The integrity of the cells' monolayer was monitored every day by measuring the transepithelial electrical resistance (TEER) by means of a volt/ohm meter (EVOM-24, World Precision Instruments Inc., Florida, FL, USA). The exposure experiment was done on Day 5 since there was a plateau of the TEER reached after four to eight days post seeding and additionally a phenol red permeability assay, performed on the experiment day, showed a penetration of phenol red of about 8% from the apical to the basolateral compartment after one hour without changing the TEER, confirming the integrity of the cell monolayer [42].

On the day of the exposure experiment, after measuring TEER and phenol red permeability, the supernatant of the basolateral and apical compartment was aspirated, and cells were washed twice with phosphate-buffered saline (PBS). Afterwards, 500 µL of incubation solution (100 µM linoleic acid, 100 µM 13-HpODE or 100 µM hexanal in DMEM, respectively) were added to the apical compartment of the Snapwell insert, whereas 3 mL of medium (DMEM without FBS) was added to the basolateral compartment prior to sealing of the Snapwell plate. These concentrations for incubations were chosen for further assays because they represent oil-characteristic amounts of the test substances [39,43–45]. After an incubation of linoleic acid and 13-HpODE for six hours and an exposure to hexanal for 0.5 h (37 °C, 5% CO_2, 95% humidity), which mimics relevant gastric retention time of food lipids, the supernatants from the basolateral and apical compartment were collected. Afterwards, pyrogallol (2 mM) was added to each sample and covered with argon, to avoid further oxidation, the samples were stored at −80 °C for further experiments.

After collection of the samples, the cells were lyzed. For that, the membrane of the apical compartment was washed with 500 µL PBS and 200 µM trypsin were added. After five minutes of trypsination (37 °C, 5% CO_2, 95% humidity), 200 µL medium were added and mixed well. After transferring the samples into reaction tubes, the Snapwell membrane was rinsed thrice with 400 µL H_2O (double-distilled), which was collected in the same reaction tube. Cell lysis was performed by means of three consecutive freeze-thaw cycles (30 s in liquid nitrogen then 3 min in heat block at 90 °C). Afterwards, centrifugation was done at 16,100× g at 4 °C for 10 min, the supernatants were collected, mixed with pyrogallol [2 mM] and covered with argon. The samples were stored at −80 °C for further experiments.

4.4. Quantitation of Linoleic Acid Peroxidation Products

The samples, obtained from the exposure assay after linoleic acid and 13-HpODE incubation, were spiked with $^{13}C_{18}$-13-HpODE, filtered through a nylon filter (0.2 µm, Phenomenex, Aschaffenburg, Germany) and subjected to quantitative analysis of 13-HpODE and 13-HODE according to Pignitter et al. [39] with slight modifications. Briefly, the samples were analyzed with a quadrupole liquid chromatography-mass spectrometry system (LCMS-8040, Shimadzu, Vienna, Austria) using a C18 column (Kinetex EVO, 150 mm × 4.6 mm; 5 µm diameter, Phenomenex, Aschaffenburg, Germany). The mobile phase was composed of 0.1% formic acid in methanol and double-distilled water. A flow rate of 1.0 mL/min was used, and the gradient elution program started with 60% methanol/40% double-distilled water, reaching a plateau of 90% methanol/10% double-distilled water after 10 min and returned to starting conditions after 30 min. The following MS settings were used for recording multiple reaction mode (MRM(-)): nebulizing gas flow, 3 L/min; drying gas flow, 10 L/min; desolvation line temperature, 250 °C; heat block temperature, 150 °C; CID gas, argon; collision energy, 15 V; MRM (precursor ion *m/z* -> product ion *m/z*), 13-HpODE (311 -> 113), $[^{13}C_{18}]$-13-HpODE (329 -> 120), 13-HODE (295 -> 277). Quantitation of 13-HpODE was done in MRM(-) mode using stable isotope dilution assay selecting transition ions *m/z* 113 for 13-HpODE and *m/z* 120 for $^{13}C_{18}$-13-HpODE, as internal standard [46]. 13-HODE was quantitated with external calibration using the transition ion *m/z* 277 [47]. Limit of detection (LOD = 0.1 µg/mL) was determined by a signal-to-noise ratio of 3, while the limit of quantitation (LOQ) was calculated based on a signal-to-noise ratio of 9.

4.5. Quantitation of Hexanal

After adding hexanal-d_{12} as an internal standard, the sample, obtained from the exposure assay, was extracted with 2 × 500 µL hexane, filtered through a nylon filter (0.2 µm, Phenomenex, Aschaffenburg, Germany) and subjected to quantitative analysis as reported previously, with minor modifications [21]. In short, the sample was injected using splitless mode and analysis was performed using a GC-MS (GCMS -QP 2010 Ultra, Shimadzu, Vienna, Austria) with a capillary column (ZB-WAX Zebron, 30 m × 0.25 mm i.d., 0.25 µm film thickness) for separation. Measurement parameters are shown in greater detail by Pignitter et al. [21]. Stable isotope dilution analysis was applied as quantitation method selecting fragment ions m/z 72 for hexanal and m/z 80 for hexanal-d_{12} in SIM mode. Limit of detection (LOD = 0.02 µg/mL) was determined by a signal-to-noise ratio of 3, while the limit of quantitation was calculated based on a signal-to-noise ratio of 9.

4.6. Genome Analysis of RNA Regulation

Effects of linoleic acid, 13-HpODE and hexanal on genomic level in HGT-1 cells were examined using a customized cDNA microarray as described before [48]. Briefly, after cells were seeded in a density of 10×10^5 cells/cm^2, on the next day cells were incubated with 100 µM linoleic acid, 100 µM 13-HpODE or 100 µM hexanal in DMEM (+ 1%FBS) for six hours at standardized conditions (37 °C, 5% CO$_2$, 95% humidity), respectively, and subsequently RNA was isolated using the RNeasy Mini Kit (Qiagen, Hilden, Germany). cDNA reverse transcription was carried out with Cy3-labeled primers (Tebu Bio, Offenbach, Germany) according to Ouellet et al. [49] After hybridization at 42 °C for 20 h, the microarrays were scanned with an Axon GenePix 4400 A microarray scanner (Molecular Devices, Sunnyvale, CA, USA). Fluorescence intensities were analyzed using NimbleScan 2.1 software and for normalization, the background intensities were corrected using robust multichip analysis (RMA). Moreover, four non-regulated reference genes (PPIA, GAPDH, TBP, UBC) were used for normalization of the fluorescence intensities of each microarray chip.

4.7. Metabolomic Analysis

To examine whether linoleic acid, 13-HpODE, and hexanal have an effect on metabolomic level in HGT-1 cells, targeted metabolomic analysis was performed using the AbsoluteIDQ p150 Kit (BIOCRATES Life Sciences AG, Innsbruck, Austria). Cells were seeded in a density of 6×10^5 cells/cm^2 and cultivated for one day. Afterwards, cells were washed with 500 µL PBS and incubated with 100 µM linoleic acid, 100 µM 13-HpODE, or 100 µM hexanal in DMEM (+ 1% FBS) for 6 h (37 °C, 5% CO$_2$, 95% humidity), respectively. After incubation, cells were washed twice with 500 µL ice-cold PBS and 300 µL H$_2$O (double-distilled) were added. Cell lysis was carried out by means of three consecutive freeze-thaw cycles (30 s in liquid nitrogen <-> 3 min in heat block at 90 °C). After collection of the cell lysates with a cell scraper and rinsing the well with 200 µL H$_2$O (double-distilled), cells were centrifuged at 20,000× g for 10 min at 4 °C. Finally, the supernatant was removed, covered with argon and stored at −80 °C until analysis.

Quantitative analysis was done according to the manufacturer's manual, which allows the determination of 163 endogenous metabolites from 5 compound classes (amino acids, acylcarnitines, sphingolipids, phospholipids, and the sum of hexoses). Analysis was carried out with a HPLC using electrospray-flow injection analysis followed by tandem mass spectrometry and isotope-labeled internal standards were used for quantitation. For validation of the kit and calculation of the metabolites' concentrations (µM) the MetIDQ Boron 5.4.8. software (BIOCRATES Life Sciences AG, Innsbruck, Austria) was used [50].

4.8. Statistical Analysis

Analyses were carried out with three to four independent biological replicates and one to two technical replicates. Statistical analyses were performed using SigmaPlot 11 (Systat Software Inc., Chicago, IL, USA). Two-way ANOVA followed by the Holm–Sidak post hoc test was used for determining statistical differences. P-values below 0.05 indicate statistical significance. Data are shown as mean ± standard deviation (SD) unless stated otherwise.

5. Conclusions

Stored and heat-treated oils are considerable sources for dietary lipid peroxidation products. There is evidence that dietary primary and secondary peroxidation products of linoleic acid, the most abundant polyunsaturated fatty acid in vegetable oils, are degraded in the stomach before being absorbed into the body. The research here presented results confirming this hypothesis, strongly suggesting that linoleic acid and its primary peroxidation product, 13-HpODE, were degraded into the corresponding hydroxy fatty acid, 13-HODE, which is very likely absorbed by gastric cells. However, it has to be noted that the degradation pathway(s) of 13-HpODE leading to the formation of 13-HODE were not investigated but need to be unraveled in future studies by means of, e.g., the CAMOLA technique [51]. For hexanal, a secondary peroxidation product of linoleic acid, results also indicated absorption by the gastric cells. Moreover, an integrated pathway analysis revealed that the pathways related to amino acid biosynthesis were mainly affected by linoleic acid primary and secondary peroxidation products in gastric cells. Elevated levels of branched-chain amino acids are supposed to improve lipid metabolism, potentially indicating beneficial effects of peroxidized lipids on metabolic health. To fully elucidate and verify effects of individual primary and secondary linoleic acid peroxidation products on cellular pathways related to amino acid metabolism, further in vivo studies are warranted.

Supplementary Materials: The following are available online at http://www.mdpi.com/1420-3049/24/22/4111/s1, Table S1: Annotation clusters from microarray probes, Table S2: Metabolic pathway analysis.

Author Contributions: Conceptualization, M.P. and V.S.; methodology, M.Z., M.P., M.M.S. and V.S.; investigation, M.Z., M.P., A.K., C.K., C.H., N.K.; resources, V.S.; writing—original draft preparation, M.Z. and M.P.; writing—review and editing, M.P., M.M.S. and V.S.

Funding: Open access funding by the University of Vienna, the Faculty of Chemistry of the University of Vienna, and the Austrian Science Fund (grant FWF P27275) are gratefully acknowledged.

Conflicts of Interest: The authors declare no conflict of interest.

References

1. Qi, Q.; Chu, A.Y.; Kang, J.H.; Huang, J.; Rose, L.M.; Jensen, M.K.; Liang, L.; Curhan, G.C.; Pasquale, L.R.; Wiggs, J.L.; et al. Fried food consumption, genetic risk, and body mass index: Gene-diet interaction analysis in three US cohort studies. *BMJ* **2014**, *348*, g1610. [CrossRef] [PubMed]

2. Guallar-Castillon, P.; Rodriguez-Artalejo, F.; Fornes, N.S.; Banegas, J.R.; Etxezarreta, P.A.; Ardanaz, E.; Barricarte, A.; Chirlaque, M.D.; Iraeta, M.D.; Larranaga, N.L.; et al. Intake of fried foods is associated with obesity in the cohort of Spanish adults from the European Prospective Investigation into Cancer and Nutrition. *Am. J. Clin. Nutr.* **2007**, *86*, 198–205. [CrossRef] [PubMed]

3. Zhang, Q.; Saleh, A.S.M.; Chen, J.; Shen, Q. Chemical alterations taken place during deep-fat frying based on certain reaction products: A review. *Chem. Phys. Lipids* **2012**, *165*, 662–681. [CrossRef] [PubMed]

4. Choe, E.; Min, D.B. Mechanisms and factors for edible oil oxidation. *Compr. Rev. Food Sci. Food Safe.* **2006**, *5*, 169–186. [CrossRef]

5. Shahidi, F.; Zhong, Y. Lipid oxidation and improving the oxidative stability. *Chem. Soc. Rev.* **2010**, *39*, 4067–4079. [CrossRef]

6. Lim, P.K.; Jinap, S.; Sanny, M.; Tan, C.P.; Khatib, A. The Influence of Deep Frying Using Various Vegetable Oils on Acrylamide Formation in Sweet Potato (Ipomoea batatas L. Lam) Chips. *J. Food Sci.* **2014**, *79*, T115–T121. [CrossRef]

7. Barison, A.; da Silva, C.W.P.; Campos, F.R.; Simonelli, F.; Lenz, C.A.; Ferreira, A.G. A simple methodology for the determination of fatty acid composition in edible oils through H-1 NMR spectroscopy. *Magn. Reson. Chem.* **2010**, *48*, 642–650. [CrossRef]

8. Li, D.; Hu, X. Fatty Acid Content of Commonly Available Nuts and Seeds. In *Nuts and Seeds in Health and Disease Prevention*, 1st ed.; Preedy, V.R., Watson, R.R., Patel, V.B., Eds.; Academic Press: Cambridge, MA, USA, 2011; pp. 35–42.

9. Shiozawa, S.; Tanaka, M.; Ohno, K.; Nagao, Y.; Yamada, T. Re-evaluation of peroxide value as an indicator of the quality of edible oils. *J. Food Hyg. Soc. Jpn.* **2007**, *48*, 51–57. [CrossRef]

10. Kanazawa, K.; Ashida, H. Dietary hydroperoxides of linoleic acid decompose to aldehydes in stomach before being absorbed into the body. *Biochim. Biophys. Acta* **1998**, *1393*, 349–361. [CrossRef]

11. Billek, G. Health aspects of thermoxidized oils and fats. *Eur. J. Lipid Sci. Technol.* **2000**, *102*, 587–593. [CrossRef]

12. Ghidurus, M.; Turtoi, M.; Boskou, G.; Niculita, P.; Stan, V. Nutritional and health aspects related to frying (II). *Rom. Biotechnol. Lett.* **2011**, *16*, 6467–6472.

13. Nourooz-Zadeh, J.; Tajaddini-Sarmadi, J.; Ling, K.L.E.; Wolff, S.P. Low-density lipoprotein is the major carrier of lipid hydroperoxides in plasma - Relevance to determination of total plasma lipid hydroperoxide concentrations. *Biochem. J.* **1996**, *313*, 781–786. [CrossRef] [PubMed]

14. Ringseis, R.; Eder, K. Regulation of genes involved in lipid metabolism by dietary oxidized fat. *Mol. Nutr. Food Res.* **2011**, *55*, 109–121. [CrossRef] [PubMed]

15. Kanazawa, K.; Ashida, H. Catabolic fate of dietary trilinoleoylglycerol hydroperoxides in rat gastrointestines. *Biochim. Biophys. Acta* **1998**, *1393*, 336–348. [CrossRef]

16. Kanner, J.; Lapidot, T. The stomach as a bioreactor: Dietary lipid peroxidation in the gastric fluid and the effects of plant-derived antioxidants. *Free Radic. Biol. Med.* **2001**, *31*, 1388–1395. [CrossRef]

17. Sack, M.; Hölz, K.; Holik, A.K.; Kretschy, N.; Somoza, V.; Stengele, K.P.; Somoza, M.M. Express photolithographic DNA microarray synthesis with optimized chemistry and high-efficiency photolabile groups. *J. Nanobiotechnol.* **2016**, *14*, 14. [CrossRef]

18. Grigoryev, D.N.; Ma, S.-F.; Irizarry, R.A.; Ye, S.Q.; Quackenbush, J.; Garcia, J.G.N. Orthologous gene-expression profiling in multi-species models: Search for candidate genes. *Genome Biol.* **2004**, *5*, R34. [CrossRef]

19. Huang, D.W.; Sherman, B.T.; Lempicki, R.A. Systematic and integrative analysis of large gene lists using DAVID bioinformatics resources. *Nat. Protoc.* **2009**, *4*, 44–57. [CrossRef]

20. Xia, J.; Wishart, D.S. Using MetaboAnalyst 3.0 for Comprehensive Metabolomics Data Analysis. *Curr. Protoc. Bioinform.* **2016**, *55*, 14.10.11–14.10.91. [CrossRef]

21. Pignitter, M.; Stolze, K.; Gartner, S.; Dumhart, B.; Stoll, C.; Steiger, G.; Kraemer, K.; Somoza, V. Cold fluorescent light as major inducer of lipid oxidation in soybean oil stored at household conditions for eight weeks. *J. Agric. Food Chem.* **2014**, *62*, 2297–2305. [CrossRef]

22. Benini, L.; Brighenti, F.; Castellani, G.; Brentegani, M.T.; Casiraghi, M.C.; Ruzzenente, O.; Sembenini, C.; Pellegrini, N.; Caliari, S.; Porrini, M.; et al. Gastric-Emptying of Solids Is Markedly Delayed When Meals Are Fried. *Dig. Dis. Sci.* **1994**, *39*, 2288–2294. [CrossRef] [PubMed]

23. Kong, F.; Singh, R.P. Disintegration of solid foods in human stomach. *J. Food Sci.* **2008**, *73*, R67–R80. [CrossRef] [PubMed]

24. Meyer, J.H.; Mayer, E.A.; Jehn, D.; Gu, Y.; Fink, A.S.; Fried, M. Gastric Processing and Emptying of Fat. *Gastroenterology* **1986**, *90*, 1176–1187. [CrossRef]

25. Wingler, K.; Muller, C.; Schmehl, K.; Florian, S.; Brigelius-Flohe, R. Gastrointestinal glutathione peroxidase prevents transport of lipid hydroperoxides in CaCo-2 cells. *Gastroenterology* **2000**, *119*, 420–430. [CrossRef] [PubMed]

26. Schieberle, P.; Grosch, W. Decomposition of linoleic acid hydroperoxides. II. Breakdown of Methyl 13-hydroperoxy-cis-9-trans-11-octadecadienoate by Radicals or Copper-II Ions. *Z. Lebensm. Unters.* **1981**, *173*, 192–198. [CrossRef]

27. Ramsden, C.E.; Ringel, A.; Feldstein, A.E.; Taha, A.Y.; MacIntosh, B.A.; Hibbeln, J.R.; Majchrzak-Hong, S.F.; Faurot, K.R.; Rapoport, S.I.; Cheon, Y.; et al. Lowering dietary linoleic acid reduces bioactive oxidized linoleic acid metabolites in humans. *Prostaglandins Leukot. Essent. Fat. Acids* **2012**, *87*, 135–141. [CrossRef]

28. Kanazawa, K.; Kanazawa, E.; Natake, M. Uptake of Secondary Autoxidation Products of Linoleic-Acid by the Rat. *Lipids* **1985**, *20*, 412–419. [CrossRef]

29. Feldstein, A.E.; Lopez, R.; Tamimi, T.A.R.; Yerian, L.; Chung, Y.M.; Berk, M.; Zhang, R.L.; McIntyre, T.M.; Hazen, S.L. Mass spectrometric profiling of oxidized lipid products in human nonalcoholic fatty liver disease and nonalcoholic steatohepatitis. *J. Lipid Res.* **2010**, *51*, 3046–3054. [CrossRef]

30. Yoshida, Y.; Umeno, A.; Akazawa, Y.; Shichiri, M.; Murotomi, K.; Hone, M. Chemistry of Lipid Peroxidation Products and Their Use as Biomarkers in Early Detection of Diseases. *J. Oleo Sci.* **2015**, *64*, 347–356. [CrossRef]

31. Guadagni, R.; Miraglia, N.; Simonelli, A.; Silvestre, A.; Lamberti, M.; Feola, D.; Acampora, A.; Sannolo, N. Solid-phase microextraction-gas chromatography-mass spectrometry method validation for the determination of endogenous substances: Urinary hexanal and heptanal as lung tumor biomarkers. *Anal. Chim. Acta* **2011**, *701*, 29–36. [CrossRef]

32. Li, J.; Peng, Y.L.; Liu, Y.; Li, W.W.; Jin, Y.; Tang, Z.T.; Duan, Y.X. Investigation of potential breath biomarkers for the early diagnosis of breast cancer using gas chromatography-mass spectrometry. *Clin. Chim. Acta* **2014**, *436*, 59–67. [CrossRef] [PubMed]

33. Paushkin, S.V.; Patel, M.; Furia, B.S.; Peltz, S.W.; Trotta, C.R. Identification of a Human Endonuclease Complex Reveals a Link between tRNA Splicing and Pre-mRNA 3′ End Formation. *Cell* **2004**, *117*, 311–321. [CrossRef]

34. Monirujjaman, M.; Ferdouse, A. Metabolic and physiological roles of branched-chain amino acids. *Adv. Mol. Biol.* **2014**, *2014*. [CrossRef]

35. Yoshizawa, F. Regulation of protein synthesis by branched-chain amino acids in vivo. *Biochem. Biophys. Res. Commun.* **2004**, *313*, 417–422. [CrossRef]

36. Zhang, Z.M.; Xu, W.T.; Tang, R.; Li, L.; Refaey, M.M.; Li, D.P. Thermally processed diet greatly affects profiles of amino acids rather than fatty acids in the muscle of carnivorous Silurus meridionalis. *Food Chem.* **2018**, *256*, 244–251. [CrossRef]

37. Jiao, J.; Han, S.F.; Zhang, W.; Xu, J.Y.; Tong, X.; Yin, X.B.; Yuan, L.X.; Qin, L.Q. Chronic leucine supplementation improves lipid metabolism in C57BL/6J mice fed with a high-fat/cholesterol diet. *Food Nut. Res.* **2016**, *60*, 31304. [CrossRef]

38. Yao, K.; Duan, Y.; Li, F.; Tan, B.; Hou, Y.; Wu, G.; Yin, Y. Leucine in Obesity: Therapeutic Prospects. *Trends Pharmacol. Sci.* **2016**, *37*, 714–727. [CrossRef]

39. Pignitter, M.; Zaunschirm, M.; Lach, J.; Unterberger, L.; Kopic, A.; Kessler, C.; Kienesberger, J.; Pischetsrieder, M.; Eggersdorfer, M.; Riegger, C.; et al. Regioisomeric distribution of 9- and 13-hydroperoxy linoleic acid in vegetable oils during storage and heating. *J. Sci. Food Agric.* **2017**, *98*, 1240–1247. [CrossRef]

40. Laboisse, C.L.; Augeron, C.; Couturier-Turpin, M.H.; Gespach, C.; Cheret, A.M.; Potet, F. Characterization of a newly established human gastric cancer cell line HGT-1 bearing histamine H2-receptors. *Cancer Res.* **1982**, *42*, 1541–1548.

41. Liszt, K.; Ley, J.P.; Lieder, B.; Behrens, M.; Stöger, V.; Reiner, A.; Hochkogler, C.M.; Köck, E.; Marchiori, A.; Hans, J.; et al. Caffeine induces gastric acid secretion via bitter taste signaling in gastric parietal cells. *Proc. Natl. Acad. Sci. USA* **2017**, *114*, E6260–E6269. [CrossRef]

42. Maznah, I., Jr. The use of Caco-2 cells as an in vitro method to study bioavailability of iron. *Malays. J. Nutr.* **1999**, *5*, 31–45. [PubMed]

43. Procida, G.; Cichelli, A.; Lagazio, C.; Conte, L.S. Relationships between volatile compounds and sensory characteristics in virgin olive oil by analytical and chemometric approaches. *J. Sci. Food Agric.* **2015**, *96*, 311–318. [CrossRef] [PubMed]

44. Romero, I.; Garcia-Gonzalez, D.L.; Aparicio-Ruiz, R.; Morales, M.T. Validation of SPME-GCMS method for the analysis of virgin olive oil volatiles responsible for sensory defects. *Talanta* **2015**, *134*, 394–401. [CrossRef] [PubMed]

45. Zhu, H.J.; Li, X.Q.; Shoemaker, C.F.; Wang, S.C. Ultrahigh Performance Liquid Chromatography Analysis of Volatile Carbonyl Compounds in Virgin Olive Oils. *J. Agric. Food Chem.* **2013**, *61*, 12253–12259. [CrossRef] [PubMed]

46. Dufour, C.; Loonis, M. Regio- and stereoselective oxidation of linoleic acid bound to serum albumin: Identification by ESI-mass spectrometry and NMR of the oxidation products. *Chem. Phys. Lipids* **2005**, *138*, 60–68. [CrossRef] [PubMed]

47. Hui, S.; Haihong, W.; Zhiming, G.; Chong, S.; Shuang, R.; Daoying, W.; Muhan, Z.; Fang, L.; Weimin, X. Simultaneous Determination of 13-HODE, 9,10-DHODE, and 9,10,13-THODE in Cured Meat Products by LC-MS/MS. *Food Anal. Methods* **2016**, *9*, 2832–2841. [CrossRef]

48. Holik, A.K.; Rohm, B.; Somoza, M.M.; Somoza, V. N(ε)-Carboxymethyllysine (CML), a Maillard reaction product, stimulates serotonin release and activates the receptor for advanced glycation end products (RAGE) in SH-SY5Y cells. *Food Funct.* **2013**, *4*, 1111–1120. [CrossRef]

49. Ouellet, M.; Adams, P.D.; Keasling, J.D.; Mukhopadhyay, A. A rapid and inexpensive labeling method for microarray gene expression analysis. *BMC Biotechnol.* **2009**, *9*. [CrossRef]

50. Hochkogler, C.M.; Lieder, B.; Rust, P.; Berry, D.; Meier, S.M.; Pignitter, M.; Riva, A.; Leitinger, A.; Bruk, A.; Wagner, S.; et al. 12-week intervention with nonivamide, a TRPV1 agonist, prevents a dietary-induced body fat gain and increases peripheral serotonin in moderately overweight subjects. *Mol. Nutr. Food Res.* **2017**, *61*. [CrossRef]

51. Schieberle, P. The carbon module labeling (CAMOLA) technique: A useful tool for identifying transient intermediates in the formation of maillard-type target molecules. *Ann. Ny Acad. Sci.* **2005**, *1043*, 236–248. [CrossRef]

Sample Availability: Samples of the compounds are not available from the authors.

 *molecules*

Article

UHPLC-HR-MS/MS-Guided Recovery of Bioactive Flavonol Compounds from Greco di Tufo Vine Leaves

Simona Piccolella [1], **Giuseppina Crescente** [1], **Maria Grazia Volpe** [2], **Marina Paolucci** [2,3] and **Severina Pacifico** [1,*]

1   Department of Environmental Biological and Pharmaceutical Sciences and Technologies, University of Campania "Luigi Vanvitelli", Via Vivaldi 43, 81100 Caserta, Italy; simona.piccolella@unicampania.it (S.P.); giuseppina.crescente@unicampania.it (G.C.)
2   Istituto di Scienze dell'Alimentazione, Consiglio Nazionale delle Ricerche (CNR), 83100 Avellino, Italy; mgvolpe@isa.cnr.it (M.G.V.); paolucci@unisannio.it (M.P.)
3   Dipartimento di Scienze e Tecnologie, via De Sanctis, snc, 82100 Benevento, Italy
*   Correspondence: severina.pacifico@unicampania.it; Tel.: +39-0823-274578

Received: 20 September 2019; Accepted: 6 October 2019; Published: 8 October 2019

Abstract: Leaves of *Vitis vinifera* cv. Greco di Tufo, a precious waste made in the Campania Region (Italy), after vintage harvest, underwent reduction, lyophilization, and ultrasound-assisted maceration in ethanol. The alcoholic extract, as evidenced by a preliminary UHPLC-HR-MS analysis, showed a high metabolic complexity. Thus, the extract was fractionated, obtaining, among others, a fraction enriched in flavonol glycosides and glycuronides. Myricetin, quercetin, kaempferol, and isorhamnetin derivatives were tentatively identified based on their relative retention time and TOF-MS^2 data. As the localization of saccharidic moiety in glycuronide compounds proved to be difficult due to the lack of well-established fragmentation pattern and/or the absence of characteristic key fragments, to obtain useful MS information and to eliminate matrix effect redundancies, the isolation of the most abundant extract's compound was achieved. HR-MS/MS spectra of the compound, quercetin-3-O-glucuronide, allowed us to thoroughly rationalize its fragmentation pattern, and to unravel the main differences between MS/MS behavior of flavonol glycosides and glycuronides. Furthermore, cytotoxicity assessment on the (poly)phenol rich fraction and the pure isolated compound was carried out using central nervous system cell lines. The chemoprotective effect of both the (poly)phenol fraction and quercetin-3-O-glucuronide was evaluated.

Keywords: food waste recovery; grape leaves; UHPLC-HR-MS/MS analysis; flavonol glycuronides recovery

---

1. Introduction

Food by-products and waste exploitation practices are gaining a lot of attention as these materials are an untapped but rich source for the recovery of bioactive compounds, favorably relevant for other food and feed scopes [1–4]. In fact, a consistent and recent literature highlights that valorizing agrofood wastes is not only a considerable alternative to composting, but also, and above all, a highly sustainable opportunity for obtaining added-value molecules, which could be efficaciously exploited in the nutraceutical and/or cosmeceutical sector, through an integrated approach involving multiple actors for an ecofriendly industrial development [5]. Phenols, carotenoids, and some other beneficial phytochemicals, together with pectin, are just a few examples of bioactives in agrofood wastes [6]. In particular, phenols and polyphenols, commonly found in high amounts in fruit and vegetable waste products, are broadly hypothesized to be used as natural food and drink preservatives, thanks to their ability to extend the expiration date of a product, thus delaying its rancidity and/or avoiding alteration of taste or other organoleptic characteristics [7]. Moreover, the pectin advantageous recovery makes the molecule largely exploitable as a gelling agent in pastry or as a fat replacement in meat products, or a binder in animal feed [8,9].

Considering that grape cultivation is one of the main agro-economic activities worldwide, with over 60 million tons produced globally every year [10], and that the entire wine production chain includes not only the production of grapes, their processing, and marketing, but also a large amount of wastes, it is reasonable to hypothesize fruitful recycle processes that go far beyond those already involved in the transformation of a grape waste part (such as that destined for distilleries) [11]. In fact, although during the wine-making process, different by-products are generated and found to be valuable for alternative use in a new production cycle (e.g., stems, grape-marc, and grapeseed) [12], waste and effluents, normally rich in sugars, proteins, fibres, and lipids as well as vitamins and other bioactive compounds, are also generated. Thus, they could represent an ideal source for obtaining chemicals and pharmaceuticals with high added value, as well as for creating biomaterials and substrates that can be used in different biological processes [13].

However, at the national and international level, an integrated approach, that allows a complete recovery of oenological wastes through the development of efficient and sustainable technologies from an economic and environmental point of view, leading to the final production of different products with standard features and demonstrating applications in specific sectors, is still lacking. Although marc and grapeseed already have an acclaimed use, even the leaves of the vine can be considered as the stalk, a curious waste to be recovered in the vine and wine industry [14]. Leaves, which unlike dregs, marc (skins and grape seeds), and stalks, are not included in the Italian Ministerial Decree from 27 November 2008 as renewable by-products, are waste material whose disposal during fruit harvesting is massive, not only in the early stages of winemaking, but mainly in destemming. Pre-bloom leaf removal, which consists of removing all or part of the leaves from the fruitful area in the period from spring to late season, is a practice commonly used with the aim of improving the quality of the harvest. On the other hand, countries such as Florida, Greece, and many Middle Eastern countries use the cultivation of vines also for the production of leaves used in kitchens for the preparation of typical dishes (e.g., the Arabian warak enab dish) in the knowledge that, like fruits, they contain numerous substances beneficial to humans such as organic acids, vitamins, and stilbenes [15]. The vine leaves are also rich in anthocyanins and tannins with vasoactive and vasoprotective properties and which have the ability to stimulate vascularization [16].

In this context and with the aim to exploit the recovery of bioactive molecules from wine production wastes, leaves of *Vitis vinifera* cv. Greco di Tufo, collected in the Campania Region (Italy), were considered. The production of this wine with the denomination of controlled and guaranteed origin represents a great resource of the territory. This ancient vine, whose name derives from the characteristic shape of the bunches, with grapes grouped in pairs, was introduced by the Greeks along the Tyrrhenian coasts. The geographical environment favors the production of this prized wine with a characteristic flavor.

Leaves were extracted through maceration and the alcoholic extract obtained was further fractionated. The phytoextract chemical composition was unravelled by UHPLC-HR-MS/MS analysis. The powerful analytical tool was thoroughly investigated for achieving suitable and valid information on the spectral behavior of Greco di Tufo leaf metabolites. The extract, and the most abundant compound isolated therefrom, were further evaluated for their potential cytotoxicity in SH-SY5Y neuroblastoma and U-251 MG glioblastoma cell lines.

2. Results and Discussion

The valorization of Greco di Tufo vine leaves, an unexplored source of bioactive molecules, took advantage of the design of a green and sustainable extraction method, pursued using ethanol, the most common biosolvent. The fraction GrM was broadly analyzed for its bioactives using UHPLC-HR-MS/MS tools, and for its cytotoxic effects. The lack of toxic effects and its ability to inhibit acetylcholinesterase enzyme, together with the observation of its richness in glucuronidated flavonols, with dissimilar fragmentation pattern in respect to that of the most common glycosides, laid the foundation for the phytochemical investigation of the extract and the purification of the most abundant compound, quercetin-*O*-glucuronide (GrM$_1$). The phytochemical approach represented a useful strategy to define the flavonol glucuronides' MS/MS chemical behavior for the rapid identification of these compounds.

2.1. GrM Chemical Composition

The alcoholic extract from the leaves of *Vitis vinifera* cv. Greco di Tufo, as evidenced by a preliminary UHPLC-HR-MS analysis, showed a high metabolic complexity. The extract was rich in (poly)phenol, alkylphenol, glycerolipid and glycerophospholipid components (Figure 1A).

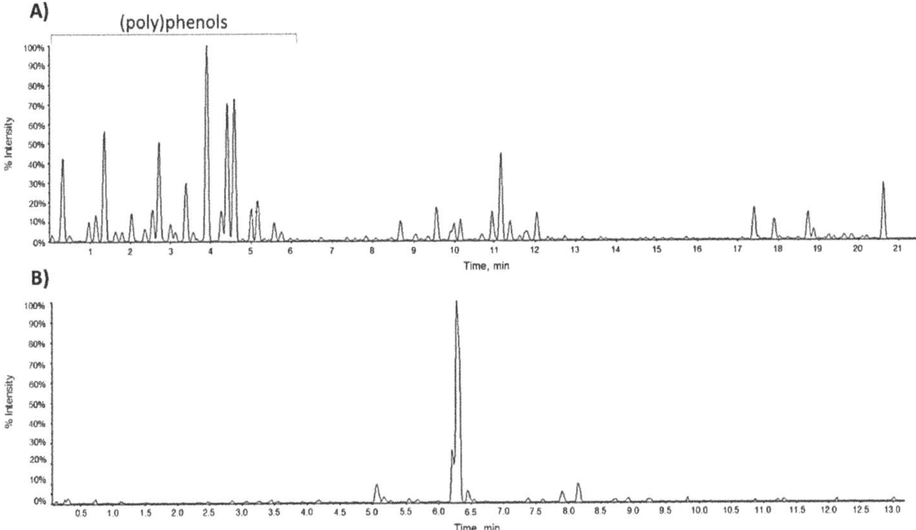

Figure 1. TIC (Total Ion Chromatogram) of (**A**) EtOH extract and (**B**) GrM fraction from Greco di Tufo vine leaves.

The parental extract was further fractionated by normal-phase column chromatography, using three solvents with increasing polarity. Among the fractions obtained, the alcoholic one, named GrM, was peculiarly enriched in flavonol glycosides and glycuronides (Table 1; Figure 1B). Flavonol hexuronides, not massively produced in the plant environment, are commonly described as bioconversion products of the phytochemicals taken with the diet or introduced by supplementation with less toxicity [17]. Indeed, their presence is not negligible in plants with common analytical techniques. In fact, these compounds, whose chemical structure was deeply elucidated by NMR spectroscopy, were also isolated from *Vitis* × *Labruscana* cv. 'Isabella' leaf methanol crude extract [18]; recently, their presence was suggested as part of the minor components in hemp seed oil [19].

Based on the relative retention time and the TOF-MS^2 data, five derivatives of myricetin (**4–7, 9**), three derivatives of quercetin (**12–14**), two derivatives of kaempferol (**15,16**) and two derivatives of isorhamnetin (**17,18**) have been tentatively identified (Table 1). The neutral loss of 162.05, 176.03 and 308.11 Da was in accordance with hexosyl, hexuronidyl and disaccharidic derivatives of the four flavonols. In particular, the neutral loss of 308.11 Da allowed us to hypothesize, for the metabolites **12, 15** and **17**, a deoxyhexose and hexose moiety, which on the basis of the relative intensity of the radical aglycone ion ([aglycone–H]$^{\bullet-}$) and [aglycone–H]$^-$, was linked to the –OH function in C-3 of the flavonolic nucleus in **12**, and to the phenolic function in C-7 in **15** and **17** (Figure 2). The identity of the flavonol glycoside **12** as rutin (rutinosyl derivative of quercetin) was further estimated by comparing the retention time and fragmentation pattern with that of the reference pure compound.

Table 1. Metabolites tentatively identified in the GrM extract.

	Rt (min)	Tentative Assignment	Formula	[M − H]⁻ Found (m/z)	[M − H]⁻ calc. (m/z)	Error (ppm)	RDB	MS/MS Fragment Ions (m/z)
1	0.308	Gallic acid hexoside	$C_{13}H_{16}O_{10}$	331.0666	331.0671	−1.3	6	169.014; 125.0248
2	3.129	Benzyl O-[arabinofuranosyl-(1-6)-glucoside]	$C_{18}H_{26}O_{10}$	401.1460 437.1232 [M + Cl]⁻	401.1453	1.7	6	269.1030; 193.0486; 101.0238; 85.0279
3	4.204	Lariciresinol hexoside	$C_{26}H_{34}O_{11}$	521.2043	521.2028	2.8	10	359.1507; 344.1273; 329.1035; 313.1078; 299.0936; 255.0657; 255.0687; 241.0509
4	5.086	Myricetin derivative	$C_{23}H_{16}O_{14}$	515.0456	515.0467	−2.2	16	339.0125; 317.0300; 316.0225; 271.0225; 178.9976; 151.0018
5	5.091	Myricetin hexoside 1	$C_{21}H_{20}O_{13}$	479.0840	479.0831	1.8	12	479.0829; 317.0310; 316.0235; 287.0209; 271.0263; 270.0187; 259.0268; 214.0280
6	5.094	Myricetin hexuronide	$C_{21}H_{18}O_{14}$	493.0630	493.0624	1.3	13	493.0665; 317.0310 (178.9980; 151.0042; 137.0243; 109.0291); 299.0203; 271.0253; 227.0347; 178.9989; 151.0039; 137.0246
7	5.224	Myricetin hexoside 2	$C_{21}H_{20}O_{13}$	479.0841	479.0831	2.1	12	317.0310; 316.0228; 287.0202; 271.0253; 270.0185; 259.0248; 242.0219; 214.0291; 178.9968; 151.0043
8	5.323	Furan/piran linalool oxide pentosyl hexoside	$C_{21}H_{36}O_{11}$	463.2198 509.2253 [M + FA]⁻	463.2185	2.8	4	463.0896; 331.1754; 161.0447; 101.0240; 85.0289
9	5.424	Myricetin hexosyl hexuronide	$C_{27}H_{28}O_{19}$	655.1143	655.1152	−1.4	14	655.1182; 493.0629; 479.0824; 317.0288
10	5.580	Caffeic acid derivative	$C_{22}H_{36}O_{12}$	491.2190	491.2134	5.9	5	329.1618; 227.1272; 101.0242
11	5.731	Hydroxygeraniol pentosyl hexoside	$C_{21}H_{38}O_{11}$	465.2353 501.2113 [M + Cl]⁻	465.2341	2.5	3	251.0794; 191.0578; 149.0447; 131.0352; 101.0241; 89.0246
12	6.270	Rutin	$C_{27}H_{30}O_{16}$	609.1486	609.1461	4.1	13	609.1498; 301.0343; 300.0268; 271.0227; 255.0285

Table 1. *Cont.*

	Rt (min)	Tentative Assignment	Formula	[M − H]⁻ Found (*m/z*)	[M − H]⁻ calc. (*m/z*)	Error (ppm)	RDB	MS/MS Fragment Ions (*m/z*)
13	6.288	Quercetin hexuronide	$C_{21}H_{18}O_{13}$	477.0685 955.1420 [2M − H]⁻	477.0675	3.2	13	477.0690; 301.0351; 283.0244; 273.0399; 255.0293; 245.0450; 211.0395; 178.9979; 151.0029; 121.0297; 107.0136
14	6.494	Quercetin hexoside	$C_{21}H_{20}O_{12}$	463.0896	463.0882	3	12	463.0896; 301.0345; 300.0268; 271.0236; 255.0281; 243.0281; 227.0327; 178.9926; 151.0017
15	7.998	Kaempferol rutinoside	$C_{27}H_{30}O_{15}$	593.1534	593.1512	3.7	13	593.1523; 285.0383; 284.0305; 255.0278; 229.0497; 227.0326
16	8.187	Kaempferol hexuronide	$C_{21}H_{18}O_{12}$	461.0738	461.0725	2.7	13	461.0757; 285.0406; 257.0452; 229.0503
17	8.722	Isorhamnetin rutinoside	$C_{28}H_{32}O_{16}$	623.1620	623.1618	0.4	13	623.1643; 315.0492; 314.0421; 300.0258; 271.0233
18	9.305	Isorhamnetin hexuronide	$C_{22}H_{20}O_{13}$	491.0845	491.0831	2.8	13	491.0834; 315.0482; 301.0333; 300.0250; 271.0221; 255.0271; 243.0271

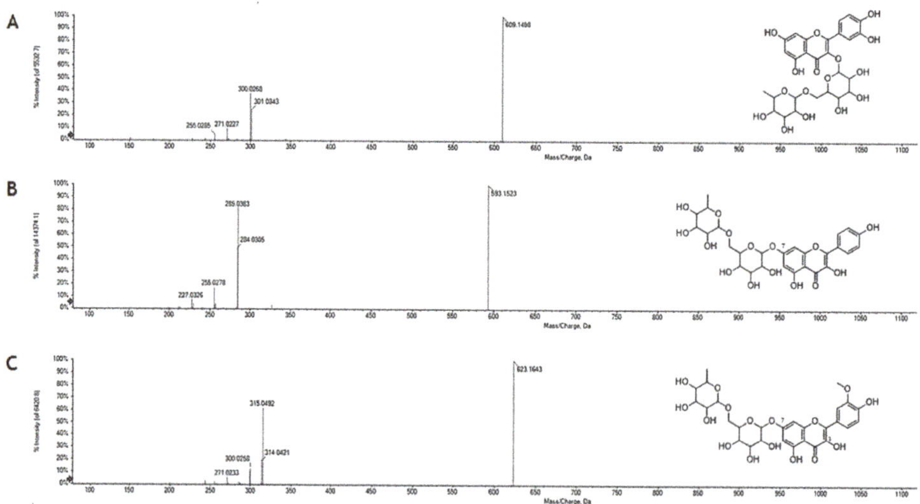

Figure 2. Flavonol rutinosyl derivatives; TOF-MS^2 spectrum of metabolites (**A**) **12**; (**B**) **15**; and (**C**) **17**.

The neutral loss of 176.03 Da, corresponding to a dehydrated hexuronic acid, characterized the MS/MS spectra of metabolites **6**, **13**, and **16**, whose deprotonated molecular ion dissociated providing the product ion [aglycone–H]$^-$ as base peak; the only exception was represented by compound **18**, for which the most favourable CH$_3$$^\bullet$ loss gave an abundant ion at *m/z* 300.0250 (Figure 3).

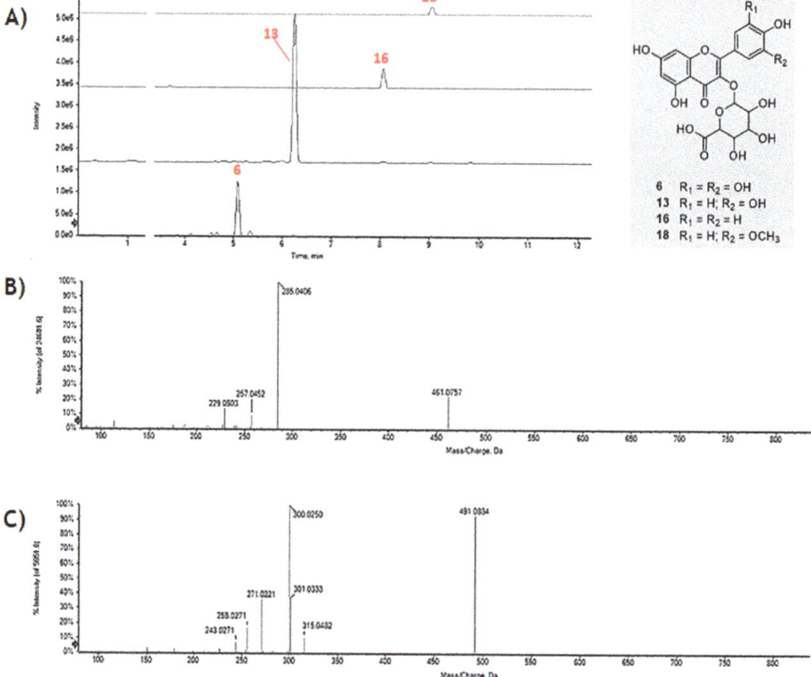

Figure 3. (**A**) Extracted ion chromatograms (XICs) of hexuronyl flavonols, whose structure is depicted in the grey box; TOF-MS^2 spectra of (**B**) **16** and (**C**) **18**.

The lack of well-established fragmentation pattern and/or the absence of characteristic key fragments make difficult the localization of the hexuronyl moiety. In fact, the main fragment detected, for example, for the abundant metabolite **13** was that corresponding to the deprotonated aglycone quercetin, as well as other characteristic ions of the flavonol such as those at *m/z* 273.0399, due to CO loss ([aglycone-28]$^-$), and at *m/z* 255.0293, which showed a relative intensity of only 5.1% and 8.5%, respectively. Other characteristic fragments of quercetin were identified in the ions at *m/z* 151.0029 and 178.9979 corresponding, respectively, to the deprotonated A ring, released by a retro-Diels Alder mechanism, and to the product of retrocyclization on bonds 1 and 2 [20]. A similar behavior was evident for the other hexuronyl derivatives and in particular, for those of myricetin. In Figure 4, the TOF-MS^2 spectra of myricetin derivatives **5**, **6** and **9** are reported; they were tentatively identified as myricetin hexoside, hexuronide and hexosyl hexuronide, respectively. It is evident that the presence of an hexuronyl moiety massively influences the fragmentation of the aglycone, impoverishing in intensity its characteristic ions.

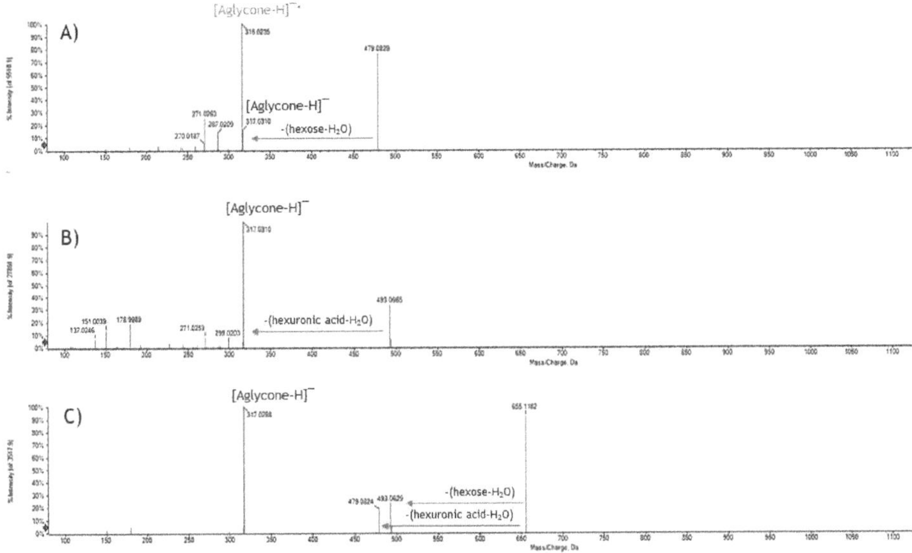

Figure 4. TOF-MS^2 spectra of metabolites (**A**) **5**; (**B**) **6**; and (**C**) **9**.

2.2. GrM$_1$ Purification

To obtain useful MS information and to eliminate matrix effect redundancies, a GrM fraction aliquot underwent thin-layer chromatography, which yielded, among others, the metabolite GrM$_1$ (the compound **13** of the GrM mixture). The UV–Vis spectrum of the molecule confirmed the presence of a flavonol skeleton molecule (Figure 5). In fact, flavonols, like flavones, present two major absorption peaks (λ_{max}) in the region between 240–280 nm (commonly referred to as band II) and between 300–380 nm (band I). This latter band favors the distinction of the two classes of flavonoids since the flavone λ_{max} is between 304–350 nm, while that of the flavonols is between 352–385 nm. Band I is associated with the absorption of the cinnamoyl system and band II with that of the benzoyl system (ring A). Literature evidence suggests that when glucuronidation occurs at the C-3 position, a hypsochromic shift of band I of about 14–29 nm is observed, whereas the glucuronidation at the phenolic function in C-7 does not lead to variations if not minimal or void [21]. GrM$_1$ spectrum showed, compared to the standard quercetin, a blue shift of the band I of 16 nm and a red shift of band II of 2 nm according to the quercetin glucuronidate in C-3.

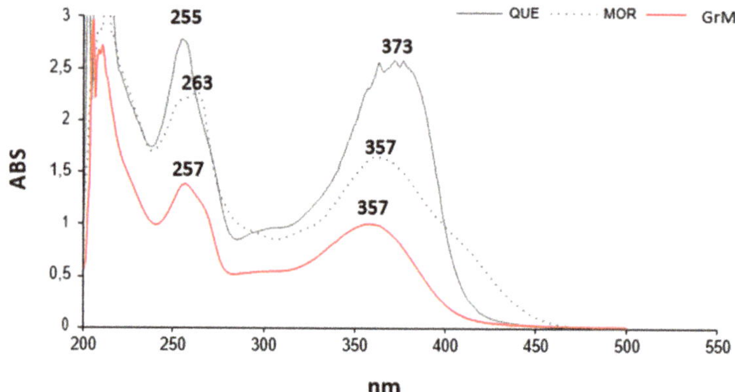

Figure 5. Comparison of the UV–Vis spectra of the metabolite GrM$_1$ and of quercetin (QUE) and morin (MOR) flavonol isomers.

The TOF mass spectrum of the molecule is completely superimposable to that recorded for peak **13** of the mixture (Figure 6) with the deprotonated molecular ion at *m/z* 477.0685, the ion [2M − H]⁻ at *m/z* 955.1420 and, again, the ion [aglycone − H]⁻ at *m/z* 301.0355. The TOF-MS^2 spectrum of the ion at *m/z* 477.0685 provided the ion [aglycone − H]⁻ as base peak and the ions [aglycone − H$_2$O − H]⁻ (*m/z* 283.0244), [aglycone − CO − H]⁻ (*m/z* 273.0399), [aglycone − CO − H$_2$O − H]⁻ (*m/z* 255.0293) and [aglycon e− 2CO − H]⁻ (*m/z* 245.0450), all with an intensity lower than 10%.

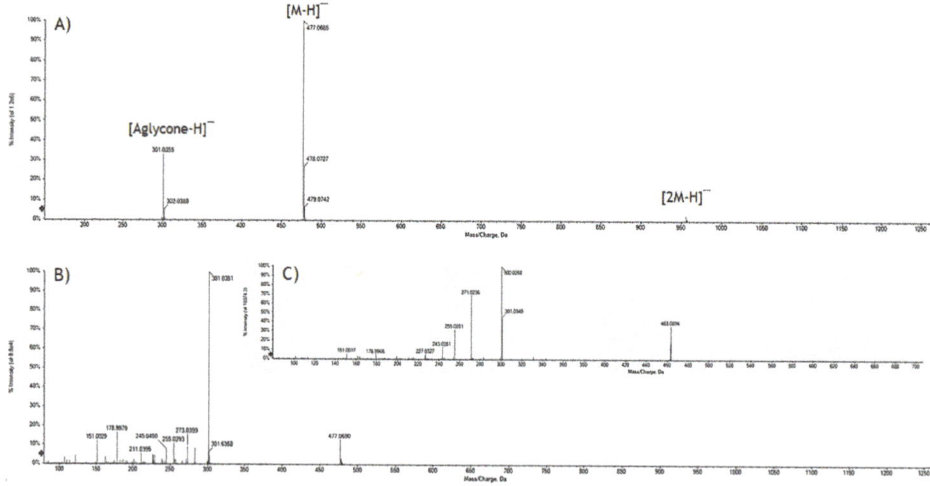

Figure 6. (**A**) TOF-MS and (**B**) TOF-MS^2 spectra of the quercetin-3-*O*-glucuronide. (**C**) Quercetin-3-*O*-glucoside TOF-MS^2 spectrum.

When [aglycone − H]⁻ ion dissociated (spectrum not shown), it provided, with relative abundance of 20%, the ion at *m/z* 151.0039 (calcd. 151.0337). The latter is the result of a loss of CO from the ion at *m/z* 178.9988 (calcd. 178.9986), whose presence, together with that of the ion at *m/z* 121.0300 (calcd. 121.0295), confirmed the presence of the flavonol quercetin. In fact, the two ions were attributable to the retrocyclization that is realized between the bonds 1 and 2 of the flavonol nucleus with formation of the fragments ^{1,2}A⁻ and ^{1,2}B⁻ (Figure 7). The loss of a CO$_2$ unit from the ion at *m/z* 151.0039 provided the ion at *m/z* 107.0134 (calcd. 107.0139). Comparing the spectrum of the standard quercetin with that obtained by GrM_1 deprotonated aglycone dissociation, the two realities were fully superimposable.

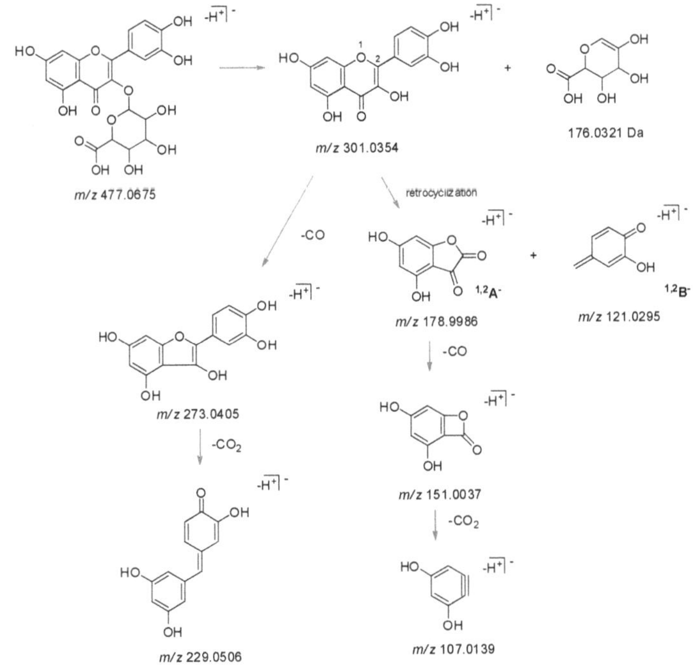

Figure 7. Proposed fragmentation pathway of GrM_1 [aglycone − H]⁻ ion. Theoretical *m/z* values are reported.

The [aglycone−H_2O−H]⁻ ion (*m/z* 283.0244), detected in GrM_1 TOF-MS² spectrum, could also represent a characteristic fragment defining the localization of the hexuronic acid in C-3 of the aglycone (Figure 8). The electronic delocalization on oxygen in C-4 could favor the formation of an enolic function, the proton abstraction with the formation of a good leaving group, whose detachment defines the formation of an anion in which there is a charge separation.

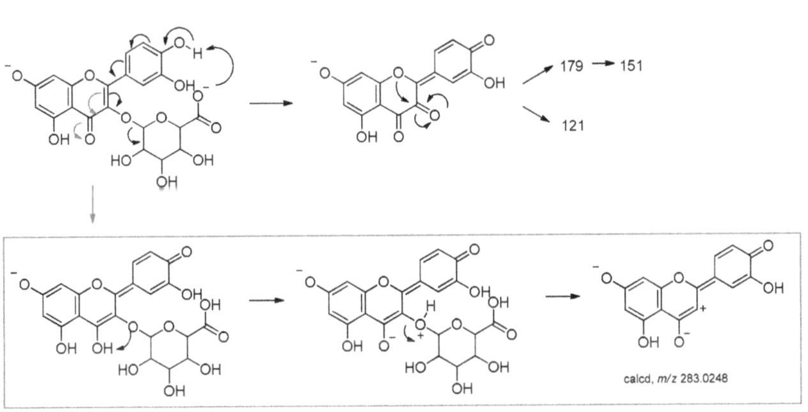

Figure 8. Proposed mechanism for the formation of the ion at *m/z* 283.0249. The calculated *m/z* value is reported (error < 5 ppm).

2.3. Cytotoxicity of GrM

Glucuronidated flavonoids display important health properties [22,23]. Baicalein 7-*O*-β-glucuronide was observed to promote wound healing and to exert antitumor activity [24,25]; quercetin 3-*O*-β-glucuronide is anti-inflammatory and neuroprotective, whereas 3-methoxyflavonol-4′-*O*-glucuronide is anti-allergenic and epicatechin glucuronide promotes vascular function. Glucuronidation greatly affects flavonoids' physiological properties, frequently their solubility and thus bioavailability. Bioactivity is differently influenced by glucuronidation and glucuronate moiety localization as it could be increased or decreased, whereas intra- and extra-cellular transport, understood as excretion, commonly increases [16]. In particular, it was reported that the oral administration of a blend of vine supplements is effective in protecting against neuropathologies and cognitive impairment that occurs with aging. Based on this evidence, the cytotoxicity of the GrM extract on the SH-SY5Y cell line was preliminarly evaluated. The choice of this cell line is based on its common use in in-vitro studies related to neurotoxicity and neurodegenerative diseases. Indeed, it is evident that primary cultures would be the best choice for this kind of investigation, as they are able to mimic the properties of neuronal cells in vivo. However, the preparation and culture of primary cells is much more challenging, especially for neuronal cells [26].

The 3-(4,5-dimethyl-2-thiazolyl)-2,5-diphenyl-2H-tetrazolium bromide (MTT) test was used for this purpose. It is able to measure the capacity of the mitochondrial dehydrogenases to reduce the tetrazolium ring of MTT, yellow colored, generating a chromogenic compound, the purple formazan. Obviously, this conversion is possible only in metabolically active cells. The results of the MTT assay suggested that the extract at doses ranging from 15.6 to 62.5 µg/mL did not massively influence the activity of mitochondrial dehydrogenases and a weak inhibitory effect on mitochondrial redox activity, equal to 31% and 38%, is recorded at the doses of 93.75 and 125 µg/mL, respectively (Figure 9).

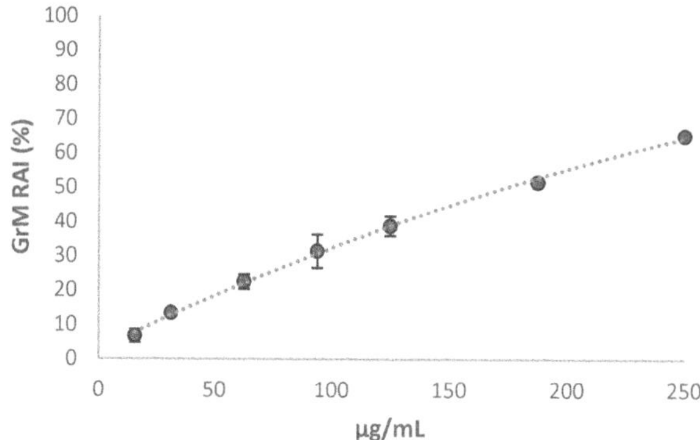

Figure 9. Redox mitochondrial inhibition (RAI%) of GrM extract towards SH-SY5Y cell line at 48 h exposure time. Values are reported as mean ± SD of measurements carried out on 3 samples (n = 3) analyzed 12 times.

The inhibition resulted was dose and time dependent and reached 62.1% of redox activity inhibition (RAI) for exposure to the highest tested dose. The lack or weak toxicity of the extract, especially at low doses, together with the chemical constitution that sees the co-presence of notoriously antioxidant molecules, led us to undertake studies evaluating the inhibitory properties of acetylcholinesterase. Acetylcholinesterase (AChE) plays an important biological role in the termination of the nerve impulse at the level of cholinergic synapses by rapid hydrolysis of its substrate, acetylcholine. Our data, although preliminary, show that the phytocomplex at 25 µg/mL inhibits the activity of the enzyme (16.8 ± 3.2%) similar to donepezil (DP; 21.7 ± 1.8), one of the drugs most commonly used to increase memory function in patients with Alzheimer's disease. The drug was tested, based on literature data, at the concentration of 3 µM, and was shown to exert an inhibition of cell proliferation of about 30% in SH-SY5Y cells [27].

2.4. Cytotoxicity of GrM₁

In order to verify the effects of the pure metabolite GrM_1, SH-SY5Y cells were exposed to the molecule and their cell viability was evaluated by the MTT test (Figure 10A). The molecule, tested at 25, 50 and 100 µM doses, did not exert toxic effects on the activity of mitochondrial dehydrogenases. These cytoprotective effects were further confirmed in U-251 MG, one of the immortalized glial cell lines which could be used instead of primary culture systems as a model for neural cells, based on the assumption that they are more homogenous, thus providing more useful tools. Indeed, glial cells are known as supportive elements of the nervous system, providing an optimal microenvironment for neurons [28].

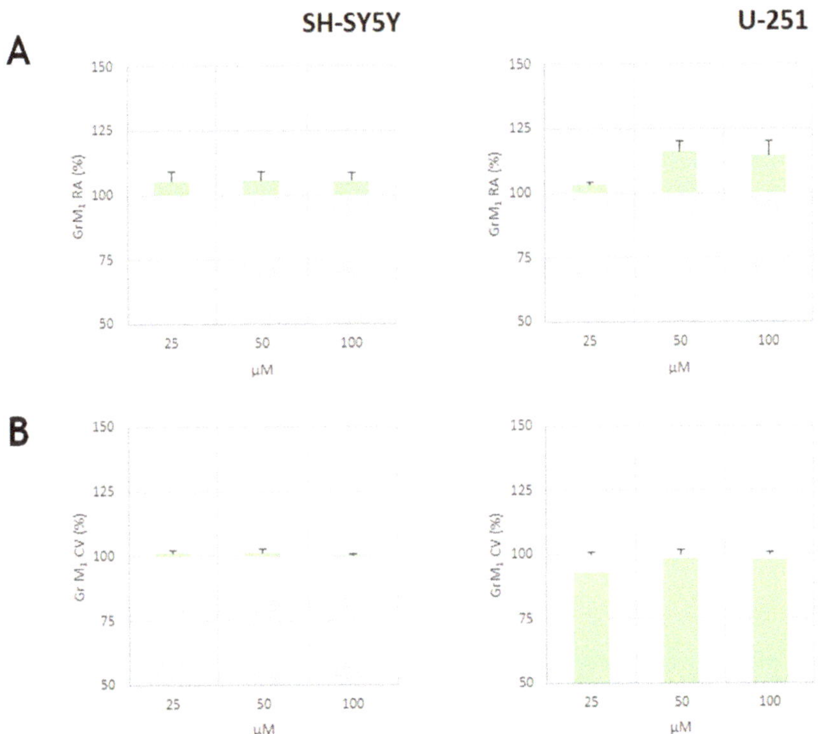

Figure 10. GrM_1 cytotoxicity on SH-SY5Y cells and U-251 MG after 48 h exposure time. Mitochondrial redox activity (RA, %) was from MTT test data (**A**), whereas cell viability (CV, %) was from sulforhodamine B (SRB) test (**B**). Values are reported as mean ± SD of measurements carried out on 3 samples (n = 3) analyzed 12 times.

The SRB (sulforhodamine B) test confirmed the absence of cytotoxicity (Figure 10D) and the weak proliferative effect was also verified by microscopic morphological change analysis (Figure 11).

Furthermore, after treating SH-SY5Y cells with GrM_1 at 50 µM concentration, in a preliminary cell metabolomics scenario, the extraction of the cell pellet with a solution of $MeCN:H_2O$ (1:1, *v:v*), after appropriate quenching and extraction operation, highlighted a peak at *m/z* 477.0696, whose TOF-MS² spectrum was super-imposable to that of GrM_1 (Figure 12).

The data obtained were in agreement with the results previously reported in literature, according to which the bioconversion of quercetin and rutin in the glucuronidate derivatives is accompanied, in the HL-60 leukemic cells [29], by a complete elimination of the toxic effect commonly ascribed to the most common flavonols and preservation of their structural entity in the intracellular environment.

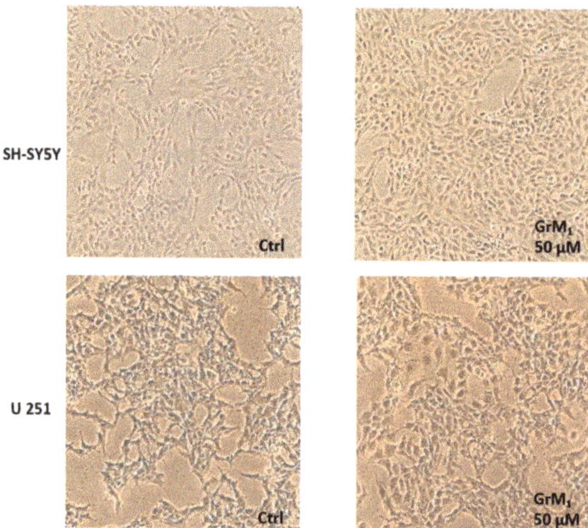

Figure 11. Morphological analysis of cells treated with GrM$_1$ pure compound after 48 h exposure time.

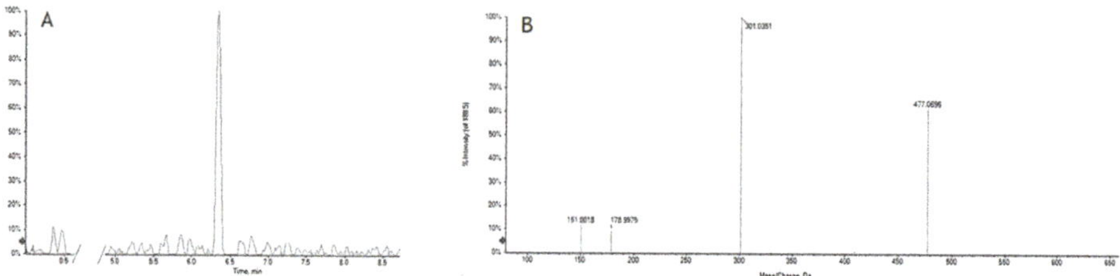

Figure 12. XIC (Extracted ion chromatogram) and TOF-MS^2 spectrum of the ion at *m/z* 477.0696 from cell pellet extract.

3. Materials and Methods

3.1. Plant Extraction and Fractionation

Leaves of *Vitis vinifera* cv. Greco di Tufo were collected in Montefusco (Avellino, Italy) in October 2017; the leaves were freeze-dried for 3 days using the FTS-System Flex-dry™ instrument (SP Scientific, Stone Ridge, NY, USA). Cryo-dried leaves were pulverized, using a rotary knife homogenizer and a sample (386.8 g) underwent solid-liquid extraction by maceration using ethanol as extracting solvent. Three extraction cycles (24 h each) were performed at 4 °C in the complete absence of light in order to obtain complete recovery of the metabolic content from *Vitis vinifera* cv. Greco di Tufo leaves. At the end of each cycle, the sample was filtered and the extraction solvent was removed using a rotary evaporator (Heidolph Hei-VAP Advantage, Schwabach, Germany). Ethanol extract was further fractionated by flash column chromatography (FCC) on Merck Kieselgel 60 (40–63 μm) silica gel under pure N$_2$ pressure (h = 20 cm, Ø = 4.0 cm), eluting first with CHCl$_3$, then with a CHCl$_3$:EtOAc solution (1:1, *v/v*), subsequently with pure EtOAc, and finally with pure MeOH. The alcoholic fraction (20.63 g) was further chromatographed on Amberlite XAD-4 (h = 70 cm, Ø = 4.0 cm) eluting with water first (GrW) and then with MeOH (GrM). GrM fraction was analyzed by UHPLC-HR-MS.

An aliquot of GrM fraction (36 mg) was chromatographed by thin-layer chromatography (TLC) using a precoated silica gel 60 F254 (20 × 20 cm, 1 mm, Merck, Darmstadt, Germany) and eluting with EtOAc:MeOH:H$_2$O:HCOOH (16:2:1:1) solution. GrM$_1$ (14.6 mg) was thus obtained and analyzed by UV–Vis and UHPLC-HR-MS. The fractionation scheme is depicted in Figure 13.

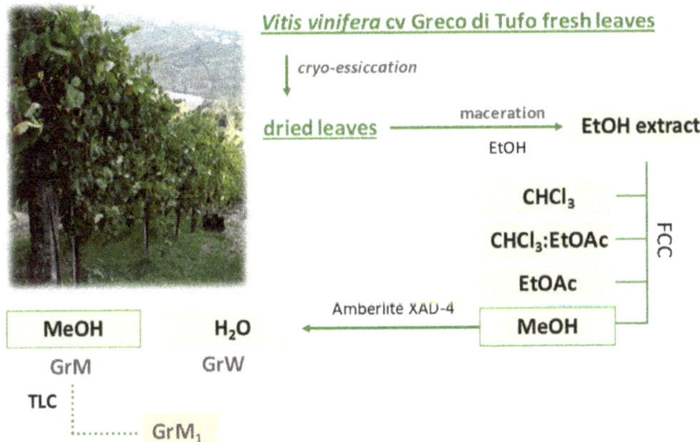

Figure 13. Extraction and fractionation scheme (FCC = flash column chromatography; TLC = thin-layer chromatography).

3.2. UHPLC-HR-MS and UV–Vis Analyses

GrM and GrM_1 fractions, placed in vials at a concentration of 10 mg/mL in pure methanol UHPLC grade, were analysed by the Shimadzu NEXERA UHPLC system and the Omega Luna C18 column (50×2.1 mm, 1.6 µm). The mobile phase consisted of a binary solution A: 0.1% formic acid in water and B: 0.1% formic acid in acetonitrile. A linear gradient was used for the analysis: 0–5 min, $5 \rightarrow 15\%$ B; 5–10 min, 15% B; 10–12 min, $15 \rightarrow 17.5\%$ B; 12–15 min, $17.5 \rightarrow 45\%$ B; 15–16.50 min, 45% B; 16.50–16.51 min, $45 \rightarrow 5\%$ B; 16.51–18.00 min, 5% B. The injection volume was 2.0 µL and the flow was set at 0.5 mL/min. MS analysis was performed using the AB SCIEX TripleTOF 4600 (AB Sciex, Concord, ON, Canada) system with a DuoSpray ion source operating in negative electrospray ionization. The APCI (Atmospheric Pressure Chemical Ionization) probe of the source was used for fully automatic mass calibration using the calibrant delivery system (CDS). CDS injects a calibration solution matching polarity of ionization and calibrates the mass axis of the TripleTOF® system in all the scan functions used (MS or MS/MS). Data were collected by information-dependent acquisition (IDA) using a TOF-MS survey scan of 100–1500 Da (250 ms accumulation time) and eight dependent TOF-MS/MS scans of 80–1300 Da (100 ms accumulation time), using a collision energy (CE) of 35 V with a collision energy spread (CES) of 25 V. The following parameter settings were also used: declustering potential (DP), 70 V; ion-spray voltage, –4500 V; ion source heater, 600 °C; curtain gas, 35 psi; ion source gas, 60 psi. Data processing was performed using the PeakView®-Analyst® TF 1.7 Software.

UV–Vis spectrum of GrM_1, as well as those of the pure reference compounds quercetin and morin, were acquired in the range 200–600 nm by a Shimadzu UV-1700 double beam spectrophotometer (Kyoto, Japan).

3.3. Cell Culture and Cytotoxicity Assessment

Tests assessing cell viability and mitochondrial activity were performed to monitor the cytotoxic potential of the GrM fraction from *Vitis vinifera* cv. Greco di Tufo and of the purified GrM_1. For this purpose, a stock solution of the two samples was prepared. Recorded activities were compared to an untreated blank arranged in parallel to the samples. Results are the mean ± SD values.

Human neuroblastoma cell line SH-SY5Y and glioma cell line U-251 MG were cultured in DMEM (Dulbecco's Modified Eagle Medium) medium supplemented with 10% fetal bovine serum, 50.0 U/mL of penicillin and 100.0 µg/mL of streptomycin at 37 °C in a humidified atmosphere containing 5% CO_2.

3.3.1. MTT (3-(4, 5-Dimethylthiazolyl-2)-2,5-Diphenyltetrazolium Bromide) Cell Viability Test

Cells were seeded in 96-multiwell plates at a density of 1.0×10^5 cells/well. After 24 h, cells were treated with different doses of the GrM extract (15.625, 31.25, 62.5, 93.75, 125.0, 187.5 and 250.0 µg/mL) and pure GrM_1 metabolite (25, 50 and 100 µM) in a culture medium for 48 h. At the end of incubation, MTT (150 µL; 0.50 mg/mL in culture medium) was added. After 1 h at 37 °C in a 5% CO_2 humidified atmosphere, MTT solution was removed and formazan was dissolved in dimethyl sulfoxide (DMSO) (100 µL). The absorbance at 570 nm was determined using a Tecan Spectra Fluor fluorescence and absorbance reader. Mitochondrial redox activity inhibition (RAI, %) was calculated using the following formula: $[(Abs_{untreated\ cells}) - (Abs_{treated\ cells})/(Abs_{untreated\ cells})] \times 100$, where Abs stands for absorbance [26].

3.3.2. SRB (Sulforhodamine B) Cell Viability Test

Cells were seeded in 96-multiwell plates at a density of 1.0×10^5 cells/well. After 24 h, cells were treated with pure GrM_1 metabolite (25, 50 and 100 µM) for 48 h. At the end of incubation, cells were fixed with ice-cold trichloroacetic acid (TCA) (10% *w/v*, 40 µL) for 1 h at 4 °C. The plates were washed five times in distilled water and allowed to dry. Then, sulforhodamine B (SBR; 50 µL, 0.4% *w/v* in 1% aqueous acetic acid) was added to each well and incubated at room temperature for 30 min. In order to remove unbound dye, the plates were quickly washed with 1% aqueous acetic acid and dried subsequently. The bound SRB was solubilized by adding 100 µL of 10 mM unbuffered Tris base (pH 10.5) to each well and shaking for 5 min on a shaker platform. Finally, the absorbance at 570 nm of each well was measured using a Tecan SpectraFluor fluorescence and absorbance reader. Cell viability inhibition (CVI, %) was determined using the following formula: $[(Abs_{untreated\ cells}) - (Abs_{treated\ cells})/(Abs_{untreated\ cells})] \times 100$, where Abs stands for absorbance [26,30].

3.3.3. Anti-Acetylcholinesterase (AChE) Activity Assay

SH-SY5Y cells were treated with GrM at a dose level equal to 25 µg/mL and then a colorimetric test was carried out using an acetylcholinesterase assay kit (Colorimetric) (Abcam, UK) in accordance with the manufacturer's instructions. The reaction was followed spectrophotometrically by the increase in absorbance at 412 nm. Donepezil (3 µM) was used as a positive standard [31].

3.3.4. Cell Metabolomic Analysis

The SH-SY5Y and U-251 MG cells (2.5×10^6 cells in Petri dish) were treated with the metabolite GrM_1 for 48 h. At the end of the exposure period, the cells were first quenched in NaCl 0.9% in order to stop the metabolic activities. The cells were scraped and collected in Eppendorf tubes, centrifuged and extracted using 1.0 mL of a cold $MeCN:H_2O$ (1:1, *v:v*) solution. The extract was dried and reconstituted for UHPLC-HR-MS analysis [32].

4. Conclusions

In the present work the possibility of recovering bioactive components from the leaves of *Vitis vinifera* cv. Greco di Tufo was investigated. The employed extractive and fractionation procedures favored the preparation of an extract enriched in flavonol hexuronides variously oxygenated at the B-ring level.

The UHPLC-HR-MS/MS characterization of these molecules showed that the analytical protocol can be used for their rapid identification. The absence of cytotoxicity of the GrM and GrM_1 extracts opens up new investigations at the cellular level aimed at ascertaining their neuroprotective potential for the functional recovery of a precious waste made in the Campania Region (Italy).

Author Contributions: Conceptualization, M.P. and S.P. (Severina Pacifico); methodology, S.P. (Severina Pacifico), S.P. (Simona Piccolella), and G.C.; formal analysis, S.P. (Simona Piccolella) and G.C.; data curation, S.P. (Simona Piccolella) and S.P. (Severina Pacifico); writing—original draft preparation, M.G.V., M.P., S.P. (Simona Piccolella), and S.P. (Severina Pacifico); writing—review and editing, S.P. (Severina Pacifico) and S.P. (Simona Piccolella); supervision, S.P. (Severina Pacifico).

Funding: This research received no external funding.

Conflicts of Interest: The authors declare no conflicts of interest.

References

1. Parfitt, J.; Barthel, M.; Macnaughton, S. Food waste within food supply chains: Quantification and potential for change to 2050. *Philos. Trans. Royal Soc. B* **2010**, *465*, 3065–3081. [CrossRef] [PubMed]
2. Piccolella, S.; Bianco, A.; Crescente, G.; Santillo, A.; Chieffi Baccari, G.; Pacifico, S. Recovering *Cucurbita pepo* cv. 'Lungo Fiorentino' wastes: UHPLC-HRMS/MS metabolic profile, the basis for establishing their nutra- and cosmeceutical valorisation. *Molecules* **2019**, *24*, 1479. [CrossRef] [PubMed]
3. Koutinas, A.; Vlysidis, A.; Pleissner, D.; Kopsahelis, N.; Lopez Garcia, I.; Kookos, I.K.; Papanikolaou, S.; Kwanb, T.H.; Sze Ki Lin, C. Valorization of industrial waste and by-product streams via fermentation for the production of chemicals and biopolymers. *Chem. Soc. Rev.* **2014**, *43*, 2587–2627. [CrossRef] [PubMed]
4. Varricchio, E.; Coccia, E.; Orso, G.; Lombardi, V.; Imperatore, R.; Vito, P.; Paolucci, M. Influence of polyphenols from olive mill wastewater on the gastrointestinal tract, alveolar macrophages and blood leukocytes of pigs. *Ital. J. Anim. Sci.* **2019**, *18*, 574–586. [CrossRef]
5. Parrillo, L.; Coccia, E.; Volpe, M.G.; Siano, F.; Pagliarulo, C.; Scioscia, E.; Varricchio, E.; Safari, O.; Eroldogan, T.; Paolucci, M. Olive mill wastewater-enriched diet positively affects growth, oxidative and immune status and intestinal microbiota in the crayfish, *Astacus leptodactylus*. *Aquaculture* **2017**, *473*, 161–168. [CrossRef]
6. Michalska, A.; Wojdyło, A.; Majerska, J.; Lech, K.; Brzezowska, J. Qualitative and quantitative evaluation of heat-induced changes in polyphenols and antioxidant capacity in *Prunus domestica* L. by-products. *Molecules* **2019**, *24*, 3008. [CrossRef] [PubMed]
7. Ramón-Gonçalves, M.; Gómez-Mejía, E.; Rosales-Conrado, N.; León-González, M.E.; Madrid, Y. Extraction, identification and quantification of polyphenols from spent coffee grounds by chromatographic methods and chemometric analyses. *Waste Manag.* **2019**, *96*, 15–24. [CrossRef] [PubMed]
8. Galanakis, C.M. Recovery of high added-value components from food wastes: Conventional, emerging technologies and commercialized applications. *Food. Sci. Technol.* **2012**, *26*, 68–87. [CrossRef]
9. Volpe, M.G.; Santagata, G.; Coccia, E.; Di Stasio, M.; Malinconico, M.; Paolucci, M. Pectin based pellets for crayfish aquaculture: Structural and functional characteristics and effects on redclaw *Cherax quadricarinatus* performances. *Aquac. Nutr.* **2014**, *21*, 814–823. [CrossRef]
10. Brezoiu, A.M.; Matei, C.; Deaconu, M.; Stanciuc, A.M.; Trifan, A.; Gaspar-Pintiliescu, A.; Berger, D. Polyphenols extract from grape pomace. Characterization and valorisation through encapsulation into mesoporous silica-type matrices. *Food Chem. Toxicol.* **2019**, *133*, 110787. [CrossRef] [PubMed]
11. Zhang, L.; Zhu, M.; Shi, T.; Guo, C.; Huang, Y.; Chen, Y.; Xie, M. Recovery of dietary fiber and polyphenol from grape juice pomace and evaluation of their functional properties and polyphenol compositions. *Food Funct.* **2017**, *8*, 341–351. [CrossRef] [PubMed]
12. Federici, F.; Fava, F.; Kalogerakis, N.; Mantzavinos, D. Valorisation of agro- industrial by- products, effluents and waste: Concept, opportunities and the case of olive mill wastewaters. *J. Chem. Technol. Biot.* **2009**, *84*, 895–900. [CrossRef]
13. Rondeau, P.; Gambier, F.; Jolibert, F.; Brosse, N. Compositions and chemical variability of grape pomaces from French vineyard. *Ind. Crop. Prod.* **2013**, *43*, 251–254. [CrossRef]
14. Torres, J.L.; Varela, B.; García, M.T.; Carilla, J.; Matito, C.; Centelles, J.J.; Cascante, M.; Sort, X.; Bobet, R. Valorization of grape (*Vitis vinifera*) byproducts. Antioxidant and biological properties of polyphenolic fractions differing in procyanidin composition and flavonol content. *J. Agric. Food Chem.* **2002**, *50*, 7548–7555. [CrossRef] [PubMed]
15. Cosme, F.; Pinto, T.; Vilela, A. Oenology in the kitchen: The sensory experience offered by culinary dishes cooked with alcoholic drinks, grapes and grape leaves. *Beverages* **2017**, *3*, 42. [CrossRef]
16. Bell, D.R.; Gochenaur, K. Direct vasoactive and vasoprotective properties of anthocyanin-rich extracts. *J. Appl. Physiol. (1985)* **2006**, *100*, 1164–1170. [CrossRef] [PubMed]
17. Wu, B.; Xu, B.; Hu, M. Regioselective glucuronidation of flavonols by six human UGT1A isoforms. *Pharm. Res.* **2011**, *28*, 1905–1918. [CrossRef]

18. Pacifico, S.; D'Abrosca, B.; Scognamiglio, M.; Gallicchio, M.; Galasso, S.; Monaco, P.; Fiorentino, A. Antioxidant polyphenolic constituents of *Vitis* × *labruscana* cv. 'Isabella'. *Open Nat. Prod. J.* **2013**, *6*, 5–11. [CrossRef]

19. Faugno, S.; Piccolella, S.; Sannino, M.; Principio, L.; Crescente, G.; Baldi, G.M.; Fiorentino, N.; Pacifico, S. Can agronomic practices and cold-pressing extraction parameters affect phenols and polyphenols content in hempseed oils? *Ind. Crop. Prod.* **2019**, *130*, 511–519. [CrossRef]

20. Fabre, N.; Rustan, I.; de Hoffmann, E.; Quetin-Leclercq, J. Determination of flavone, flavonol, and flavanone aglycones by negative ion liquid chromatography electrospray ion trap mass spectrometry. *J. Am. Soc. Mass Spectrom.* **2001**, *12*, 707–715. [CrossRef]

21. Singh, R.; Wu, B.; Tang, L.; Liu, Z.; Hu, M. Identification of the position of mono-*O*-glucuronide of flavones and flavonols by analyzing shift in online UV spectrum (lambdamax) generated from an online diode array detector. *J. Agric Food Chem.* **2010**, *58*, 9384–9395. [CrossRef] [PubMed]

22. Docampo, M.; Olubu, A.; Wang, X.; Pasinetti, G.; Dixon, R.A. Glucuronidated flavonoids in neurological protection: Structural analysis and approaches for chemical and biological synthesis. *J. Agric. Food Chem.* **2017**, *65*, 7607–7623. [CrossRef] [PubMed]

23. Dueñas, M.; Mingo-Chornet, H.; Pérez-Alonso, J.J.; Di Paola-Naranjo, R.; González-Paramás, A.M.; Santos-Buelga, C. Preparation of quercetin glucuronides and characterization by HPLC–DAD–ESI/MS. *Eur. Food Res. Technol.* **2008**, *227*, 1069–1076. [CrossRef]

24. Gao, C.; Zhou, Y.; Li, H.; Cong, X.; Jiang, Z.; Wang, X.; Cao, R.; Tian, W. Antitumor effects of baicalin on ovarian cancer cells through induction of cell apoptosis and inhibition of cell migration in vitro. *Mol. Med. Rep.* **2017**, *16*, 8729–8734. [CrossRef] [PubMed]

25. Chen, H.; Gao, Y.; Wu, J.; Chen, Y.; Chen, B.; Hu, J.; Zhou, J. Exploring therapeutic potentials of baicalin and its aglycone baicalein for hematological malignancies. *Cancer Lett.* **2014**, *354*, 5–11. [CrossRef] [PubMed]

26. Piccolella, S.; Nocera, P.; Carillo, P.; Woodrow, P.; Greco, V.; Manti, L.; Fiorentino, A.; Pacifico, S. An apolar *Pistacia lentiscus* L. leaf extract: GC-MS metabolic profiling and evaluation of cytotoxicity and apoptosis inducing effects on SH-SY5Y and SK-N-BE(2)C cell lines. *Food Chem. Toxicol.* **2016**, *95*, 64–74. [CrossRef]

27. Pacifico, S.; Piccolella, S.; Lettieri, A.; Nocera, P.; Bollino, F.; Catauro, M. A metabolic profiling approach to an Italian sage leaf extract (SoA541) defines its antioxidant and anti-acetylcholinesterase properties. *J. Funct. Foods* **2017**, *29*, 1–9. [CrossRef]

28. Sak, K.; Illes, P. Neuronal and glial cell lines as model systems for studying P2Y receptor pharmacology. *Neurochem. Int.* **2005**, *47*, 401–412. [CrossRef]

29. Araújo, K.C.F.; Costa, E.M.M.B.; Pazini, F.; Valadares, M.C.; Oliveira, V. Bioconversion of quercetin and rutin and the cytotoxicity activities of the transformed products. *Food Chem. Toxicol.* **2013**, *51*, 93–96. [CrossRef]

30. Pacifico, S.; Piccolella, S.; Papale, F.; Nocera, P.; Lettieri, A.; Catauro, M. A polyphenol complex from *Thymus vulgaris* L. plants cultivated in the Campania Region (Italy): New perspectives against neuroblastoma. *J. Funct. Foods* **2016**, *20*, 253–266. [CrossRef]

31. Pacifico, S.; Gallicchio, M.; Lorenz, P.; Duckstein, S.M.; Potenza, N.; Galasso, S.; Marciano, S.; Fiorentino, A.; Stintzing, F.C.; Monaco, P. Neuroprotective potential of *Laurus nobilis* antioxidant polyphenol enriched leaf extracts. *Chem. Res. Toxicol.* **2014**, *27*, 611–626. [CrossRef] [PubMed]

32. Dietmar, S.; Timmins, N.E.; Gray, P.P.; Nielsen, L.K.; Kromer, J.O. Towards quantitative metabolomics of mammalian cells: Development of a metabolite extraction protocol. *Anal. Biochem.* **2010**, *404*, 155–164. [CrossRef] [PubMed]

Sample Availability: Samples of the compounds are available from the authors.

Article

In vitro Fermentation of Polysaccharides from *Aloe vera* and the Evaluation of Antioxidant Activity and Production of Short Chain Fatty Acids

Antonio Tornero-Martínez [1], Rubén Cruz-Ortiz [1], María Eugenia Jaramillo-Flores [1], Perla Osorio-Díaz [2], Sandra Victoria Ávila-Reyes [2,3], Guadalupe Monserrat Alvarado-Jasso [2] and Rosalva Mora-Escobedo [1,*]

[1]   Instituto Politécnico Nacional, ENCB, Campus Zacatenco. Miguel Othón de Mendizábal 699, Alcaldía G.A. Madero, Ciudad de México C.P. 07360, CDMX, Mexico; torneroayora@gmail.com (A.T.-M.); rub66.63@gmail.com (R.C.-O.); Jaramillo_flores@hotmail.com (M.E.J.-F.)

[2]   Instituto Politécnico Nacional, CEPROBI, Carretera Yautepec-Jojutla, Km. 6, Yautepec C.P. 62731, Morelos, Mexico; posoriod@gmail.com (P.O.-D.); Sandra_victory@yahoo.com (S.V.Á.-R.); gpemonserratjasso@gmail.com (G.M.A.-J.)

[3]   CONACYT—Instituto Politécnico Nacional, CEPROBI, Carretera Yautepec-Jojutla, Km. 6, Yautepec C.P. 62731, Morelos, Mexico

*   Correspondence: rosalmorae@gmail.com; Tel.: +52-5555-57296000 (ext. 57872)

Received: 24 August 2019; Accepted: 3 October 2019; Published: 7 October 2019

Abstract: Soluble or fermentable fibre has prebiotic effects that can be used in the food industry to modify the composition of microbiota species to benefit human health. Prebiotics mostly target Bifidobacterium and Lactobacillus strains, among others, which can fight against chronic diseases since colonic fermentation produces short chain fatty acids (SCFAs). The present work studied the changes produced in the fibre and polyphenolic compounds during in vitro digestion of gel (AV) and a polysaccharide extract (AP) from *Aloe vera*, after which, these fractions were subjected to in vitro colonic fermentation to evaluate the changes in antioxidant capacity and SCFAs production during the fermentation. The results showed that the phenolic compounds increased during digestion, but were reduced in fermentation, as a consequence, the antioxidant activity increased significantly in AV and AP after the digestion. On the other hand, during in vitro colon fermentation, the unfermented fibre of AV and AP responded as lactulose and the total volume of gas produced, which indicates the possible use of *Aloe vera* and polysaccharide extract as prebiotics.

Keywords: *Aloe vera*; *Aloe vera* polysaccharides; in vitro fermentation; SCFAs; antioxidant capacity

1. Introduction

Plants are a source of a wide spectrum of compounds such as polyphenols, carotenoids, glucosinolates, and lignans, among others. These phytochemicals provide potential beneficial properties to each plant matrix [1]. *Aloe vera* has been long used thanks to its curative and therapeutic properties. It has been reported that only the pulp has more than 75 bioactive compounds [2]. *Aloe vera*, originally from Africa, belongs to the genus *Aloe*, and it is a perennial, succulent xerophyte grown in temperate and sub-tropical regions of the world. *Aloe vera* or *Aloe barbadensis* is part of the *Asphodelaceae* family, of which there are over 360 known species. There are several species under the genus *Aloe*, including *Aloe vera*, *Aloe barbadensis*, *Aloe ferox*, *Aloe chinensis*, *Aloe indica*, *Aloe peyrii* etc. Amongst these, *Aloe vera Linn syn. Alo barbadensis Miller* is accepted unanimously as the ideal botanical source of *Aloe* [3]. The pulp or gel of *Aloe vera* is the part of the plant that is of great commercial, pharmacological, alimentary, industrial, and cosmetic importance [3,4]. *Aloe vera* gel contains more

coumaroyl ester, which are anthraquinones reported as powerful stabilizers in oxidative stress [17]. Aloesin is a natural polyphenol originated from *Aloe* plants. Aloesin and/or *Aloe* polysaccharides can reduce systemic oxidative stress by acting directly as potent anti-oxidants and also indirectly by regulating the production of adiponectin and gene expression pathways related to insulin sensitivity, glucose transportation, and fatty acid biosynthesis [18].

Aloin A and B are some of the main components of *Aloe vera* gel, which are not only laxatives but are also known to have antioxidant properties at low concentrations [17]. Aloinoside compounds are also anthraquinones, along with aloesin, which exhibits great antioxidant activity and is used to lower oxidative stress indices in diseases such as diabetes [17].

Table 2. Phenolic compounds identified in the aqueous extract of *Aloe vera* by UPLC-MS.

PEAK	Rt [a] (min)	*m/z* [M-H]⁻	*m* [M]	Tentative Identification
1	1.52	393.18	394.18	Aloesin
2	3.15	433.19	434.19	10-Hydroxyaloin
3	3.15	561.36	562.36	Aloinoside
4	4.45	417.19	418.19	Aloin A
5	4.50	555.25	556.25	Aloenin-2′-*p*-coumaroyl ester
6	4.68	417.19	418.19	Aloin B

[a] Retention time.

2.3. Total Fibre Content of Non-Digestible Fibre Fractions AV-TNDF and AP-TNDF and of Their Soluble (AV-SNDF and AP-SNDF) and Insoluble (AV-INDF and AP-INDF) Fractions

After the digestion process insoluble non-digestible fraction of AV (AV-INDF), insoluble non-digestible fraction of AP (AP-INDF), soluble fraction of the non-digestible fibre of AV (AV-SNDF), soluble fraction of non-digestible fibre of AP (AP-SNDF), were obtained.

The fibre contents of the total non-digestible fibre fraction (TNDF) and its soluble (SNDF) and insoluble (INDF) fractions are shown in Table 3. The INDF content was approximately 25% higher in AP than in AV. The content obtained for AP was comparable to these of nopal cactus (57.7 g/100 g) [19]. The SNDF proportion was 52% higher in AP than in AV, indicating that AP may have a higher amount of partially acetylated mannans as the primary polysaccharide of the gel, although it also contains pectins, some hemicelluloses, mucilages, and gums [4]. Acetylated mannan molecules are mainly responsible for the thick, mucilage-like properties of raw *Aloe vera* gel [4,6]. These results are interesting because *Aloe vera* polysaccharides have been considered the main component responsible for most of the beneficial properties attributed to the *Aloe vera* plant [15]. They are dietary carbohydrates resistant to digestion in the upper GI tract. Due to the human physiological effect of these polysaccharides, they could be labelled as prebiotics, selectively fermented ingredients that allow for specific changes in the composition and/or activity in the gastrointestinal microflora that confer benefits to the hosts wellbeing and health [20].

Table 3. Non-digestible fraction of *Aloe vera* gel (AV) and its polysaccharide extract (AP).

Sample	SNDF (g/100g)	INDF (g/100g)	TNDF (g/100g)
AV	20.57 [a] ± 0.83	26.57 [a] ± 0.02	47.14
AP	39.34 [b] ± 0.55	53.05 [b] ± 0.40	92.39

SNDF = Soluble non digestible fraction; INDF = Insoluble non digestible fraction; TNDF = Total non-digestible fraction. The results are the average of three repetitions ± SD. Means between rows with different letters (a and b) are significantly different for a significance level of ($p \leq 0.05$) by t-student test.

The high fermentability of SNDF produces SCFAs (acetic, propionic, and butyric acids) that contribute to the proper functioning of the large intestine and the prevention of pathologies through their actions in the lumen, colonic and vascular musculature as well as in the colonocyte metabolism [21].

Furthermore, SCFAs reduce intestinal pH and increase water and salt absorption in the large intestine [22]. On the other hand, the fibre contents in AV-INDF and AP-INDF were 26.57 g/100 g and 57.7 g/100 g, respectively. The high proportion of INDF, could increase the stool water content provides bulky/soft/easy-to-pass stools [23].

2.4. In Vitro Fermentation

2.4.1. Changes in pH

Fermentation of TNDF in the large intestine is associated with reduced levels of pH and the consequent proliferation of beneficial microbes such as bifidobacteria [22]. Therefore, changes in pH are often used as indicators of the fermentability of non-digestible matter. Figure 2A shows that the pH decreased with reaction time in both samples: from 7.09 (h0) to 6.76 (h24) in AV and from 7.13 (h0) to 6.79 (h24) in AP. However, the change was gradual when compared to that of the lactulose used as a control. Lactulose has high fermentability and promotes a greater reduction in pH. A significant difference ($p < 0.05$) was observed in the pH of AV and AP with respect to the pH of the lactulose control during the reaction time and up to 24 h of incubation. This behaviour occurs because the control is a disaccharide of fast fermentation [23]. In contrast, AV and AP are matrices mainly composed of non-digestible oligosaccharides or polysaccharides of slow fermentation that produce lower pH [24] or buffer or antacid effects, as Al-Madboly et al. have reported [25].

2.4.2. Volume of gas produced

Gases are produced by microbiota as by-products of their metabolic activities. These microorganisms of the gastrointestinal tract satisfy their energy requirements largely through non-digested carbohydrate fermentation and the subsequent production of SCFAs and certain gas species that include carbon dioxide (CO_2), methane (CH_4), and hydrogen (H_2). The latter is the main by-product of hydrogen fermentation and is generated by hydrogen producers of the phyla Firmicutes and Proteobacteria [26]. Hydrogen gas can be generated from pyruvate cleavage or reoxidation of the reduced pyridine and flavin nucleotides.

Figure 2B shows the volume of gas produced by the microbiota feeding on different components during in vitro fermentation. In AV, the increase in the volume of gas was slow, reaching over 12 mL at 24 h. The volumes indicate that, in the first four hours, fermentation is very slow since, in addition to fermentable fibres of AV, other compounds hamper the beginning of the process, such as insoluble fibres and polyphenols. On the other hand, the AP fraction immediately begins with active fermentation and progressively increases the volume of gas until reaching 15.8 mL (24 h). In this case, the available compounds are mostly soluble polysaccharides, which are rapidly fermentable when compared against the control. The AP extract clearly produced more gases from the beginning, and at 24 h, there was no significant difference with that produced by lactulose (17.7 mL). These results were in agreement with the fibre proportion since the total non-digestible fibre proportion (TNDF) in AP was approximately 50% more than in AV. These results indicate that glucooligosaccharides from AP could be the best carbon source for different microorganisms since they show specific preferences for defined substrates, as reported by Gullon et al. [21]. Gas production was constant throughout the fermentation period, showing a significant difference ($p < 0.05$) against the control at 24 h of fermentation.

2.4.3. Unfermented fibre (UF)

The TNDF that passes to the large intestine is composed of soluble fibre that is mostly fermented and insoluble fibre, which is mostly not fermented by microorganisms. The latter are residual compounds or UF in the expelled faeces. Figure 2C shows that the UF residues in AV and AP were scarcely fermented, as evidenced by 29.8% and 39.2% of the TNDF for AV and AP, respectively, results which match what was reported by Mudgil and Barak [27].

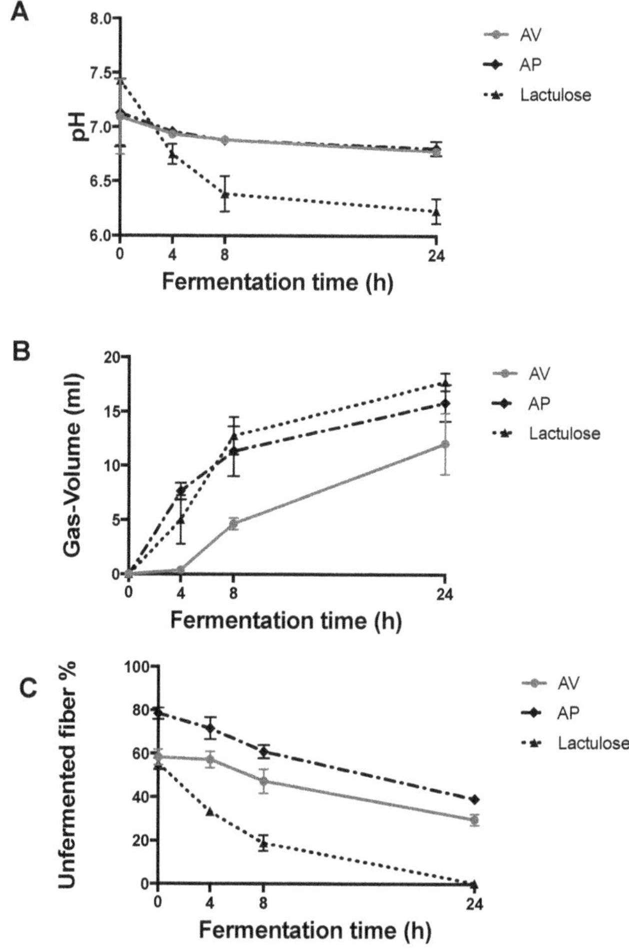

Figure 2. Fermentation kinetics of *Aloe vera* gel (AV) and its polysaccharide extract (AP) during 24 h. (**A**) pH changes during fermentation, (**B**) Volume of gas produced during fermentation, and (**C**) Unfermented residue of samples AV and AP.

2.4.4. Quantification of SCFAs

Table 4 shows the changes in the proportion of SCFAs in AV and AP during fermentation until 24 h, using faeces from healthy volunteers. The SCFAs were quantified as acetate, propionate, and butyrate and were compared with the SCFAs produced by lactulose fermentation, which works as a control given its high fermentability and its consideration as a fermentation standard [21]. Acetate was the main SCFA quantified in both samples, and no significant difference ($p < 0.05$) was found with that produced by lactulose fermentation. This compound is the most common compound in the large intestine [26] and enters the peripheral circulation through peripheral tissues to be metabolized. It has been reported that acetic acid is the main product of bifidobacterial fermentation in a human faecal environment [28]. Acetate is produced by most anaerobes, including acetogens that are able to perform reductive acetogenesis from formate or hydrogen combined with CO_2 [29]. Acetate plays an important role in controlling inflammation and in combating pathogen invasion [30,31].

Table 4. Concentration of SCFA (μmol/mg) produce during the fermentation of *Aloe vera* gel (AV), its polysaccharide extract (AP) and lactulose as a carbon source

Fermentation Time	Acetic Acid			Propionic Acid			Butyric Acid			Total SCFA		
	AV	AP	Lactulose	AV	AP	Lactulose	AV	AP	Lactulose	AV	AP	Lactulose
0	2.16 ± 1.07 [b]	0.93 ± 0.36 [a]	2.16 ± 0.08 [b]	0.09 ± 0.05 [a]	0.14 ± 0.03 [b]	0.09 ± 0.03 [a]	0.09 ± 0.04 [b]	0.14 ± 0.01 [c]	0.05 ± 0.01 [a]	2.33 = 1.08 [ab]	1.21 ± 0.39 [a]	2.34 ± 0.11 [b]
4	3.15 ± 0.55 [a]	3.20 ± 0.64 [a]	4.50 ± 0.73 [a]	1.32 ± 0.26 [a]	1.30 ± 0.14 [a]	1.07 ± 0.06 [a]	0.31 ± 0.05 [a]	0.63 ± 0.04 [c]	0.47 ± 0.02 [b]	4.77 ± 0.84 [a]	5.13 ± 0.80 [a]	5.15 ± 0.93 [a]
8	6.12 ± 0.66 [a]	7.00 ± 0.26 [a]	9.75 ± 1.92 [b]	2.50 ± 0.27 [ab]	2.16 ± 0.32 [a]	3.42 ± 0.90 [b]	0.79 ± 0.04 [a]	1.02 ± 0.13 [b]	2.09 ± 0.03 [c]	9.41±0.96 [a]	10.18 ± 0.70 [a]	12.51 ± 2.79 [a]
24	10.67 ± 0.3 [a]	10.93 ± 1.33 [a]	11.43 ± 1.57 [a]	3.68 ± 0.24 [b]	3.12 ± 0.14 [a]	4.44 ± 0.69 [b]	1.92 ± 0.10 [a]	1.79 ± 0.14 [a]	4.25 ± 1.12 [b]	16.27 ± 0.60 [a]	15.83 ± 1.47 [a]	17.03 ± 1.51 [a]

Standard deviation for n = 3. Different letters (a and b) indicate significant differences ($p < 0.05$) between columns for the same acid.

Propionate is another key SCFA in large intestine fermentation. At 8 h of fermentation, it was produced in similar amounts in the three substrates. However, at 24 h, AV and the control were the substrates that produced the most amount of propionate (3.68 and 4.44 µmol/mg, respectively), which was significantly higher ($p < 0.05$) than that produced by AP (3.12 µmol/mg) at 24 h of fermentation. Propionate, a gluconeogenerator, has been shown to inhibit cholesterol synthesis. On the other hand, after absorption, acetate has been shown to increase cholesterol synthesis [26]; therefore, the higher propionate proportion obtained from AV fermentation could decrease the acetate: propionate ratio and reduce serum lipids and possibly cardiovascular disease risk. The proportion of acetate:propionate in the three samples at 24 h was AV = 2.90, AP = 3.50, and lactulose = 2.57. This indicates that AV and lactulose fermentation produce the highest concentration of propionate. In AV, the proportion was lower than that reported by Gullón et al. [32]. This difference could be due to seasonal changes and geographic locations that lead to significant variations in gel polysaccharides [13]. Producers of propionate largely belong to the phylum Bacteroidetes but also include some Firmicutes.

Finally, AV (1.92 µmol/mg) and AP (1.79 µmol/mg) butyrate yields at 24 h were lower than that of lactulose (4.25 µmol/mg). Butyrate formation occurs in certain Firmicutes bacteria, either via butyrate kinase (in many *Clostridium* and *Coprococcus* species) or via butyric acetate CoA transferase [23]. Butyrate has been studied for its role in nourishing the colonic mucosa and in the prevention of cancer of the colon by promoting cell differentiation, cell-cycle arrest and apoptosis of transformed colonocytes, inhibiting the enzyme histone deacetylase and decreasing the transformation of primary bile acids to secondary bile acids as a result of colonic acidification [26]. Butyric acid exerts a positive influence on the colonic mucosa and affects cell differentiation. It has anti-inflammatory properties and reduces the incidence of colon cancer [29]. Butyrate irrigation (enema) has also been suggested in the treatment of colitis.

Therefore, SCFA production and their potential delivery to the distal colon due to AV and AP fermentation may result in a protective effect. First, the human large intestine is colonized by dense microbial communities that utilize both diet-derived and host-derived energy sources for growth, predominantly through fermentative metabolism. This highly diverse community has the capacity to perform an extraordinary range of biochemical transformations that go well beyond those encoded by the host genome, and these activities exert an important influence upon many aspects of human health [25].

2.5. Quantification of Phenolic Compounds in AV and AP and Identification of AV, AV-TNDF, AP, AP-TNDF by HPTLC

2.5.1. Quantification of Phenolic Compounds in AV and AP

Figure 3A shows the amount of phenolic compounds from AV and AP before and after undergoing an in vitro enzymatic process of digestion simulating the passing of these samples throughout the digestive tract. AV had a higher proportion of phenolic compounds than the extract (AP). In the same figure, a significant increase ($p < 0.05$) of 66% in the amount of phenolic compounds of AV-TNDF with respect to AV was observed, and a significant increase in the amount of phenolic compounds in AP-TNDF compared to AP was also observed, with an increase greater than 100%. However, it is important to note that even when the number of phenolic compounds of AP was lower compared to AV, the sample AP-TNDF had a proportion of phenolic compounds similar to that of AV-TNDF. This may be due to a major release of insoluble phenolic compounds with the acid, alkaline, and enzymatic digestion process, as Anokwuru, et al., have reported [7]. This increase in phenolic compounds, also, could be the result of the opening and breakdown of polysaccharide chains and lattices, liberating most of the phenols when the AV and AP samples undergo the fermentation process [29].

Figure 3B shows the change in the total phenolic compounds before and during fermentation. Fermented products of AV had a stable amount of phenolic compounds during the first 4 h (243.65 to 239.18 meq aloin/100 g), which was reduced at 24 h until 25.72 meq aloin/100 g. For AP-TNDF the fall was higher, starting with a concentration of 207.48 meq aloin/100 g at 0 h fermentation time to 4.58

meq aloin/100 g at 24 h. In humans, insoluble phenolic compounds are released from the matrix in the colon during the fermentation of the ingested material. The release of these phenolic compounds has been identified as beneficial against colon cancer [33]. Free and conjugated phenolic compounds are known as extractable phenolic compounds while bound or insoluble phenolic compounds are known as non-extractable [34]. It is observed that they disappear with fermentation and part of them are degraded, although they remain a constant and significant amount due to the release of insoluble phenolic compounds. There are studies of phenolic compounds that show how the microbiota populations and the phenolic compounds metabolized by this microbiota evolve and change. Cueva et al. (2012) [35] observed a great degradation of phenolic compounds in grape seed extract during in vitro fermentation. Similar results to those obtained in this work.

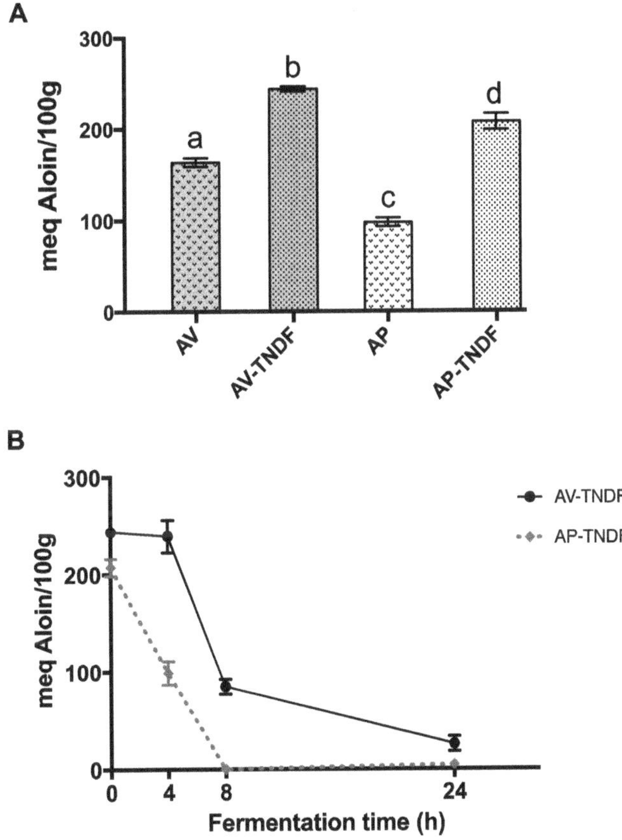

Figure 3. Total phenolic compounds. (A) Total phenolic compounds before and after digestion of *Aloe vera* gel (AV) and its polysaccharide extract (AP). Where AV and AP correspond to samples before digestion; AV-TNDF and AP-TNDF correspond to samples after digestion. (B) Changes in the total content of phenolic compounds during AV-TNDF and AP-TNDF fermentation.

2.5.2. Identification of phenolic compounds by HPTLC

Taking into account the results obtained for total phenolic compounds before and after digestion of AV and AP, it was assessed the identification of phenolic compounds in AV, AV-TNDF, AP, and AP-TNDF by HPTLC (Table 5). The standards used were vanillic acid, *p*-hydroxybenzaldehyde, *p*-hydroxybenzoic acid, *p*-cresol and ferulic acid.

Table 5. Identification of phenolic compounds of AV, AV-TNDF, AP, AP-TNDF and their fermented fractions obtained by HPTLC.

Phenolic Compounds	Colour	Rf	AV	AV-TNDF	AP	AP-TNDF	FAV4	FAV8	FAV24	FPA4	FPA8	FPA24
Vanillinic acid	Dark blue	0.54	-	-	-	-	-	-	-	-	-	-
p-Hydroxybenzaldehyde	Dark blue	0.65	-	-	-	-	-	-	-	-	-	-
p-Hydroxybenzoic acid	Dark blue	0.45	-	-	-	-	-	-	-	-	-	-
p-Cresol	Dark blue	0.72	-	-	-	-	-	-	-	-	-	-
p-Creosol	Dark blue	0.86	-	-	-	-	-	-	-	-	-	-
Ferulic acid	Light blue	0.55	+	+	+	+	-	-	-	-	-	-
Unidentified	Violet	0.64	-	-	-	-	++	++	++	-	-	-
Unidentified	Yellow	0.57	-	-	-	-	++	++	++	-	-	-
Unidentified	Yellow	0.12	-	-	-	-	++	++	++	+	+	+
Unidentified	Red	0.06	++	+	-	-	-	-	-	-	-	-
Unidentified	Light blue	0.04	+	+	++	+++	+	+	+	+	+	+

AV = *Aloe vera*, AV-TNDF = non-digestible fraction of AV, AP= AV polysaccharides, AP-TNDF= non-digestible fraction of AP and their fermented fractions obtained by HPTLC. Samples obtained at 4, 8 and 24 (FAV4, FAV8, FAV24, FPA4, FPA8, FPA24). The symbols indicate: +++ highly concentrated compound, ++ compound concentrate, + little concentrated compound, - no appreciable concentration of the compound. HPTLC, remission at 366 nm. CAMAG.

In the samples AV, AP, AV-TNDF and AP-TNDF, the presence of ferulic acid and an unidentified compound with Rf's of 0.05 and 0.04, respectively, was detected. However, with the fermentation process for AV, in the samples FAV4, FAV8 and FAV24, ferulic acid disappeared, and other components appeared with Rf's of 0.64, 0.57, 0.12 and 0.04. For FPA4, FPA8, and FPA24, only two unidentified compounds with Rf's of 0.12 and 0.04 appear in small quantities.

Unidentified components can appear when fibre and phenolic compounds are fermented in the intestine. Lee, et al., [36] reported that some bacterial phenolic metabolites were modified in colonic fermentation since specific intestinal bacteria metabolize phenolic compounds to different extents and produce different aromatic metabolites. Changes in intestinal phenolic levels may influence microflora composition. Phenolic compounds are likely to benefit the host by inhibiting pathogen growth and regulating commensal bacteria, including probiotics, and could therefore be considered as prebiotics.

2.6. Antioxidant Activity of AV-TNDF and AP-TNDF

2.6.1. Reducing Power

To establish the appropriate concentration to determine the antioxidant activity of the AV-TNDF and AP-TNDF samples, several concentrations from 0.78 mg/mL to 12.5 mg/mL were evaluated. It was found that the samples had a concentration-dependent behaviour, with 12.5 mg/mL being the one with the highest activity ($p < 0.05$) (data not shown); therefore, the tests of reducing power and radical ·OH sequestration correspond only to this concentration. Figure 4A shows the changes in the reducing power during 24 h of the fermentation kinetics at 12.5 mg/mL of the fraction of non-digestible *Aloe vera* (AV-TNDF) and the non-digestible fraction of *Aloe vera* polysaccharide extract (AP-TNDF). The highest activity of reducing power of AV-TNDF was observed at 4 h (AVF4). This indicated that, in general, fractions derived from AV show a greater reducing power than fractions from AP. A possible explanation is that AV-TNDF had higher concentrations of inactive polyphenols in their matrices that can interact with the medium after gastrointestinal digestion. Acids and enzymes can open the fibres containing the polyphenols, which in turn are more bioavailable at the end of digestion. These results agree with the identification of the phenolic compounds shown in Table 5. The appearance of the unidentified compounds with Rf's of 0.64, 0.57 and 0.12 may have greater reducing power than those identified in AP-TNDF.

The fermentation samples of AP-TNDF (AP0-24) had antioxidant capacities and reducing powers that experience limited change over time. As shown in Table 5, during fermentation, AP-TNDF lost unidentified phenolic compounds, which leads to a decrease in reducing power, and a correlation between the antioxidant capacity and the reducing power was seen, as Alugoju et al. have reported for certain bioactive compounds [37]. The reducing power of 0.2 of inhibition was reported with methanolic extracts from non-digested fresh samples [28]. Cueva et al. [35] demonstrated the appearance of new phenolic compounds that were not present or were present in small amounts before fermentation such as hydroxyphenylacetic acid, phenylpropionic acid or phenylacetic acid with colonic in vitro fermentation of grape seed extract.

2.6.2. OH Radical Sequestration

The statistical analysis of ·OH radical sequestration revealed that all extracts, except AP, were dose dependent. The elimination of hydroxyl radicals from fractionated polysaccharides and AV gel extracts has been reported by Chun-Hui et al. [38]. Figure 4B shows the efficacy of 12.5 mg/mL solutions of AV (35.47%), AV-TNDF (93.54%), AP (3.07%) and AP-TNDF (94.78%) to sequester hydroxyl radicals. The results of AV are even lower than assays reported by authors such as Ray et al. [39], who report values of up to 48.01% using methanolic extracts of fresh *Aloe vera* gel without a digestion process. This difference may be due to the age of the plant and edaphoclimatic conditions as Sánchez-Machado et al., have reported [40]. The higher efficacy of AV-TNDF and AP-TNDF could be due to the increase in phenolic compounds, because fresh samples have large concentrations of inactive polyphenols in their matrices that can interact with the medium after a digestion process such as gastro intestinal digestion. Additionally, it could be the result of the opening and breakdown of polysaccharide chains and lattices, liberating most of the phenols when the AV and AP samples undergo the fermentation process [29].

Figure 4B shows the ·OH sequestration percentage of the synthesized or released compounds in solutions of fermented AV (AVF) and the fermented polysaccharide extract (APF) at 12.5 mg/mL. At this concentration, the ferments exhibited a decrease in antioxidant activity during the first 4 h of fermentation in the two samples up to 1.85% (AVF4) and 11.52% (APF4). The AVF antioxidant activity experienced a slight increase until 24 h of fermentation with 7.76% effectiveness observed for AVF24, and the APF antioxidant activity remained low and constant, reaching 5.74% effectiveness at 24 h.

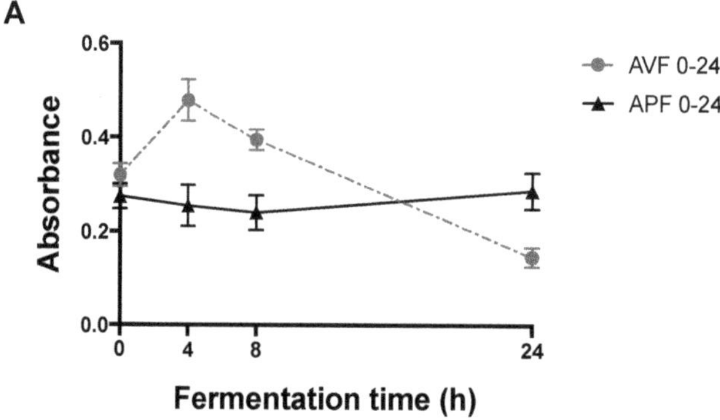

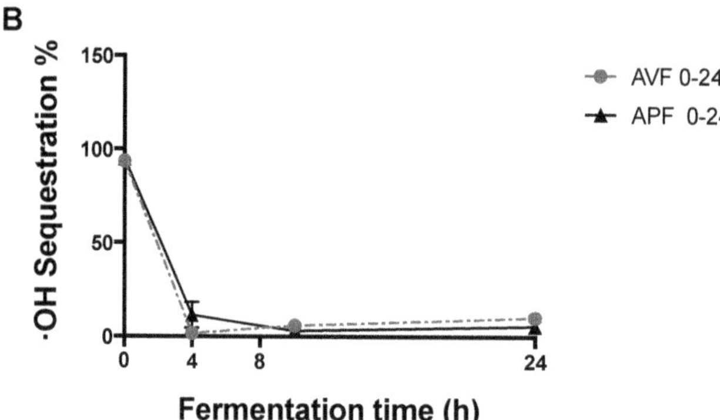

Figure 4. Changes in antioxidant activity during colonic fermentation of *Aloe vera* gel (AV) and its polysaccharide extract (AP). (**A**) Changes in reducing power (Abs 700 nm), due to colonic fermentation. (**B**) Changes in % of sequestration of hydroxyl radical ·OH due to colonic fermentation. (**B**) The results are expressed as the ME ± DS, n = 3.

In contrast to the reducing power, in this case, the percent of ·OH sequestration drops dramatically over time within the first few hours of fermentation. In this case, the phenolic or other compounds that have effective ·OH sequestration in the undigested fractions AV-TNDF and AP-TNDF are rapidly degraded by the enzymes, bacteria or fermentation conditions, drastically lowering the antioxidant activity. Lee et al., have reported a decrease in concentration after incubation with a faecal bacteria homogenate in compounds present in the tea extract, including epicatechin, catechin, 3-O-methyl gallic acid, gallic acid, and caffeic acid [36]. The difference in the behavior of reducing power and ·OH radicals are because they have different ways of sequestering Reactive Oxygen Species (ROS). ·OH radicals react with biomolecules such as lipid, protein and DNA [41].

3. Materials and Methods

3.1. Raw Material

Leaves of *Aloe vera* plants (aged 3–4 years) were obtained from a market in Mexico City. The leaves were washed with soap and water, followed by disinfection with ethanol. The gel was extracted by removing the cortex and was frozen and lyophilized until use.

3.2. Physicochemical Characterization of Aloe Vera (AV)

In the raw gel of *Aloe vera*, the acidity was determined using AOAC method 939.05 (2015) [42], while the pH was determined using a potentiometer (pH-120; Conductronic, Puebla, Pue, Mexico). Soluble solids (°Brix) were assessed using a digital refractometer (HSR-500; Atago, Kobe, Hyogo-ken, Japan) at a 0–32 °Brix scale considering that 1 °Brix = 1 g soluble solid in 100 g of solution [43]. Moisture was determined according to method 2005.02, raw protein by the Kjeldahl method (2005.06), raw fat by a Soxhlet method (920.39c), and ash by method 923.03 (AOAC, 2015) [42]. Neutral detergent fibre was assessed in the lyophilized sample of the gel, according to the 2002:04/ISO technique (16472:2005) from AOAC [42]. Total dietary fibre was determined through the enzymatic-gravimetric method (32–05) outlined by AOAC (2015) [42].

3.3. Identification of phenolic Compounds by Ultra-Performance Liquid Chromatography-Tandem Mass Spectrometry (UPLC-MS) in Aloe Vera (AV)

The identification of phenolic compounds in the gel of *Aloe vera* (AV) was performed with ultra-performance liquid chromatography (UPLC) by utilizing an Acquity UPLC H-Class (Waters Corp, Milford, MA, USA) coupled with a Waters Xevo G2-X2 Tof system with an electrospray ionization (ESI) interface. The conditions for the chromatographic analysis were as follows: the mobile phase consisted of 0.1% formic acid in water/acetonitrile at a flow rate of 0.3 mL/min and 35 °C. A BEH C18 Column (2.1 mm X 100 mm) was used and the chromatograms were registered through detection of the Total Ion Current. The MS conditions of analysis in negative-ion mode were as follows: drying gas (nitrogen) flow rate, 8 L/min; gas temperature, 180 °C; scan range, *m/z* 50–3000; capillary voltage, 4500 V; and nebulizer pressure, 2.5 bar [44]. A solution of 8 mg/mL of lyophilized AV was prepared, homogenized, and passed through a 0.22 µm cellulose filter prior to analysis.

3.4. Extraction of Polysaccharides (AP) from Aloe Vera Gel

The extraction of *Aloe vera* gel polysaccharides (AP) followed a modification of the technique reported by Ni et al. [4]. One gram of lyophilized *Aloe vera* gel was suspended in 80 mL of a 95% water/methanol solution, homogenized with a Polytron, and centrifuged at 936× *g* (Allegra 64R; Beckman, Pasadena, CA, USA) for 10 min. The first precipitate was stored, and the supernatant was centrifuged one more time at 10,285× *g* for 15 min. The first and second precipitates were combined, re-suspended in deionized water and lyophilized. The sum of these two fractions was the polysaccharide extract (AP). The first precipitate corresponds to the cell wall fibre, and the second precipitate corresponds to micro-particles from cell organelles of *Aloe vera* [4].

3.5. Indigestible Fraction and in Vitro Fermentation

3.5.1. Indigestible Fermentation of Aloe Vera Gel (AV) and Polysaccharides (AP)

Figure 5 shows the flowchart of the isolation of the indigestible fibre fractions of AV and AP The indigestible fraction (IF) of (AV) and (AP) (Figure 5) was determined following the method proposed by Saura-Calixto et al. [45] with some modifications. Briefly, 300 mg of both fractions were suspended in HCl-KCl buffer (pH 1.5). Then, 0.2 mL of a pepsin solution at a concentration of 300 mg/mL in HCl-KCl buffer (>250 units/mg, P7000; Sigma Aldrich, Toluca, Edo. Mex. Mexico) was added and incubated at 40 °C for 1 h with constant stirring to simulate the digestive process.

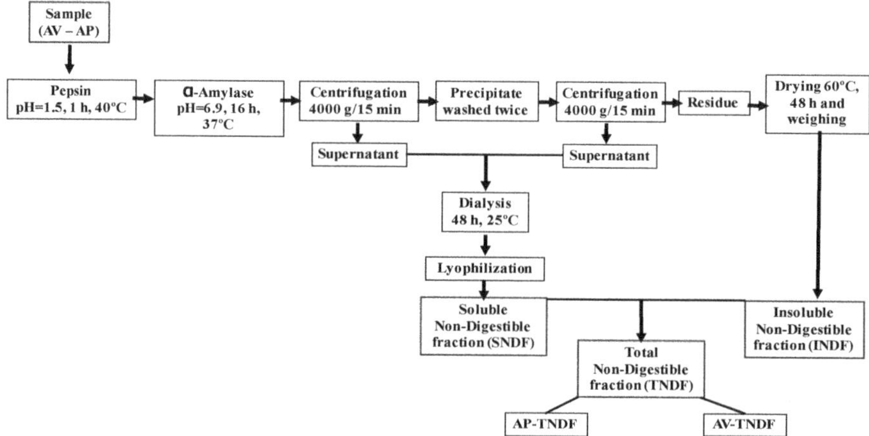

Figure 5. Indigestible fraction. The obtaining of indigestible fractions of AV and AP. AV= *Aloe vera gel*, AP= *Aloe* polysaccharides; AV-NDF = Insoluble non-digestible fraction of AV, AP-NDF = Insoluble non digestible fraction of AP, AV-SNDF = Soluble fraction of the non-digestible fibre of AV, AP-SNDF = Soluble fraction of non-digestible fibre of AP.

To simulate small intestine digestion, pH was adjusted to 6.9, and 9 mL of tris-maleate buffer (0.1 M) was added, followed by 1 mL of α-amylase solution (3480 units/mL, A3176; Sigma Aldrich), at a concentration of 120 mg/mL in tris-maleate buffer. The mix was incubated at 37 °C for 16 h at constant stirring. Lipid digestion enzymes such as lipase or biliary juices were omitted given that the amount of fat (lower than 4%) was not significant. After the reaction time, the samples were centrifuged at 4000× *g* for 15 min, and the supernatants were saved. The precipitate was washed twice with 10 mL of distilled water and centrifuged; supernatants were collected in 80 mL beakers. The tubes containing the solid residue were placed in an oven at 60 °C for 48 h, and the residue was then weighed as the amount of insoluble non-digestible fibre in AV (AV-INDF) and AP (AP-INDF). Both insoluble non-digestible fibre fractions were stored at 4°C until the fermentation experiments.

The supernatant was placed on dialysis membranes previously treated with boiling water for 20 min. The supernatant was dialyzed (cellulose membrane 12000–14000 Dalton MWCO-18) at a constant water flow of 7 L/h (30 mL/15 sec) at 25 °C for 48 h. Finally, the content of the membranes was lyophilized and weighed to obtain the soluble fraction of the non-digestible fibre of AV (AV-SNDF) and non-digestible fibre of AP (AP-SNDF); these fractions were stored at 5 °C until use in the fermentation process [46].

3.5.2. In Vitro Fermentation of AV-TNDF and AP-TNDF

The samples obtained in the digestion, which were the soluble fraction (SNDF) and the insoluble fraction (INDF), were combined as the total non-digestible fraction (TNDF) of AV and AP to be fermented. In vitro fermentation was assessed in the AV-TNDF and AP-TNDF fractions. Fermentation was prepared according to a modified version of Martín-Carrón and Goñi [46]. The samples were placed in a culture system under strict anaerobic conditions for 24 h. Human faeces were used as inocula, and anaerobic conditions were maintained using oxygen-free carbon dioxide.

Inoculum

To obtain the 10% *w/v* inoculum, the faeces of four volunteers were weighed and placed in beakers containing a sterile and anaerobic fermentation medium. The inocula were mixed in a Stomacher 80 Lab blender (Seward Medical, London, UK for 10 min and then filtered (1-mm mesh) before use.

Medium

A micromineral solution was prepared using $CaCl_2$-$2H_2O$, $MnCl_2$-$4H_2O$, $CoCl_2$-$6H_2O$, and $FeCl_3$-$6H_2O$ mixed in a buffer solution ((NH_4)HCO_3 and $NaHCO_3$) in distilled water. A macromineral solution containing Na_2HPO_4, KH_2PO_4, and $MgSO_4$-$7H_2O$ was made in distilled water, as was a reducing solution containing cysteine hydrochloride (C-1276, Sigma-Aldrich, Toluca, Edo. Mex., Mexico).

Procedure

Substrate (100 mg) of the total non-digestible fraction (AV-TNDF or AP-TNDF) was weighed in a 60 mL serum vial (Supelco; Bellefonte, PA, USA), and 8 mL of fresh medium and 2 mL of inoculum were added. Vials were sealed with rubber stoppers and placed in a water bath with constant stirring at 37 °C. The fermentation kinetics was monitored at 0, 4, 8 and 24 h to obtain fermented samples at each time point (AVF4, AVF8 and AVF24 or APF4, APF8 and APF24). A control for each time containing lactulose (L-7877, Sigma-Aldrich, Toluca, Edo. Mex., Mexico) and a target without substrate were included as controls.

Gas volume production and pH were measured at 0, 4, 8 and 24 h. Fermentation was stopped by adding 2.5 mL of1 M NaOH. Samples were centrifuged at 2500× *g* for 10 min, and 3 mL of supernatant was taken in duplicate for SCFA determination by gas chromatography.

Unfermented Fibre

The precipitate of each tube from the fermentation kinetics experiment was homogenized in 50 mL of 0.9% NaCl for 3 min. All precipitates were filtered using Dacron filters (pore size 50 μm), which were dried to a constant weight before use. Unfiltered residues were washed twice with 50 mL of 0.9% NaCl and washed twice with 5 mL of acetone to eliminate mineral and inorganic residues in addition to lipids and other hydrophobic compounds. Filter papers were dried at 60 °C to constant weight, assessing the unfermented fibre (UF) [47] that was calculated by the following formula:

$$UF = \text{weight (filter + washed residue)} - \text{initial filter weight.} \qquad (1)$$

Then, UF(AV-TNDF) and UF(AP-TNDF) were obtained.

3.5.3. Quantification of SCFAs in AV-TNDF and AP-TNDF

SCFA quantification was carried out in AV-TNDF, AP-TNDF and lactulose fermented according the methodology reported by Saura-Calixto et al., [45] and adapted from Zhao et al., [48]. Briefly, 500 μL of supernatant of each fermentation time sample was mixed with 400 mL of 2-methylvaleric acid as an internal standard (10987-8; Sigma, Toluca, Edo. Mex., Mexico) and 100 mL of $HClO_4$ to maintain a constant pH in the samples. The mixture was centrifuged at 10,000 at 4 °C for 15 min, and the supernatant was placed in gas chromatography vials. A Clarus 500/580 GC gas chromatographer (Perkin-Elmer, Inc. Shelton, CO., USA) was used with a TG-WAXMS-A GC column (Thermo Fisher Scientific, CDMX, Mexico) and a flame ionization detector with an injector temperature of 270 °C and a detector temperature of 300 °C. Glacial acetic acid, propionic acid, butyric acid, and 4-methylvaleric acid (Sigma, 320099, 94425, 19215 and 10987-8) were used as standards to obtain calibration curves and assess retention times. The oven was heated to 95 °C at 2 min and increased at 120 °C/min to 240 °C; the carrier flow was 1 mL/min, and the carrier gas flow was 20 mL/min.

3.5.4. Quantification of Phenolic Compounds

Total phenolic compounds were quantified using the Folin–Ciocalteu method [49] in the AV, AP, AV-TNDF, AP-TNDF, AVF4, AVF8, AVF24, APF4, APF8 and APF24 samples. The samples were homogenized in 0.5 mL methanol, added Folin-Ciocalteu reagent, and saturated sodium carbonate solution and water. After 60 min, the absorbance was measured at 760 nm. A calibration curve was performed with aloin, because it is the most representative phenolic compound in *Aloe vera* [50]. The concentration of total phenolic compounds was expressed as milligram equivalents of aloin/100 g sample.

3.5.5. Identification of Phenolic Compounds by HPTLC

HPTLC was accomplished according to a modified version of the methodology published by Paillat, et al., [51]. At the beginning volumes of standard solutions of ferulic acid, *p*-hydroxybenzaldehyde (PHB), *p*-hydroxybenzoic acid (APHB), *p*-cresol, *p*-creosol, and vanillic acid (722820, 54590, 54630, 61030, 41340, and 68654, Sigma-Aldrich, Química S.A. de C.V.) were applied at a concentration of 3 mg/mL. After that, were applied 15 µL of AV, AV-TNDF, AP, AP-TNDF and the kinetic fermentation samples (AVF4, AVF8, AVF24 and APF4, APF8, APF24) at a concentration of 12.5 mg/mL. Samples were placed on TLC silica gel 60 F254 plates (E. Merck, Darmstadt, Germany) using an ATS 4 TLC sampler (CAMAG, Muttenz, Switzerland) at a constant application rate of 120 nL s^{-1} and developed in a CAMAG automated developing chamber ADC2 (47% moisture) saturated and preconditioned for 5 min to a 50-mm distance with an *n*-hexane: chloroform:methanol:acetic acid solvent system (5:36:4:0.5 *v/v/v/v*). Plates were scanned at 254 nm and 366 nm in a CAMAG TLC III scanner (slit size 4 mm × 0.3 mm) at a scanning speed of 10 mm s^{-1} and a data step resolution of 50 µm.

3.6. Antioxidant Activity

3.6.1. Reducing Power and Hydroxyl Radical Scavenging Activity

The reducing power of AV, AV-TNDF, AP, and AP-TNDF was assessed by the Oyaizu method [52]. Various concentrations of methanolic extracts (12.5, 6.25, 3.12, 1.56, and 0.78 mg/mL) were mixed with 2.5 mL 200 mmol/L sodium phosphate buffer (pH 6.6) and 2.5 mL 1% potassium ferricyanide. The mixture was incubated at 50 °C for 20 min. After 2.5 mL of 10% trichloroacetic acid (*w/v*) were added, the mixture was centrifuged at 650 rpm for 10 min. The upper layer (5 mL) was mixed with 5 mL deionised water and 1 mL 0.1% of ferric chloride; the absorbance was measured at 700 nm, a higher absorbance indicating a higher reducing power. The assays were carried out in triplicate and the results were expressed as mean values ± standard deviations. Finally, the reducing power in the kinetic fermentation samples (AVF4, AVF8, AVF24, APF4, APF8 and APF24) was determined but only with a concentration of 12.5 mg/mL. This was the concentration that gave the best results in the previous samples and was chosen for this second part of the work.

3.6.2. ·OH Radical Sequestration

To determine the scavenging activity of the hydroxyl radical (·OH), the method described by Li et al. [53] was applied. The antioxidant activity of AV, AV-TNDF, AP and AP-TNDF was assessed at 12.5, 6.25, 3.12, 1.56 and 0.78 mg/mL. Both 1,10-phenanthroline (0.75 mM) and FeSO4 (0.75 mM) were dissolved in phosphate buffer (pH 7.4) and mixed thoroughly. To start the reaction, H_2O_2 (0.01%) and samples were added. The mixture was incubated at 37 °C for 60 min and the absorbance was measured at 536 nm. Finally, the scavenging activity of the hydroxyl radical in the kinetic fermentation samples (AVF4, AVF8, AVF24, APF4, APF8 and APF24) was determined but only with a concentration of 12.5 mg/mL his was the concentration that gave the best result in the previous samples and was chosen for this second part of the work.

3.7. Statistical Analysis

Differences between experimental groups were analysed by one-way ANOVA and Tukey'post hoc test for repeated samples. To compare the maximum and minimum values of the total SCFA concentration, the data were analysed by a paired one-tailed Student's t-test. Data were processed using GraphPad Prism 6.0 (GraphPad Software Inc., San Diego, CA, USA). Values of $p \leq 0.05$ were considered statistically significant.

4. Conclusions

Aloe vera gel (AV) and polysaccharide extract of *Aloe vera* (AP) are a matrix mainly composed of non-digestible oligosaccharides or polysaccharides of slow fermentation, since that produces a lower pH. The behavior of AV and AP during in vitro colon fermentation was similar to that of lactulose, what indicates the possibility of using *Aloe vera* and polysaccharide extracts as prebiotics. The SCFA production and their potentially delivery to the distal colon due to AV and AP digestion and fermentation process, may result in a protective effect, firstly of human large intestine, since this could be colonised by dense microbial communities that utilized both diet-derived and host-derived energy sources for growth predominantly through fermentative metabolism. In the same way, the antioxidant activity also increases significantly in both the reducing power test and the ·OH radical sequestration when going from AV and AP to its indigestible fraction AV-TNDF and AP-TNDF. Finally, we can conclude that there were no significant differences during the digestion and fermentation of *Aloe vera* and its extract despite the fact that the content of dietary fibre in AP was significantly higher than that of *Aloe vera* gel. During in vitro colon fermentation, the unfermented fibre of AV and AP had a similar response to that of lactulose, as well as the total volume of gas produced, which indicates that *Aloe vera* and polysaccharide extract can possibly be used as prebiotics.

Author Contributions: Conceptualization, A.T.-M., R.M.-E. and M.E.J.-F.; Methodology, A.T.-M.; P.O.-D., S.V.Á.-R.; G.M.A.-J.; Validation, A.T.-M. and R.C.-O.; Formal analysis, A.T.-M.; Investigation, A.T.-M.; Resources, R.M.-E.; writing A.T.-M., R.M.-E., Supervision, R.M.-E., M.E.J.-F. and P.O.-D.; Funding acquisition, R.M.-E. All authors accepted the final version of the manuscript.

Funding: This research was supported by grants from the Instituto Politécnico Nacional (SIP:20181718 project) and CONACyT (PY-SEP-CONACyT 242860, project), The work was also supported by a PhD scholarship to A.T.-M., from CONACyT.

Conflicts of Interest: The authors declare no conflicts of interest.

References

1. Durazzo, A.; D'Addezio, L.; Camilli, E.; Piccinelli, R.; Turrini, A.; Marletta, L.; Marconi, S.; Lucarini, M.; Lisciani, S.; Gabrielli, P.; et al. Plant compounds to botanicals and back: A current snapshot. *Molecules* **2018**, *23*, 18–44. [CrossRef] [PubMed]

2. Nejatzadeh-Barandozi, F. Antibacterial activities and antioxidant capacity of *Aloe vera*. *Org. Med. Chem. Lett.* **2013**, *3*, 2–8. [CrossRef] [PubMed]

3. Sharma, P.; Kharkwal, A.C.; Kharkwal, H.; Abdin, M.Z.; Varma, A. A Review on pharmacological properties of *Aloe vera*. *Int. J. Pharm. Sci. Rev. Res.* **2014**, *29*, 31–37.

4. Ni, Y.; Turner, D.; Yates, K.M. Isolation and characterization of structural components of *Aloe vera* L. leaf pulp. *Int. Immunopharmacol.* **2004**, *4*, 1745–1755. [CrossRef]

5. Wang, P.G.; Zhou, W.; Wamer, W.G.; Krynitsky, A.J.; Rader, J.I. Simultaneous determination of aloin A and aloe emodin in products containing *Aloe vera* by ultra–performance liquid chromatography with tandem mass spectrometry. *Anal. Methods* **2012**, *4*, 3612–3619. [CrossRef]

6. Hamman, J.H. Composition and applications of *Aloe vera* leaf gel. *Molecules.* **2008**, *13*, 1599. [CrossRef]

7. Anokwuru, C.; Sigidi, M.; Boukandou, M.; Tshisikhawe, P.; Traore, A.; Potgieter, N. Antioxidant Activity and spectroscopic characteristics of extractable and non–extractable phenolics from *Terminalia sericea* Burch. ex DC. *Molecules* **2018**, *23*, 1303. [CrossRef]

8. Rajasekaran, S.; Sivagnanam, K.; Subramanian, S. Antioxidant effect of *Aloe vera* gel extract in streptozotocin–induced diabetes in rats. *Pharmacol. Rep.* **2005**, *57*, 90–96.

9. Ravi, S.; Kabilar, P.; Velmurugan, S.; Ashok, K.R.; Gayathiri, M. Spectroscopy studies on the status of aloin in *Aloe vera* and commercial samples. *J. Exper. Sci.* **2011**, *2*, 10–13.

10. Patel, D.K.; Patel, K.; Dhanabal, S.P. Phytochemical standardization of *Aloe vera* extract by HPTLC techniques. *J. Acute Dis.* **2012**, *1*, 47–50. [CrossRef]

11. Gibson, G.R.; Roberfroid, M.B. Dietary modulation of the human colonic microbiota: Introducing the concept of prebiotics. *J. Nutr.* **1995**, *125*, 1401–1412. [CrossRef]

12. Topping, D.L.; Clifton, P.M. Short–Chain fatty acids and human colonic function: Roles of resistant starch and nonstarch polysaccharides. *Physiol. Rev.* **2001**, *81*, 1031–1064. [CrossRef] [PubMed]

13. Calderón-Oliver, M.; Quiñones, P.M.A.; Pedraza-Chaverri, J. Health benefits of *Aloe Vera*. *Rev. Esp. Cienc. Salud.* **2011**, *14*, 53–73.

14. Loots, D.T.; van der Westhuizen, F.H.; Botes, L. *Aloe ferox* leaf gel phytochemical content, antioxidant capacity, and possible health benefits. *J. Agric. Food Chem.* **2007**, *55*, 6891–6896.

15. Minjares-Fuentes, R.; Femenia, A.; Comas-Serra, F. Compositional and structural features of the main bioactive polysaccharides present in the *Aloe vera* plant. *Plant J. AOAC Inter.* **2018**, *101*, 1711–1719. [CrossRef] [PubMed]

16. Zhao, Y.; Sun, Y.N.; Lee, M.J.; Kim, Y.H.; Lee, W.; Kim, K.H.; Kim, K.T.; Kang, J.S. Identification and discrimination of three common Aloe species by high performance liquid chromatography–tandem mass spectrometry coupled with multivariate analysis. *J. Chromat. B.* **2016**, *1031*, 163–171. [CrossRef] [PubMed]

17. Wu, X.; Ding, W.; Zhong, J.; Wan, J.; Xie, Z. Simultaneous qualitative and quantitative determination of phenolic compounds in *Aloe barbadensis* Mill by liquid chromatography–mass spectrometry–ion trap–time–of–flight and high performance liquid chromatography–diode array detector. *J. Pharm. Biomed. Anal.* **2013**, *80*, 94–106. [CrossRef] [PubMed]

18. Yimam, M.; Brownell, L.; Jia, Q. Aloesin as a medical food ingredient for systemic oxidative stress of diabetes. *World J. Diabetes* **2015**, *6*, 1097–1107. [CrossRef]

19. Reynoso-Camacho, R. Gonzalez de Mejia. Nopal (*Opuntia* spp.) and other traditional mexican plants. In *Nutraceutical Glycemic Health & Type 2 Diabetes*; Pasupulety, V.K., Anderson, J.M., Eds.; John Wiley & Sons: Hoboken, NJ, USA, 2008; Chapter 15; pp. 379–399.

20. Slavin, J. Fiber and prebiotics: Mechanisms and health benefits. *Nutrients.* **2013**, *5*, 1417–1435. [CrossRef]

21. Gullón, B.; Gullón, P.; Sanz, Y.; Alonso, J.L.; Parajó, J.C. Prebiotic potential of a refined product containing pectic oligosaccharides. *LWT–Food Sci. Technol.* **2011**, *44*, 1687–1696.

22. Romero-López, M.R.; Osorio-Díaz, P.; Flores-Morales, A.; Robledo, N.; Mora-Escobedo, R. Chemical composition, antioxidant capacity and prebiotic effect of aguamiel (Agave atrovirens) during in vitro fermentation. *Rev. Mex. Ing. Quím.* **2015**, *14*, 281–292.

23. Johnson, W.; McRorie, J.W., Jr.; Nicola, M.; McKeown. Understanding the physics of functional fibers in the gastrointestinal tract: An evidence–based approach to resolving enduring misconceptions about insoluble and soluble fiber. *J. Acad. Nutr. Diet.* **2017**, *117*, 251–264.

24. Goñi, I.; Martín, N.; Saura-Calixto, F. In vitro digestibility and intestinal fermentation of grape seed and peel. *Food Chem.* **2005**, *90*, 281–286. [CrossRef]

25. Al-Madboly, L.A.; Kabbash, A.; Yassin, A.M.; Yagi, A. Dietary cancer prevention with butyrate fermented by *Aloe vera* gel endophytic microbiotata. *J. Gastroenterol. Hepat. Res.* **2017**, *6*, 2312–2317.

26. Wong, J.M.W.; Souza, R.; Kendall, C.W.C.; Emam, A.; Jenkins, D.J.A. Colonic health: Fermentation and short chain fatty acids. *J. Clin. Gastroenterol.* **2006**, *40*, 235–243. [CrossRef]

27. Mudgil, D.; Barak, S. Composition, properties and health benefits of indigestible carbohydrate polymers as dietary fiber: A review. *Int. J. Biol. Macromol.* **2013**, *61*, 1–6. [CrossRef] [PubMed]

28. Laparra, J.M.; Sanz, Y. Interactions of gut microbiota with functional food components and nutraceuticals. *Pharmacol. Res.* **2010**, *61*, 219–225. [CrossRef]

29. Russell, W.R.; Hoyles, L.; Flint, H.J.; Dumas, M.E. Colonic bacterial metabolites and human health. *Curr. Opin. Microbiol.* **2013**, *16*, 246–254. [CrossRef]

30. Maslowski, M.K.; Kendle, M.; Vieira, T.A.; Ng, A.; Kranich, J.; Sierro, F.; Yu, D.; Schilter, C.H.; Rolph, S.M.; Mackay, F.; et al. Regulation of inflammatory responses by gut microbiota and chemoattractant receptor GPR4. *Nature* **2009**, *461*, 1282–1286. [CrossRef]

31. Fukuda, S.; Toh, H.; Hase, K.; Oshima, K.; Nakanishi, Y.; Yoshimura, K.; Tobe, T.; Clarke, M.J.; Topping, L.D.; Suzuki, T.; et al. Bifidobacteria can protect from enteropathogenic infection through production of acetate. *Nature.* **2011**, *469*, 543–547. [CrossRef]

32. Gullón, B.; Gullón, P.; Tavaria, F.; Alonso, J.L.; Pintado, M. In vitro assessment of the prebiotic potential of Aloe vera mucilage and its impact on the human microbiota. *Food Funct.* **2015**, *6*, 525.

33. Verma, B.; Hucl, P.; Chibbar, R. Phenolic acid composition and antioxidant capacity of acid and alkali hydrolysed wheat bran fractions. *Food Chem.* **2009**, *116*, 947–954.

34. Scaglioni, P.T.; de Souza, T.D.; Schmidt, C.G.; Badiale-Furlong, E. Availability of free and bound phenolic compounds in rice after hydrothermal treatment. *J. Cereal Sci.* **2014**, *60*, 526–532. [CrossRef]

35. Cueva, C.; Sanchez-Patan, F.; Monagas, M.; Walton, G.E.; Gibson, G.R.; Martın-Alvarez, P.J.; Bartolome, B.; Moreno-Arribas, M.V. In vitro fermentation of grape seed flavan–3–ol fractions by human faecal microbiota: Changes in microbial groups and phenolic metabolites. *FEMS Microbiol. Ecol.* **2013**, *83*, 792–805. [CrossRef] [PubMed]

36. Lee, H.C.; Jenner, A.M.; Low, C.S.; Lee, Y.K. Effect of tea phenolics and their aromatic fecal bacterial metabolites on intestinal microbiota. *Res. Microbiol.* **2006**, *157*, 876–884. [CrossRef] [PubMed]

37. Alugoju, P.; Dinesh, B.J.; Latha, P. Free radicals: Properties, sources, targets, and their implication in various diseases. *Ind. J. Clin. Biochem.* **2015**, *30*, 11–26.

38. Chun-Hui, L.; Chang-Hai, W.; Zhi-Liang, X.; Yi, W. Isolation, chemical characterization and antioxidant activities of two polysaccharides from the gel and the skin of *Aloe barbadensis* Miller irrigated with sea water. *Process Biochem.* **2007**, *42*, 961–970. [CrossRef]

39. Ray, A.; Gupta, S.D.; Ghosh, S. Evaluation of anti–oxidative activity and UV absorption potential of the extracts of *Aloe vera* L. gel from different growth periods of plants. *Ind. Crops Prod.* **2013**, *49*, 712–719. [CrossRef]

40. Sánchez-Machado, D.I.; López-Cervantes, J.; Sendón, R.; Sanches-Silva, A. *Aloe vera*: Ancient knowledge with new frontiers. *Trends Food Sci. Technol.* **2017**, *61*, 94–102. [CrossRef]

41. Kunwar, A.; Priyadarsini, K.I. Free radicals, oxidative stress and importance of antioxidants in human health. *J. Med. Allied Sci.* **2011**, *1*, 53–60.

42. *AOAC Official Methods of Analysis (2015)*; AOAC International: Rockville, MD, USA, 2015.

43. Cassani, J.; Ferreyra-Cruz, O.A.; Dorantes-Barrón, A.M.; Villaseñor, R.M.; Arrieta-Baez, D.; Estrada-Reyes, R. Antioxidant, hepatoprotective, and antidepression effects of *Rumex tingitanus* extracts and identification of a novel bioactive compound. *J. Ethnopharmacol.* **2015**, *171*, 295–306. [CrossRef] [PubMed]

44. Hernández-Rosas, N.A.; García-Zebadúa, J.C.; Hernández-Delgado, N.; Torres-Castillo, S.; Figueroa-Arredondo, P.; Mora-Escobedo, R. Perfil de polifenoles, capacidad antioxidante y efecto citotóxico in vitro en líneas celulares humanas de un extracto hidroalcohólico de pétalos de Calendula officinalis L. *TIP Rev. Espec. Cienc. Quím. Biol.* **2018**, *21*, 1–11.

45. Saura-Calixto, F.; García-Alonso, A.; Goñi, I.; Bravo, L. In vitro determination of the indigestible fraction in foods: An alternative to dietary fiber analysis. *J. Agric. Food Chem.* **2000**, *48*, 3342–3347. [CrossRef] [PubMed]

46. Martín-Carrón, N.; Goñi, I. Prior exposure of cecal microflora to grape pomaces does not inhibit in vitro fermentation of pectin. *Nutr. Res.* **1998**, *46*, 1064–1070. [CrossRef]

47. Wenzel, M.E.; Dan, M.C.T.; Cardenette, H.L.G.; Goñi, I.; Bello, L.A.; Lajolo, F.M. In vitro colonic fermentation and glycemic response of different kinds of unripe banana flour. *Plant Foods Human Nutr.* **2010**, *63*, 379–385.

48. Zhao, G.; Nyman, M.; Jönsson, J.A. Orthogonal comparison of GC–MS and H NMR spectroscopy for short chain fatty acid quantitation. *Biomed. Chromatogr.* **2006**, *20*, 674–682. [CrossRef] [PubMed]

49. Blainski, A.; Lopes, G.C.; Palazo de Mello, J.C. Application and analysis of the Folin Ciocalteu method for the determination of the total phenolic content from *Limonium brasiliense* L. *Molecules.* **2013**, *18*, 6852–6865. [CrossRef] [PubMed]

50. Flores-López, M.L.; Aloia-Romani, M.; Cerqueira, A.; Rodríguez-García, R.; Jasso de Rodríguez, D.; Vicente, A.A. Compositional features and bioactive properties of whole fraction from *Aloe vera* processing. *Ind. Crops Prod.* **2016**, *91*, 179–185. [CrossRef]

51. Paillat, L.; Périchet, C.; Lavoine, S.; Meierhenrich, U.J.; Fernandez, J. Validated high–performance thin–layer chromatography (HPTLC) method for quantification of vanillin β–D–glucoside, and four major phenolic compounds in vanilla (*Vanilla planifolia*) fruits, beans, and extracts. *J. Planar Chromat.* **2012**, *4*, 295–300. [CrossRef]

52. Oyaizu, M. Studies on products of browning reactions: Antioxidative activities of product of browning reaction prepared from glucosamine. *Jpn. J. Nutr.* **1986**, *44*, 307–315. [CrossRef]

53. Li, Y.; Jiang, B.; Zhang, T.; Mu, W.; Liu, J. Antioxidant and free radical–scavenging activities of chickpea protein hydrolysate (CPH). *Food Chem.* **2008**, *106*, 444–450. [CrossRef]

Sample Availability: Samples of the compounds are not available from the authors.

Review

Therapeutic Perspectives of Molecules from *Urtica dioica* Extracts for Cancer Treatment

Sabrina Esposito *, Alessandro Bianco, Rosita Russo, Antimo Di Maro, Carla Isernia and Paolo Vincenzo Pedone

Department of Environmental, Biological, and Pharmaceutical Sciences and Technologies, University of Campania "Luigi Vanvitelli", 81100 Caserta, Italy
* Correspondence: sabrina.esposito@unicampania.it; Tel.: +39-0823-274620

Academic Editor: Derek J. McPhee
Received: 26 June 2019; Accepted: 27 July 2019; Published: 29 July 2019

Abstract: A large range of chronic and degenerative diseases can be prevented through the use of food products and food bioactives. This study reports the health benefits and biological activities of the *Urtica dioica* (*U. dioica*) edible plant, with particular focus on its cancer chemopreventive potential. Numerous studies have attempted to investigate the most efficient anti-cancer therapy with few side effects and high toxicity on cancer cells to overcome the chemoresistance of cancer cells and the adverse effects of current therapies. In this regard, natural products from edible plants have been assessed as sources of anti-cancer agents. In this article, we review current knowledge from studies that have examined the cytotoxic, anti-tumor and anti-metastatic effects of *U. dioica* plant on several human cancers. Special attention has been dedicated to the treatment of breast cancer, the most prevalent cancer among women and one of the main causes of death worldwide. The anti-proliferative and apoptotic effects of *U. dioica* have been demonstrated on different human cancers, investigating the properties of *U. dioica* at cellular and molecular levels. The potent cytotoxicity and anti-cancer activity of the *U. dioica* extracts are due to its bioactive natural products content, including polyphenols which reportedly possess anti-oxidant, anti-mutagenic and anti-proliferative properties. The efficacy of this edible plant to prevent or mitigate human cancers has been demonstrated in laboratory conditions as well as in experimental animal models, paving the way to the development of nutraceuticals for new anti-cancer therapies.

Keywords: *Urtica dioica*; natural products bioactivity; food bioactives; nutraceuticals; cancer therapy; breast cancer

---

1. Introduction

A wide range of chronic and degenerative diseases can be prevented using food product nutraceuticals or functional foods and food bioactive molecules [1,2].

In this context, researchers have studied the *Urtica dioica* [3,4], an evergreen edible plant commonly used since ancient times in traditional medicine to treat several diseases.

U. dioica is the most common species of the Urticaceae family commonly known as Stinging nettle and one of the most studied medicinal plants worldwide. It is an herbaceous perennial plant and has a long history of usage for various kinds of health problems [3,4]. The plant grows in tropical and temperate wasteland areas around the world, and well tolerates all environments. The name *Urtica* comes from the Latin verb urere, namely 'to burn,' attributed to its stinging hairs. The most common species *dioica* is so defined because the plant generally contains either female or male flowers [5]. The leaves are oval, long petiolate, elongated with toothed margins, the flowers are dioecious, the fruit is a small oval and greenish-yellow achene. The plant has stinging hairs with a tuft of hair at the apex. The leaves and stems contain abundant non-stinging hairs, with touch sensitive tips, needles that will inject chemicals including serotonin, histamine, acetylcholine, moroidin, leukotrienes and possibly formic acid into the skin. The irritant compounds provoke pain, wheals or a stinging sensation [6]. However, this edible plant will lose its irritant powers during cooking, the burning property of the juice is, indeed, dissipated by heat and the young shoots may be used for culinary purposes.

The medicinal properties of *U. dioica* are linked to its anti-inflammatory, anti-asthmatic, astringent, depurative, galactogogue, diuretic, nutritive and stimulating effects. The powered leaf's extract has been used as an anti-haemorrhagic agent to reduce excessive menstrual flow and nose bleedings. The roots and herbs are used in different ways: the roots for benign prostate hyperplasia, the herbs for urinary tract disorders and rheumatic conditions, while fresh freeze-dried leaves are used to treat allergies [3,4]. Several studies have also reported its analgesic potential and its role as anti-aggregating factor, as well as describing its favorable effects on cardiovascular and smooth-muscle activity as a hypotensive agent [4]. Indeed, the uses of the plant are extended to different fields including the dye industry, veterinary medicine, the textile industry, cosmeceutics for hair loss lotions and anti-dandruff products and also for culinary usage in the preparation of common dishes [7]. In the popular tradition *U. dioica* leaves are eaten, both raw and blanched, gently fried or steamed in many foodstuffs such as pesto, quiches, soups, purées, sauces, cookies, gelatines and jams. Dried herb is processed for capsules, tablets and teas, and other preparations. Freeze-dried herbs are commonly prepared in capsules. Formulations from fresh plant material include homeopathic products, juice and liquid extracts. Targeted studies favored the reuse of *U. dioica* to produce curd for fresh cheese [8]. The proteins (3.7%), dietary fibers (6.4%) and low total calories contents (45.7 kcal/100 g) offer *U. dioica* shoots as a nutritional valuable source and a valid contribution in vitamins (A and C), calcium, iron, sodium and fatty acids [9]. Essential fatty acids, such as linoleic and α-linolenic acid, account for 20.2% and 12.4% of total fatty acids, respectively [10]. Although, the polyunsaturated fatty acids (PUFAs)/saturated fatty acids ratio was reported to be comparably contained, among PUFAs, fatty acids of n-3 and n-6 series were more abundant than monounsaturated fatty acids. This finding was in contrast with a recent investigation regarding fatty acids in *U. dioica* leaves which analyzed a rapid solid–liquid extraction (Soxtherm) using petroleum ether as the solvent, revealing a favorable outcome for saturated acids compared to unsaturated acids [11]. Among monounsaturated fatty acids, a small amount of C17:1 fatty acid was detected (0.13%). Recently, oxylipins, which are bioactive lipid metabolites derived from PUFAs via cycloxygenase (COX), lipoxygenase and cytochrome P450 pathways, were isolated from a *U. dioica* hydroalcoholic extract [12].

Furthermore, *U. dioica* powder was found to be rich in proteins, three-fold higher than traditional cereals, such as rice, wheat and barley [9,13]. *U. dioica* is also rich in minerals, as is characterized by high levels of calcium (169 mg/100 g) and iron (277 mg/100 g) followed by potassium, phosphorus, magnesium, sodium and zinc. According to reported data, *U. dioica* powder is most likely one of the richest source of minerals among the plant foods. The content of carbohydrate is low (37.4%) compared to cereals, such as wheat and barley, showing that *U. dioica* powder has a low glycemic index in relation to the conventional sources of plant foods such as cereals and the potato [13].

The benefits of *U. dioica* may be linked to its diversity in secondary metabolites and its content appears to be strongly influenced by the geographic conditions and taxonomical, morphological and genetics factors [14]. In particular, a number of studies have explored and confirmed *U. dioica* as a favorable source of flavonoids and phenylpropanoids [4].

U. dioica may be considered as an infesting plant albeit possessing a longstanding history of medicinal usage. Some medicinal capacities of *U. dioica* have been confirmed by modern research and scientific evidences.

The purpose of this review is to focus, among the many biological activities, on the anti-cancer effects of *U. dioica* by examining the cellular and molecular results of *U. dioica* extract treatments on various human cancer cell lines and in-vivo animal models with special attention to breast cancer, the most prevalent cancer among women.

The study of the mechanisms of action of *U. dioica* may be of importance in the development of food bioactives with natural products for anti-cancer therapies.

PubMed, Scopus and Science Direct were used to consult literature.

2. Phytochemical Investigations and Biological Activities of *Urtica dioica*

The phytochemical composition investigation on *U. dioica* [13–28] revealed that it contains phenolic compounds (including flavonoids, tannins, coumarins and lignans), sterols, fatty acids, polysaccharides and isolectins.

The growing interest in plant phenolic compounds is illustrated by the extensive body of literature devoted to this field of study. Nowadays, it is generally accepted that the therapeutic effects of many plant species are due to the presence of antioxidative phenolics in their tissues. These compounds represent a plant defense mechanism against UV radiation, insects and microorganism, but may also act as plant pigments [29–31].

Phenols and polyphenols in dietary plants have gained considerable attention as therapeutic and prophylactic agents in the treatment of chronic and degenerative diseases [32,33]. In particular, it was observed that all the parts (roots, stalk and leaves) of *U. dioica* are a rich source of these substances and that their content is higher in wild plants than in domesticated plants [18,34]. Root samples from Mediterranean cultivar were reported to contain phenol compounds, such ferulic acid and polyphenols as naringin, ellagic acid, myricetin and rutin (Figure 1). The roots also contained lignans (secoisolariciresinol, 9,90-bisacetyl-neo-olivil and their glucosides), phytosterols (e.g., β-sitosterol), polysaccharides, isolectins (mainly *U. dioica* agglutinine), coumarins (e.g., scopoletin), simple phenols (e.g., p-hydroxy-benzaldehyde), triterpenoic acids and monoterpendiols.

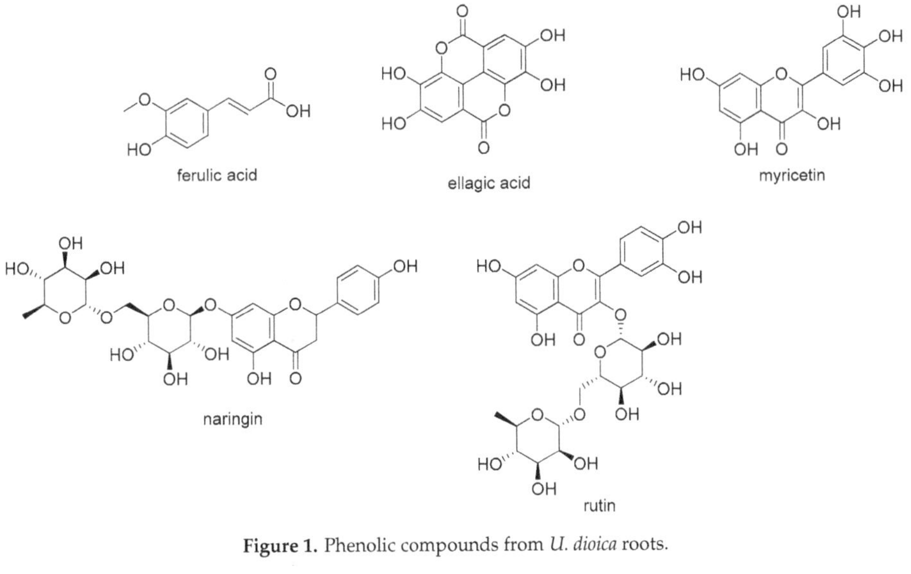

Figure 1. Phenolic compounds from *U. dioica* roots.

U. dioica leaves are also constituted by flavonoid glycosides, mainly rutinosyl flavonols, as well as by different depsides of hydroxycinnamic acids with quinic or malic acid (Figure 2). Chlorogenic acid and caffeoyl malic acid represented approximately 76.5% of total phenolic compounds, whereas rutin was the most abundant flavonol derivative [18,19]. Isorhamnetin-3-*O*-rutinoside was found, together with rutin, quercetin-3-*O*-glucoside and kaempferol-3-*O*-glucoside in methanolic extracts of *U. dioica* leaves and stalks [35]. The polyphenol profile seems to be strongly dependent on the parts of the plant investigated, but also on the harvest site and season. The quantification of *U. dioica* phenolics in different extracts, by high-performance liquid chromatography coupled with tandem mass spectrometric detection, evidenced that inflorescence extracts were the richest extracts [17]. Thus, the consumption of *U. dioica* was in line with an amelioration of phenolic compounds food intake and thus, the exploitation of these anti-oxidant and anti-inflammatory compounds defined the plant as a valuable tool towards mutagenesis and carcinogenesis [36].

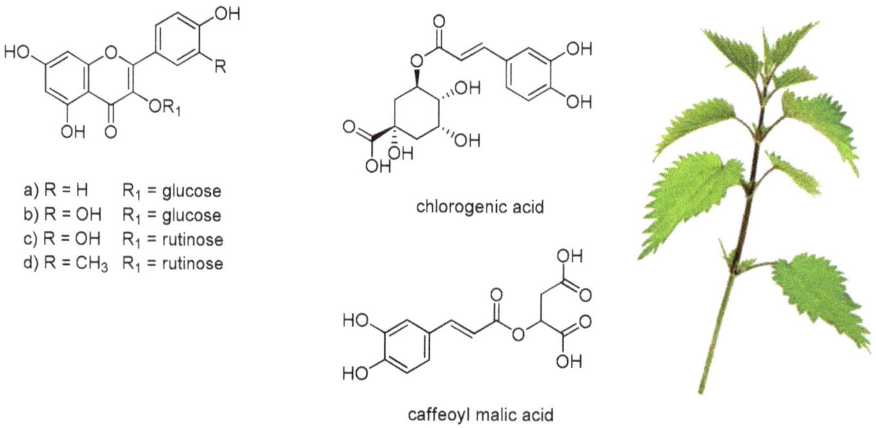

Figure 2. Phenols and polyphenols mostly detected in *U. dioica* leaves. (**a**) kaempferol-3-*O*-glucoside; (**b**) quercetin-3-*O*-glucoside; (**c**) rutin; (**d**) isorhamnetin-3-*O*-rutinoside.

Among lipid secondary metabolites, carotenoids were detected in the leaves and their total content was estimated equal to 29.6 mg/100 g dry weight [4,23].

Several studies have established that extracts of *U. dioica* possess various pharmacological effects [37–39], including anti-inflammatory [40–42], anti-oxidant [43–47], anti-microbial [46–49], anti-diabetic [39,50,51], cardiovascular [39], anti-ulcer, analgesic [36], immuno-modulatory [35,52], anti-mutagenic [44] and anti-cancer properties. Moreover, this edible plant has considerable chemopreventive capacities and disease-preventing effects on animals and humans. These health benefits of *U. dioica* may be related to the wide range of bioactive natural products present in the various parts of the plant.

The preventive activity of both polyphenols and carotenoids is associated to the health promoting effects of *U. dioica* against chronic and degenerative diseases such as cancer.

3. Anticancer Activities of *Urtica dioica*

Among the biological activities of *U. dioica*, we report in detail the studies on the anti-cancer effects, due to the induction or inhibition of the key processes in cellular metabolism and the ability to activate the apoptotic pathways. Various studies have recently demonstrated the cytotoxic and anti-cancer properties of *U. dioica*, in particular against colon, gastric, lung, prostate and breast cancers. In this section, we will review the main anti-tumour activities of *U. dioica* demonstrated against several human cancer cell lines and in animal models.

Cancer is a group of diseases in which normal cells grow uncontrollably and abnormally, invade and spread to other parts of the body. Unfortunately, it is a main cause of death worldwide and the incidence and mortality rates are still unacceptably high [53–56]. Notable progresses have been obtained in conventional therapies (as chemotherapy, radiotherapy and surgical excision), but these treatments cause many serious side effects and often may prolong life for only a few years.

Cancer chemoprevention [57,58] has become an important therapeutic option through which the battle against cancer could be possible, using natural, synthetic or biological agents able to reverse, suppress or prevent either carcinogenesis or the progression of premalignant cells towards invasive tumors. For this purpose, plants and herbs may be a promising source for adjuvant, complementary or alternative anti-cancer therapy [59–68], since some of them contain bioactive natural products and anti-cancer compounds, including polyphenols [32,33,69,70] as flavonoids, tannin etc. Today, many phytochemical compounds, normally biosynthesized and accumulated in the plants, have shown chemopreventive actions and several anti-cancer drugs, including podophilotoxins, camptotoxins, taxans, arise from herbal compounds and are successfully used for anti-cancer therapy [63,71]. Chemoprevention due to the natural plant-based bioactivity can be achieved through different biochemical and molecular mechanisms involved in cancer control and development. The plants, indeed, contain a number of bioactive molecules that are able to induce cellular protection and responses to stresses, such as anti-oxidant enzymes, apoptosis and/or cell cycle arrest [72,73]

The anti-mutagenic activity of a protein fraction from the aerial parts of *U. dioica* was demonstrated, via the Ames test in various bacteria strains, against the mutagen 2-aminoanthracene; the anti-mutagenic activity can be due to the inhibition of CYP450-isoenzymes, involved in the 2-aminoanthracene mutagen bioactivation [44].

The anti-oxidant and radical scavenging activities of *U. dioica* were reported by ABTS and superoxide-radical scavenger assays and on analysis of the changes in antioxidant enzymes [43–47]. In mice, the treatment with *U. dioica* methanolic extract from the aerial parts demonstrated hepatoprotective and nephroprotective activities against cisplatin-induced [74,75] toxicity, most likely due to increasing antioxidant defense mechanisms, in fact, this extract has been shown to increase the activities of catalase (CAT) and superoxide dismutase (SOD) enzymes and the content of glutathione (GSH) [76]. These activities may be attributed to the flavonoid content of *U. dioica*, in fact, the flavonoids were associated to anti-oxidant and radical scavenging activities [77–79].

The reactive oxygen metabolites have a well-known role in cancer pathogenesis [80,81]. The oxidative stress, with the loss of cellular redox homeostasis and elevated levels of oxygen free radicals, causes the production of mutagenic agents and can be tumorigenic, with a key role in initiation and progression of cancer [82]. Due to the presence of large quantities of compounds with anti-oxidant and free radical scavenger properties, *U. dioica*, and specifically the leaves, are able to reduce the high level of oxidative stress present in cancerous cells and exert a chemopreventive function.

Most drugs in use for cancer treatment are cytotoxic and/or cytostatic [83]. Since the rate of apoptosis was reduced during cancer in several studies, an efficient cancer treatment requires the induction of apoptosis in cancer cells, with the programmed cell death of cancerous and damaged cells, without destructive adverse effects in normal dividing cells [84]. Furthermore one of the more significant properties analyzed in putative cytotoxic anti-tumor agents is the ability to induce apoptosis and/or cell cycle arrest. The cytotoxic activity of *U. dioica* was widely tested by in vitro MTT assay and trypan blue viability exclusion dye assay (to evaluate the number of live cells). The cellular and molecular mechanisms of toxicity were analyzed using different assays. The most commonly used were DNA fragmentation assay and TUNEL test (to detect the type of cell death, apoptosis or necrosis), quantitative Real-Time PCR (qRT-PCR, to quantify the apoptosis- and metastasis- related mRNA expression levels), Western Blotting (to quantify apoptosis-related protein levels) and flow cytometry (to analyze cell cycle distribution and apoptosis). The Tables 1, 2 and 4 (Table 4 is included in the next paragraph) summarize the main findings regarding the anti-cancer properties of *U. dioica* with the plant extracts and parts used (specifying collection sites and the biologically active molecules identified), the cancer cell lines/tissues or animal models tested, the IC50 (concentration required for 50% inhibition) and the effects. The studies (first author and date) are chronologically listed for various tumoral groups. The studies on benign prostatic hyperplasia are also reported.

Table 1. Anti-cancer activities of *U. dioica*: cervical, epidermoid, colon, gastric and lung cancer.

U. dioica Extracts *	Cancer Cell Lines	IC50	Effects	References
U. dioica L. (Germany) roots aqueous extract *U. dioica* agglutinine (UDA)	HeLa human cervical cancer A431 human epidermoid carcinoma	5 µg/mL (24 h treatment) 21 µg/mL (24 h treatment)	↓ Proliferation ↓ EGF binding	Wagner 1994 [85]
U. dioica L. (Iran) roots ethanolic extract	HT29 human colon cancer MKN45 human gastric cancer	24.7 µg/mL (72 h treatment) 249.9 µg/mL (72 h treatment)	↓ Proliferation ↑ Apoptosis	Ghasemi 2016 [86]
U. dioica L. (Iran) aerial parts dichloromethane extract	HCT-116 human colon cancer	23.61 µg/mL (48 h treatment)	↓ Proliferation ↑ Apoptosis G2/M arrest	Mohammadi 2016 [87]
U. dioica L. (Italy) leaves methanolic extract, oxylipins	NSCLC H1299 human non-small cell lung cancer NSCLC A549 human non-small cell lung cancer	52.3 µg/mL (72 h treatment) 47.5 µg/mL (72 h treatment)	↓ Proliferation ↑ Apoptosis extrinsic pathway ↑ caspase 3 ↑ caspase 8 ↑ cPARP ↑ tBid ↑ GADD153 ↑ DR5 G2/M arrest	D'Abrosca 2019 [12]

* The plant extracts and parts used with specified collection sites and the biologically active molecules identified.

Table 2. Anti-cancer effects of *U. dioica*: prostate cancer, in-vitro and in-vivo studies.

U. dioica Extracts *	Cancer Cell Lines/Tissues/Animal Models	IC50	Effects	References
U. dioica L.(Germany) roots methanolic extract	Balb/c mouse model of benign prostatic hyperplasia (28 days, 5 mg oral treatment)		↓ hyperplasia 51.4 % growth inhibition	Lichius 1997 [88]
U. dioica L.(Germany) roots methanolic extract	LNCaP human prostate cancer		↓ Proliferation 30% (5 day treatment with 1 µg/mL)	Konrad 2000 [89]
U. dioica L. leaves aqueous extract	prostate tissue from prostate cancer patients	50 µg/mL (30 min treatment)	↓ ADA	Durak 2004 [90]
U. dioica L. (India) roots petroleum ether extract, β-sitosterol roots ethanolic extract, β-sitosterol and scopoletin	rat model of benign prostatic hyperplasia (28 days, 50 mg/Kg oral treatment)	0.19 mg/mL (28 day treatment) 0.12 mg/mL (28 day treatment)	↓ hyperplasia ↓ 5α-reductase	Nahata 2012 [91]
U. dioica L. (Iran) leaves dichloromethane extract	PC3 human prostate cancer	15.54 µg/mL (48 h treatment)	↓ Proliferation ↑ Apoptosis intrinsic pathway ↑ caspase 3 ↑ caspase 9 ↓ Bcl-2 G2/M arrest	Mohammadi 2016 [92]

* The plant extracts and parts used with specified collection sites and the biologically active molecules identified. ADA: adenosire deaminase.

In the study by Ghasemi et al. (2016) [86] the cytotoxic effects of an ethanolic extract of *U. dioica* roots (0–2000 µg/mL) were demonstrated on human colon (HT29) and gastric (MKN45) cancer cells. Cells were treated with increasing concentrations of *U. dioica* for 24–72 h. *U. dioica* decreased cell viability in a dose- and time- dependent manner, with IC50 values of 24.7 and 249.9 µg/mL, respectively, after 72 h exposure. In addition, *U. dioica* treatment induced apoptotic cell death, as shown by flow cytometry analysis. The different studied cell lines showed a diverse sensitivity to the *U. dioica* treatment, with more sensitive human colon cancer HT29 cells compared to human gastric cancer MKN45 cells, a cell line poorly differentiated and usually resistant to chemotherapy. Interestingly, the anti-proliferative effects of *U. dioica* treatments are comparable to those obtained with oxaliplatin, a current anti-neoplastic drug. In line with these findings, Mohammadi et al. (2016) [87], demonstrated the cytotoxic effects of a dichloromethane extract of *U. dioica* aerial parts plant on human colon cancer cell line HCT-116, with IC50 of 23.61 µg/mL (48 h treatment) and by eliciting apoptotic cell death and arresting the cell cycle at the G2/M phase.

The dichloromethane extract of *U. dioica* leaves inhibited the growth and proliferation of human prostate cancer cells (PC3), showing a IC50 concentration of 15.54 µg/mL in 48 h exposure and a cell cycle arrest in G2/M phase (Mohammadi et al. 2016) [92]. In this study, the observed increased expression levels of pro-apoptotic genes caspase 3 and 9 and reduced anti-apoptotic Bcl-2 suggested that cytotoxicity was due to apoptosis induction from intrinsic (mitochondrial) pathway. Moreover, a methanolic extract of *U. dioica* roots previously produced a significant dose- and time- dependent reduction in proliferation of human prostate carcinoma cells (LNCaP), with a 30% maximum growth inhibition after 5 day exposure with 1 µg/mL extract concentration (Konrad et al. 2000) [89].

More recently, D'Abrosca et al. (2019) [12] reported the effects of *U. dioica* leaves methanolic extract against the human non-small cell lung cancer cell lines (NSCLC). Exposure of H1299 and A549 NSCLC cells to this extract inhibited cell proliferation, with an IC50 of 52.3 and 47.5 µg/mL, respectively. NSCLC cells have a low sensitivity to cisplatin [74,75,93], a cytotoxic agent largely utilized for chemotherapy cancer cure. The co-treatment with the *U. dioica* extract and cisplatin ameliorated the cisplatin cytotoxicity, thus showing a synergistic effect. *U. dioica* extract induced arrest at G2/M cell cycle phase and apoptosis from extrinsic pathway, as demonstrated by the observed decreased levels of pro-caspase 3 and pro-caspase 8 proteins (indicating the activation and increasing of the apoptotic proteolytic enzymes caspase 3 and caspase 8) and increased levels of cPARP and tBid (substrates of caspase 3 and caspase 8, respectively). GADD153 (a marker of endoplasmic reticulum stress) [94–96] and DR5 (death receptor [97] a cell surface receptor of the TNF-receptor superfamily, which directly promotes the extrinsic apoptotic pathway) were also upregulated after *U. dioica* treatment, confirming that the extracts promoted the extrinsic apoptotic pathway. Interestingly, rutin and oxylipins (polyunsaturated oxidised fatty acids) [98–100] were identified in the *U. dioica* extract on investigation of the exact mechanism involved in cell death, via spectroscopic techniques NMR [101,102] and mass spectrometry analyses. In particular, oxylipins, including the most abundant 13-S-hydroxy-9Z, 11E, 15Z-octadecantrienoic acid, were proved to be the bioactive natural products responsible for anti-cancer activity (Figure 3 and Table 3). Oxylipins are a large and diverse family of secondary metabolites derived from the oxidation of PUFAs [103]. The oxylipin 13-S-hydroxy-9Z, 11E, 15Z-octadecantrienoic acid also possess anti-inflammatory properties in human chondrocytes [41]. It is noteworthy that a plant oxylipin, 12-oxo-phytodienoic acid, potently suppressed the proliferation of human breast cancer cell lines T47-D and MDA-MB-231, reducing the expression of the cyclin D1 and inducing arrest at G1 phase of the cell cycle [104,105]. In addition, another biologically active molecule of *U. dioica* is a rare lectin (carbohydrate-binding protein), *U. dioica* agglutinine (UDA, accession number P11218 in Protein Data Bank) [106], isolated from the *U. dioica* aqueous roots extract (Wagner et al. 1994) [85]. This molecule demonstrated in vitro anti-proliferative properties on human cervical cancer HeLa cells and human epidermoid carcinoma A431 cancer cells. On HeLa and A431 cells UDA inhibited the binding of EGF to its receptor with an IC50 of 5 and 21 µg/mL, respectively, after 24 h exposure [107].

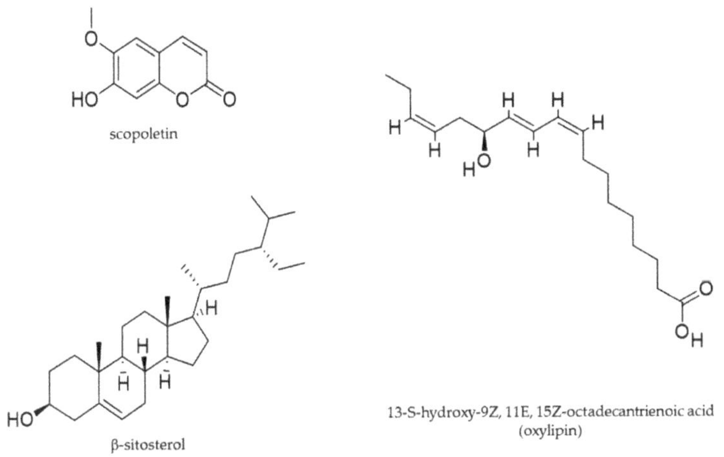

scopoletin

β-sitosterol

13-S-hydroxy-9Z, 11E, 15Z-octadecantrienoic acid (oxylipin)

Figure 3. Molecular structures of selected bioactive phytochemicals isolated from *U. dioica*.

Moreover, in-vitro studies and investigations regarding testosterone-induced rat models of prostatic hyperplasia, examined the effects of petroleum ether and ethanolic extracts of the *U. dioica* roots (28 day, 50 mg/Kg, oral treatment), Nahata et al. (2012) [91]. The results demonstrated that *U. dioica* significantly reduced the activity of 5α-reductase enzyme and the dimension of the prostatic hyperplasia. 5α-reductase [108] is a key enzyme involved in testosterone metabolism, thus in hormone-dependent prostate hyperplasia and prostate cancer. The IC50 values for 5α-reductase were of 0.19 and 0.12 mg/mL, respectively, on the 28th day of treatment. In both the extracts, via spectroscopic techniques, the β-sitosterol was isolated; in ethanolic extract scopoletin was also found (Figure 3 and Table 3). β-sitosterol is a sterol originated by a complex and multistage biosynthetic process [109]; it was jet reported to treat patients with prostate diseases [110,111]. Scopoletin is a courmarin derived by a known biosynthesis pathway [112]; it possesses anti-inflammatory properties [113] and has been reported to induce anti-proliferative and pro-apoptotic effects on the prostate cancer cell line PC3 [114]. In a previous animal study (Lichius et al. 1999) [88], the experimentally induced benign prostatic hyperplasia in Balb/c mouse was reduced, by a methanolic extract of *U. dioica* roots, orally administered for 28 days (5 mg), with an observed 51.4% growth inhibition.

Table 3. Biological activities of specific molecules isolated from *U. dioica*.

Molecules	Biological Activities	Cells	References
13-S-hydroxy-9Z, 11E, 15Z-octadecantrienoic acid (oxylipin)	Anti-proliferation ro-apoptotis Stop cell cycle Anti-inflammation	lung cancer chondrocytes	[12,41]
U. dioica agglutinine (UDA)	Anti-proliferation Anti-EGF binding	cervical cancer epidermoid carcinoma	[85,107]
β-sitosterol	Anti-proliferation Inhibition 5α-reductase	prostate	[91,110,111]
scopoletin	Anti-proliferation Inhibition 5α-reductase Anti-inflammation Pro-apoptotis	prostate prostate cancer	[91,113,114]

Figure 3 present the molecular structures of specific biologically active compounds identified from *U dioica*. Table 3 summarizes the biological activities of specific bioactive molecules isolated from the *U dioica* extracts in each anti-cancer study.

Urtica dioica and Breast Cancer

Breast cancer is the most prevalent cancer among women and one of the main causes of death worldwide. Breast cancer statistics indicate that in the US one woman in eight will suffer from breast cancer and that more than 200,000 new patients with breast cancer will be diagnosed every year [115,116]. Tumor invasion and metastasis remain the main causes of patients' mortality and still present an important therapeutic challenge. For the treatment of breast cancer, a multidisciplinary approach is currently used, involving surgery, radiotherapy, chemotherapy, hormone therapy, immunotherapy and other novel treatment strategies such as gene silencing, but they are associated with serious side effects [56,117,118]. Various new therapeutic targets for adjuvant, complementary and alternative medicines, including natural products bioactivity from plants, have been proposed by several new studies to treat breast cancer patients [60,62,119].

In this paragraph, we report investigations, via in-vitro studies and animal models, on *U. dioica* as a potential natural source of food bioactives and chemotherapeutic agent for breast cancer (Table 4).

In a study by Fattahi et al. (2013) [45], the activity of *U. dioica* leaves aqueous extract was analyzed on the human breast cancer cell line MCF-7. The *U. dioica* extract demonstrated anti-oxidant and anti-proliferative activity. After 24, 48 and 72 h of exposure to different concentrations of the *U. dioica* extract, significant cell death was observed in a dose-dependent manner, with an IC50 value of 2 mg/mL concentration after 72 h of treatment. In accordance with previous observations regarding anticancer drugs [118], the decrease of cell viability caused by *U. dioica* was due to the induction of apoptosis but not to necrosis. The treatment-induced apoptosis was demonstrated at the cellular level by morphological observation, DNA ladder formation, flow cytometry analysis and at the molecular level by measuring the increased amount of the different apoptotic-related proteins caspase 3, caspase 9, Bax (a pro-apoptotic protein), Bcl-2, calpain 1 (a calcium-dependent cytosolic cysteine protease) and calpastatin (a specific inhibitor of calpain 1). Noteworthy was the increase in the anti-apoptotic Bcl-2 protein; indeed, Bcl-2 can interact with Nur 77/TR3 and convert to Bax-like death effector, subsequently inducing apoptosis [120]. These findings were in agreement with a recent study by Fattahi et al. (2018) [121]. In this study, two human breast cancer cell lines, MCF-7 (estrogen and progesterone receptors positive; wild-type P53) and MDA-MB-231 (estrogen and progesterone receptors negative; mutated P53), were treated with *U. dioica* aqueous extract of leaves. The cytotoxicity of the treatment was confirmed, with IC50 values for both breast cancer cell lines of approximately 2 mg/mL, after 72 h exposure. *U. dioica* extract induced the apoptosis and increased the expression levels of Bax, especially in MCF-7 cells. Interestingly, *U. dioica* extract has been shown to influence the gene expression of two other proteins, adenosine deaminase (ADA) [122] and ornithine decarboxylase (ODC1) [123]. The expression level of ADA gene in MCF-7 cell line was increased in a dose-dependent manner, but did not modify in the MDA-MB-231 cell line. Alternatively, the ODC1 gene was upregulated in both cell lines. These enzymes show a regulatory role in cellular processes such as proliferation, cell growth and apoptosis [123,124]. In particular, ADA is a key enzyme in adenosine metabolism and nucleotide DNA turnover; ODC1 is the key enzyme in biosynthesis of polyamines [125,126]. Considering that *U. dioica* extracts contain phytoestrogens [127], the differences observed in these two cell lines could be due to a diverse status of hormone receptors. Moreover, it is possible that *U. dioica* in MDA-MB-231 cells may induce apoptosis via a P53-independent pathway.

Table 4. Anti-cancer effects of *U. dioica*: breast cancer, in-vitro and in-vivo studies.

U. dioica Extracts *	Cancer Cell Lines/ Animal Models	IC50	Effects	References
U. dioica L. (Jordan) leaves and stems ethanol extract	MCF-7 human breast cancer		↓ Proliferation 7% (72 h treatment with 50 µg/mL)	Abu-Dahab 2007 [128]
U. dioica, L. (Iran) leaves aqueous extract	MCF-7 human breast cancer	2 mg/mL (72 h treatment)	↓ Proliferation ↑ Apoptosis intrinsic pathway ↑ caspase 3 ↑ caspase 9 ↑ Bax ↑ Bcl-2 ↑ calpain 1 ↑ calpastatin	Fattahi 2013 [45]
U. dioica, L. (Iran) leaves dichloromethane extract	MDA-MB-468 human breast cancer	15.54 µg/mL (48 h treatment)	↓ Proliferation ↑ Apoptosis intrinsic pathway ↑ caspase 3 ↑ caspase 9 ↓ Bcl-2	Mohammadi 2016 [129]
U. dioica, L. (Iran) leaves dichloromethane extract	MDA-MB-468 human breast cancer	0.59 µM (24 h co-treatment paclitaxel + extract)	↓ Proliferation ↑ Apoptosis ↓ Migration ↓ Snail-1 ↓ ZEB1, ZEB2, twist G2/M arrest ↓ Cdc2	Mohammadi 2016 [130]
U. dioica, L. (Iran) leaves dichloromethane extract	MCF-7 human breast cancer MDA-MB-231human breast cancer 4T1 mouse breast cancer Balb/c mouse model of breast cancer (28 day, 20 mg/Kg injection treatment)	31.37 mg/mL (48 h treatment) 38.14 mg/mL (48 h treatment) 35.21 mg/mL (48 h treatment)	↓ Proliferation ↓ Migration ↓ miR-21 ↓ MMP1, MMP9, MMP13, vimentin, CXCR4 ↑ E-cadherin	Mansoori 2017 [131]
U. dioica, L. (Iran) leaves dichloromethane extract	Balb/c mouse model of breast cancer (28 day, 20 mg/Kg injection treatment)		↓ Metastasis ↑ Apoptosis intrinsic pathway ↑ caspase 3 ↑ caspase 9 ↓ Bcl-2 ↓ Ki-67	Mohammadi 2017 [132]
U. dioica, aqueous extract	rat model of breast cancer (5.5 months, 50 g/kg food treatment)		↓ Metastasis ↓ lipid peroxidation ↑ catalase	Telo 2017 [133]
U. dioica, L. (Iran) leaves aqueous extract	MCF-7 human breast cancer	2 mg/mL (72 h treatment)	↓ Proliferation ↑ Apoptosis ↑ ADA ↑ ODC1	Fattahi 2018 [121]
	MDA-MB-231 human breast cancer	2 mg/mL (72 h treatment)	↓ Proliferation ↑ Apoptosis = ADA ↑ ODC1	

* The plant extracts and parts used with specified collection sites. ADA: adenosine deaminase; ODC1: ornithine decarboxylase.

Furthermore, it is interesting to report that the *U. dioica* leaves aqueous extract in the prostate tissue of patients with prostate cancer, while, inhibited the activity of ADA, with IC50 of about 50 µg/mL (calculated, 30 min treatment) (Durak et al. 2004) [90].

Different modes of actions and effects of the *U. dioica* extracts are then possible in the various cell lines and tissues. Differences in in-vitro conditions related to patient's tissue and cell lines also to be considered.

Mohammadi et al. (2016) [129] in a previous study also demonstrated the cytotoxic and apoptotic effects of a dichloromethane extract of *U. dioica* leaves, in MDA-MB-468 cells, a human breast adenocarcinoma cell line. The dichloromethane extract of *U. dioica* induced a dose- and time-dependent anti-proliferative effect. The IC50 concentrations were of 29.46 and 15.54 µg/mL for 24 and 48 h exposure, respectively. These experimental values for IC50 demonstrate that the dichloromethane extract of *U. dioica* leaves have a more potent cytotoxic effect, on MDA-MB-468 cells. In this cell line, *U. dioica* dichloromethane extract caused cell death through apoptosis as revealed by morphological changes, TUNEL test, DNA fragmentation ladders and mRNA expression levels of apoptotic-related genes. In particular, *U. dioica* activated apoptosis through the intrinsic pathway, as revealed by the increase in the caspase 3 and caspase 9, the decrease in the Bcl2 and any significant changes in caspase 8 expression levels. Interestingly, Mohammadi et al. (2016) [130] demonstrated a synergic effect on cell death and invasion of human breast cancer MDA-MB-468 cell line, by treatment of *U. dioica* leaves dichloromethane extract in combination with the paclitaxel drug. Paclitaxel is one of the most commonly used natural drugs (derived from the bark of pacific yew tree) approved for chemotherapy in different types of cancers, such as ovary cancer, breast cancer and non-small cell lung cancer, acting as an anti-microtubule chemotherapy drug [134]. The antitumor potency of combinational therapy with paclitaxel and *U. dioica* extract was investigated on the human breast cancer cell line MDA-MB-468, demonstrating that *U. dioica* significantly increased the sensitivity of breast cancer cells to paclitaxel therapy, ameliorating its cytotoxicity. In fact, the MTT test demonstrated, in a time- and dose-dependent manner, a strong reduction of cell viability and of IC50 values for paclitaxel in the co-treatment with *U. dioica* extract: from 6.73 µM for paclitaxel alone, to 0.59 µM for co-treatment, after 24 h. The synergic effect of *U. dioica* extract and paclitaxel was also demonstrated on cell migration; in fact, by scratch test [135], a decreased invasion rate and a reduced number of migrated cells were observed. The molecular mechanism involved was elucidated studying the synergic effect of *U. dioica* and paclitaxel on the expression of snail-1 and related genes ZEB1, ZEB2 and twist. Snail-1 is a protein involved in invasion and migration of cancer cells [136–138] and is required for metastatic ability in breast cancer [139]; in fact, the silencing of snail-1 gene using specific siRNA prevented the metastasis of breast cancer cells. The observed reduction of the expression of snail-1 and related genes after the co-treatment was in agreement with its anti-metastatic potential. Moreover, a synergic effect of *U. dioica* and paclitaxel on the cell cycle arrest also revealed cell cycle arrest occurring at the G2/M phase, with a decreased Cdc2 expression. In agreement with these studies, Mansoori et al. (2017) [131], demonstrated that the dichloromethane extract of *U. dioica* leaves significantly decreased the cell proliferation of three different breast cancer cell lines, the human MCF-7 and MDA-MB-231 and mouse 4T1. The observed IC50 concentration of *U. dioica* extract was 31.37 mg/mL in MCF-7, 38.14 mg/mL in MDA-MB-231 and 35.21 mg/mL in 4T1 cells, at 48 h of treatment. Moreover, the scratch assay demonstrated an inhibitory effect of *U. dioica* on the migration of the breast cancer cell lines. Moreover, the authors investigated the signalling pathway by which *U. dioica* could inhibit the cell migration. In detail, demonstrated that *U. dioica* extract could inhibit tumor metastasis by regulating miR-21 (a crucial oncomir that is overexpressed in advanced tumors and metastasis) [140–146], the matrix metalloproteinases [147] MMP1, MMP9, MMP13, the vimentin [148], CXCR4 [149,150] and E-cadherin [151], important metastasis-related genes involved in cellular invasion by modifying adhesion junctions and the migratory capacity of cells [152]. In particular, miR-21, MMP1, MMP9, MMP13, CXCR4 and vimentin were found overexpressed in the invasive margins of breast cancer tissues of clinical samples and in the cancer cell lines; E-Cadherin, on the other hand, was decreased.

U. dioica extract treatment decreased miR-21 expression, which substantially reduced the overexpressed MMP1, MMP9, MMP13, vimentin and CXCR4, and increased E-cadherin in the treated tumor cell lines. In a previous study (Abu-Dahab and Afifi, 2007) [128] the cytotoxic effects of *U. dioica* ethanol extracts from leaves and stems were tested on human breast cancer cell lines MCF-7 and the percentage of MCF-7 cell survival after 72 hrs exposure to 50 µg/mL extract was 93.12%.

Finally, the *U. dioica* root methanolic extract inhibited, via an in-vitro test, aromatase enzyme activity in a concentration dependent manner [153]. Aromatase is a key enzyme involved in steroid hormone metabolism (mediating the conversion of androgens into estrogens) and is targeted in hormonal therapy of hormone-sensible breast cancers, thus acting on the cancer promotion.

Taken together, the various studies on cell lines demonstrated a diverse sensitivity to the *U. dioica* treatments and the extract tested showed differences in anti-cancer potency. It is important to consider that the cell survival and IC50 discrepancies observed in the various studies could be caused by differences in the habitat and parts of the plant used, in cell line types investigated and in the *U. dioica* extraction process.

The effectiveness of *U. dioica* to treat breast cancer has been proved not only in laboratory conditions but also in in-vivo experimental models.

Mohammadi et al. (2017) [132] prepared an *in vivo*-induced model of breast cancer, mice Balb/c with allograft tumors caused by injecting subcutaneously 4T1 murine breast tumor cells, and then administered *U. dioica* leaves dichloromethane extract (10 or 20 mg/kg body weight) by intraperitoneal injection for 28 days. Interestingly, *U. dioica* extract significantly reduced the tumor masses in the treated mice and significantly diminished the size and weight of the tumors removed from the treated mice. By TUNEL assay, it was shown that the *U. dioica* extract induced apoptosis in Balb/c allograft tumor models. Furthermore, the Ki-67 test demonstrated that the *U. dioica* treatment reduced tumor growth, decreasing the percentage of cell proliferation in the breast cancer tissue. Then, real-time PCR studies revealed that the intraperitoneal injection of *U. dioica* extract into the model mice was able to elicit the intrinsic pathway of apoptosis with increased expression of pro-apoptotic caspase 3 and caspase 9 and a downregulation of anti-apoptotic Bcl2. In the treated mice (4T1-induced Balb/c mouse model of breast cancer), according to the results previously reported on the breast cancer cell lines MCF-7, MDA-MB-231 and 4T1, *U. dioica* treatment induced the anti-metastatic pathway, with decreased expression of miR-21, MMP1, MMP9, MMP13, vimentin and CXCR4 and increased expression of E-cadherin (Mansoori et al. 2017) [131].

Finally, in an animal study by Telo et al. (2017) [133], the effects of *U. dioica* in N-methyl-N-nitrosourea-induced rat model of breast cancer were investigated. Aqueous extract of *U. dioica*, 50 g/kg powdered, was added into the food of rats for 5.5 months. The lipid peroxidation, the antioxidant enzyme activities and the formation of mammary gland cancer was then evaluated. *U. dioica* administration decreased the levels of lipid peroxidation and increased catalase antioxidant enzyme activity in rats generated mammary tumors. The results demonstrated, besides, a reduced rate in formation of breast cancer, with a decreased number of cancer masses.

Each *U. dioica* extract and its effects are identified using different color codes. The dashed arrows denote the results obtained with whole extracts; the solid arrows indicate the specific molecules isolated from the extracts and their effects. A generic picture of apoptosis signalling, indicating where and how molecules or extracts of *U. dioica* act on the different cellular targets, is included. By different images are specified if these results were observed in-vitro or in-vivo, on specific cancer types.

In Figure 4 we summarize the main anti-cancer studies and results reported for the various *U. dioica* extracts and cancer types.

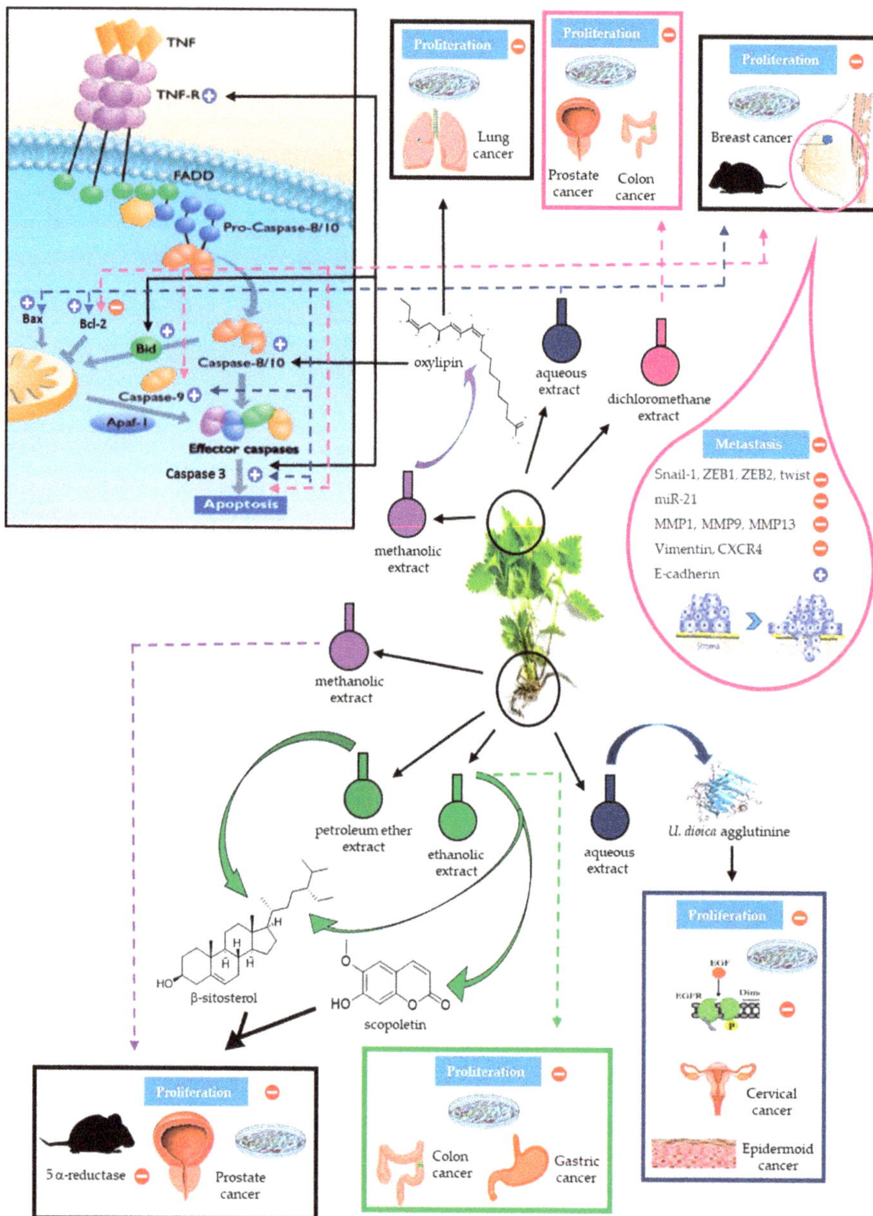

Figure 4. Schematic drawing of *U. dioica* anti-cancer effects. Each *U. dioica* extract and its effects are identified using different color codes. The dashed arrows denote the results obtained with whole extracts; the solid arrows indicate the specific molecules isolated from the extracts and their effects. A generic picture of apoptosis signalling, indicating where and how molecules or extracts of *U. dioica* act on the different cellular targets, is included. By different images are specified if these results were observed in-vitro or in-vivo, on specific cancer types.

4. Conclusions and Future Perspectives

Cancer has become the second most frequent cause of death after cardiac diseases and recent analyses provide an increase of its prevalence in the near future. Breast cancer is the second cause of cancer death among women. Currently, a variety of treatments such as chemotherapy, radiotherapy, hormone therapy and surgery, as well as newer nanotechnology and gene silencing therapy, are used in the treatment of cancer, but induce side effects. Hence, the need to develop the most effective anti-cancer therapy with few side effects and high cytotoxicity that will effectively arrest the initiation and progression of the cancer.

In recent years, many researchers have analyzed natural products and low cost drugs for cancer cure and prevent cancer development. Plants are a precious source of anti-cancer agents; the use of plants for cancer treatment is popular in many Asian cultures and today, beneficial compounds from these plants are used in the production of different modern anti-cancer drugs. Recent studies have illustrated that adjuvant therapy with natural products could help to prevent the development of cancer, as well as cure and improve the survival rate of patients.

Several studies have shown the anti-cancer properties of *U. dioica*, however, to the best of our knowledge, no previous report has reviewed the effects of *U. dioica* extracts on different cancer cell lines and animal cancer models.

Taken together, the main studies on the anti-cancer ability of *U. dioica* extracts provide a promising chance for the use of *U. dioica* as a nutraceutical food for the prevention and treatment of several cancers, including breast cancer. The various extracts of *U. dioica* tested, in fact, prevent cancerogenesis, kill human cancer cells and inhibit their migration. The extracts were not toxic and differences in the growth of the cancer cells was observed compared to the controls (untreated cancer cells) and normal cells, indicating their safety and a promising strategy to reduce adverse effects and ameliorate the efficacy of cancer chemotherapies.

U. dioica may exert biological anti-cancer activities through various mechanisms of actions, including antioxidant and anti-mutagenic properties, induction or inhibition of key processes in cellular metabolism and ability to activate the apoptotic pathways. Most anti-cancer drugs induce apoptosis, as a primary mechanism for inhibition of cell proliferation. The apoptotic effect of *U. dioica* in cancer cell lines and animal models was studied at the cellular and molecular levels. The type of cell death (apoptosis or necrosis) was investigated and the pathways involved in apoptosis induction were demonstrated, by studying the genes and proteins involved in the apoptosis process.

The *U. dioica* extracts contain varied bioactive molecules and the ability of these extracts to treat cancer, is due to those components that inhibit tumor growth and induce the apoptosis pathway.

The principal bioactivity of *U. dioica* was found in lipophilic fractions (e.g., dichloromethane extracts), suggesting that lipophilic compounds are mostly responsible for the anti-cancer actions. From the lipophilic fractions phytosterols, pentacyclic triterpenoids, coumarins, ceramides and hydroxyl fatty acids were isolated. In the hydrophobic extracts of the roots the sterols stigmast-4-en-3-one, stigmasterol and campesterol are present, which inhibited the enzyme activity of Na+, K+-ATPase in patients' tissues with benign prostatic hyperplasia and may subsequently repress prostate-cell metabolism and proliferation. The hydrophilic fractions (e.g., water, methanol, ethanol extracts) of *U. dioica* also demonstrate a high bioactivity, indicating that even polar active principles are responsible for the anti-cancer activity. The hydrophilic fractions contain isolectins and some polysaccharides. In the polar extracts of the roots the lignans as (+)-neoolivil, (-)-secoisolariciresinol, dehydrodiconiferyl alcohol, isolariciresinol, pinoresinol and 3,4-divanillyltetrahydrofuran are also present.

The most likely explanation for the considerable anti-cancer effect of *U. dioica* is the content of flavonoids and other known molecules and/or still unknown substances. Among the food bioactive molecules of *U. dioica*, the flavonoids are polyphenolic compounds that are able to induce anti-cancer effects through different mechanisms such as anti-oxidant activity, induction of apoptosis, inhibition of cell growth and cell migration. In fact, several plants rich in flavonoids possess disease preventive and therapeutic properties and, in particular, the consumption of vegetable and fruit rich in flavonoids is associated with reduced cancer risk.

Therefore, *U. dioica* may be used as a nutraceutical food bioactive in cancer treatment to prevent or reduce cancer without presenting the side effects of current anti-cancer treatments.

However, the effects observed could be caused by several molecules and, probably, molecules that have not yet been identified. Further studies are required to isolate and characterize the pure bioactive molecules in this plant to better understand its multiple anti-cancer actions and to explore these potentials in the fight against human cancers.

Author Contributions: Conceptualization, S.E.; software, R.R.; writing—original draft preparation, S.E., P.V.P. and A.B.; writing—review and editing, S.E., R.R. and A.B.; supervision, S.E., P.V.P., A.D.M. and C.I.

Funding: This research received no external funding.

Acknowledgments: This work was supported by University of Campania "Luigi Vanvitelli" and "Progetti per la ricerca oncologica della Regione Campania", Grant: I-Cure. The authors thank Angela Chambery for helpful discussions, Giuseppina Caraglia for english eding, Angela Giojelli and Marco Mammucari for careful support. This article is dedicated to the memory of Marco Esposito for his humanity and scientific competence.

Conflicts of Interest: The authors declare no conflicts of interest.

References

1. Wildman, R.E.C.; Wildman, R.; Wallace, T.C. *Handbook of Nutraceuticals and Functional Foods*; CRC Press: Boca Raton, FL, USA, 2016.
2. Angela, M.; Meireles, A. *Extracting Bioactive Compounds for Food Products: Theory and Applications*; CRC Press: Boca Raton, FL, USA, 2008.
3. Kavalali, G.M. *Urtica Therapeutic and Nutritional Aspects of Stinging Nettles*; Taylor & Francis: London, UK, 2003; Volume 37.
4. Upton, R. Stinging nettles leaf (*Urtica dioica* L.): Extraordinary vegetable medicine. *J. Herb. Med.* **2013**, *3*, 9–38. [CrossRef]
5. Ahmed, K.K.M.; Parsuraman, S. *Urtica dioica* L., (*Urticaceae*): A Stinging Nettle. *Syst. Rev. Pharm.* **2014**, *5*, 6–8. [CrossRef]
6. Oliver, F.; Amon, E.U.; Breathnach, A.; Francis, D.M.; Sarathchandra, P.; Black, A.K.; Greaves, M.W. Contact urticaria due to the common stinging nettle (*Urtica dioica*)—Histological, ultrastructural and pharmacological studies. *Clin. Exp. Dermatol.* **1991**, *16*, 1–7. [CrossRef] [PubMed]
7. Di Virgilio, N.; Papazoglou, E.; Jankauskiene, Z.; Lonardo, S.; Praczyk, M.; Wielgusz, K. The potential of stinging nettle (*Urtica dioica* L.) as a crop with multiple uses. *Ind. Crops Prod.* **2015**, *68*, 42–49. [CrossRef]
8. Law, B.A.; Tamime, A.I. *Technology of Cheese Making*, 2nd ed.; Wiley-Blackwell: London, UK, 2010; p. 512.
9. Rutto, L.K.; Xu, Y.; Ramirez, E.; Brandt, M. Mineral Properties and Dietary Value of Raw and Processed Stinging Nettle (*Urtica dioica* L.). *Int. J. Food Sci.* **2013**, *2013*, 857120. [CrossRef] [PubMed]
10. Rafajlovska, V.; Rizova, V.; Djarmati, Z.; Tesevic, V.; Cvetkov, L. Contents of fatty acids in stinging nettle extracts (*Urtica dioica* L.) obtained with supercritical carbon dioxide. *Acta Pharm.* **2001**, *51*, 45–51.
11. Đurović, S.; Zekovic, Z.; Šorgić, S.; Popov, S.; Vujanović, M.; Radojković, M. Fatty acid profile of stinging nettle leaves: Application of modern analytical procedures for sample preparation and analysis. *Anal. Methods* **2018**, *10*, 1080–1087. [CrossRef]
12. D'Abrosca, B.; Ciaramella, V.; Graziani, V.; Papaccio, F.; Della Corte, C.M.; Potenza, N.; Fiorentino, A.; Ciardiello, F.; Morgillo, F. *Urtica dioica* L. inhibits proliferation and enhances cisplatin cytotoxicity in NSCLC cells via Endoplasmic Reticulum-stress mediated apoptosis. *Sci. Rep.* **2019**, *9*, 4986. [CrossRef]
13. Adhikari, B.M.; Bajracharya, A.; Shrestha, A.K. Comparison of nutritional properties of Stinging nettle (*Urtica dioica*) flour with wheat and barley flours. *Food Sci. Nutr.* **2016**, *4*, 119–124. [CrossRef]
14. Farag, M.A.; Weigend, M.; Luebert, F.; Brokamp, G.; Wessjohann, L.A. Phytochemical, phylogenetic, and anti-inflammatory evaluation of 43 *Urtica* accessions (stinging nettle) based on UPLC-Q-TOF-MS metabolomic profiles. *Phytochemistry* **2013**, *96*, 170–183. [CrossRef]
15. Yan, X.G. New chemical constituents of roots of Urtica triangularis HAND-MASS. *Chem. Pharm. Bull.* **2008**, *56*, 1463–1465. [CrossRef] [PubMed]

16. Đurovic, S.; Pavlic, B.; Šorgic, S.; Popov, S.; Savic, S.; Pertonijevic, M.; Radojkovic, M.; Cvetanovic, A.; Zekovic, Z. Chemical composition of stinging nettle leaves obtained by different analytical approaches. *J. Funct. Food.* **2017**, *32*, 18–26. [CrossRef]

17. Orčić, D.; Francišković, M.; Bekvalac, K.; Svirčev, E.; Beara, I.; Lesjak, M.; Mimica-Dukić, N. Quantitative determination of plant phenolics in *Urtica dioica* extracts by high-performance liquid chromatography coupled with tandem mass spectrometric detection. *Food Chem.* **2014**, *143*, 48–53. [CrossRef] [PubMed]

18. Otles, S.; Yalcin, B. Phenolic compounds analysis of root, stalk, and leaves of nettle. *Sci. World J.* **2012**, *2012*, 564367. [CrossRef] [PubMed]

19. Pinelli, P.; Ieri, F.; Vignolini, P.; Bacci, L.; Baronti, S.; Romani, A. Extraction and HPLC analysis of phenolic compounds in leaves, stalks, and textile fibers of *Urtica dioica* L. *J. Agric. Food Chem.* **2008**, *56*, 9127–9132. [CrossRef] [PubMed]

20. Krauss, R.; Spitteler, G. Phenolic compounds from roots of *Urtica dioica*. *Phytochemistry* **1990**, *29*, 1653–1659. [CrossRef]

21. Grevsen, K.; Frette, X.C.; Christensen, L.P. Concentration and composition of flavonol glycosides and phenolic acids in aerial parts of stinging nettle (*Urtica dioica* L.) are affected by nitrogen fertilization and by harvest time. *Eur. J. Hortic. Sci.* **2008**, *73*, 20–27.

22. Chaurasia, N.; Wichtl, M. Flavonol Glycosides from *Urtica dioica*. *Planta Med.* **1987**, *53*, 432–434. [CrossRef]

23. Guil-Guerrero, J.L.; Rebolloso-Fuentes, M.M.; Isasa, M.E.T. Fatty acids and carotenoids from Stinging Nettle (*Urtica dioica* L.). *J. Food Compos. Anal.* **2003**, *16*, 111–119. [CrossRef]

24. Hirano, T.; Homma, M.; Oka, K. Effects of stinging nettle root extracts and their steroidal components on the Na+, K+-ATPase of the benign prostatic hyperplasia. *Planta Med.* **1994**, *60*, 30–33. [CrossRef]

25. Chaurasia, N.; Wichtl, M. Sterols and Steryl Glycosides from *Urtica dioica*. *J. Nat. Prod.* **1987**, *50*, 881–885. [CrossRef]

26. Sajfrtová, M.; Sovova, H.; Opletal, L.; Bártlová, M. Near-critical extraction of b-sitosterol and scopoletin from stinging nettle roots. *J. Supercrit. Fluid* **2005**, *35*, 111–118. [CrossRef]

27. Carvalho, A.R.; Costab, G.; Figueirinha, A.; Liberal, J.; Prior, J.A.V.; Lopes, M.C.; Cruz, M.T.; Batista, M.T. Urtica spp.: Phenolic composition, safety, antioxidant and anti-inflammatory activities. *Food Res. Int.* **2017**, *99*, 485–494. [CrossRef] [PubMed]

28. Lapinskaya, E.S.; Kopytko, Y. Composition of the lipophilic fraction of stinging nettle (*Urtica dioica* L. and *U. urens* L.) homeopathic matrix tinctures. *Pharm. Chem. J.* **2008**, *42*, 699–702. [CrossRef]

29. Di Maro, A.; Pacifico, S.; Fiorentino, A.; Galasso, S.; Gallicchio, M.; Guida, V.; Severino, V.; Monaco, P.; Parente, A. Raviscanina wild asparagus (*Asparagus acutifolius* L.): A nutritionally valuable crop with antioxidant and antiproliferative properties. *Food Res. Int.* **2013**, *53*, 180–188. [CrossRef]

30. Pacifico, S.; Galasso, S.; Piccolella, S.; Kretschmer, N.; Pan, S.; Marciano, S.; Bauer, R.; Monaco, P. Seasonal variation in phenolic composition and antioxidant and anti-inflammatory activities of *Calamintha nepeta* (L.) Savi. *Food Res. Int.* **2015**, *69*, 121–132. [CrossRef]

31. Pacifico, S.; Piccolella, S.; Nocera, P.; Tranquillo, E.; Dal Poggetto, F.; Catauro, M. New insights into phenol and polyphenol composition of Stevia rebaudiana leaves. *J. Pharm. Biomed. Anal.* **2019**, *163*, 45–57. [CrossRef] [PubMed]

32. Del Rio, D.; Rodriguez-Mateos, A.; Spencer, J.P.; Tognolini, M.; Borges, G.; Crozier, A. Dietary (poly)phenolics in human health: Structures, bioavailability, and evidence of protective effects against chronic diseases. *Antioxid. Redox Signal.* **2013**, *18*, 1818–1892. [CrossRef] [PubMed]

33. Pacifico, S.; Piccolella, S. Plant-Derived Polyphenols: A Chemopreventive and Chemoprotectant Worth-Exploring Resource in Toxicology. In *Advances in Molecular Toxicology*; Fishbein, J.C., Heilman, J.M., Eds.; Elsevier: Amsterdam, The Netherlands, 2015; pp. 161–214.

34. Spina, M.; Cuccioloni, M.; Sparapani, L.; Acciarri, S.; Eleuteri, A.M.; Fioretti, E.; Angeletti, M. Comparative evaluation of flavonoid content in assessing quality of wild and cultivated vegetables for human consumption. *J. Sci. Food Agric.* **2008**, *88*, 294–304. [CrossRef]

35. Akbay, P.; Basaran, A.A.; Undeger, U.; Basaran, N. *In vitro* immunomodulatory activity of flavonoid glycosides from *Urtica dioica* L. *Phytother. Res.* **2003**, *17*, 34–37. [CrossRef] [PubMed]

36. Gülçin, I.; Küfrevioglu, O.I.; Oktay, M.; Büyükokuroglu, M.E. Antioxidant, antimicrobial, antiulcer and analgesic activities of nettle (*Urtica dioica* L.). *J. Ethnopharmacol.* **2004**, *90*, 205–215. [CrossRef] [PubMed]

37. Chrubasik, J.E.; Roufogalis, B.D.; Wagner, H.; Chrubasik, S.A. A comprehensive review on nettle effect and efficacy profiles, Part I: Herba urticae. *Phytomedicine* **2007**, *14*, 423–435. [CrossRef] [PubMed]
38. Chrubasik, J.E.; Roufogalis, B.D.; Wagner, H.; Chrubasik, S. A comprehensive review on the stinging nettle effect and efficacy profiles. Part II: Urticae radix. *Phytomedicine* **2007**, *14*, 568–579. [CrossRef] [PubMed]
39. El Haouari, M.; Rosado, J.A. Phytochemical, Anti-diabetic and Cardiovascular Properties of *Urtica dioica* L. (*Urticaceae*): A Review. *Mini Rev. Med. Chem.* **2019**, *19*, 63–71. [CrossRef] [PubMed]
40. Zemmouri, H.; Sekiou, O.; Ammar, S.; El Feki, A.; Bouaziz, M.; Messarah, M.; Boumendjel, A. *Urtica dioica* attenuates ovalbumin-induced inflammation and lipid peroxidation of lung tissues in rat asthma model. *Pharm. Biol.* **2017**, *55*, 1561–1568. [CrossRef]
41. Schulze-Tanzil, G. Effects of the antirheumatic remedy hox alpha-a new stinging nettle leaf extract–on matrix metalloproteinases in human chondrocytes in vitro. *Histol. Histopathol.* **2002**, *17*, 477–485. [CrossRef] [PubMed]
42. Riehemann, K.; Behnke, B.; Schulze-Osthoff, K. Plant extracts from stinging nettle (*Urtica dioica*), an antirheumatic remedy, inhibit the proinflammatory transcription factor NF-kappaB. *FEBS Lett.* **1999**, *442*, 89–94. [CrossRef]
43. Bisht, R.; Joshi, B.C.; Kalia, A.N.; Prakash, A. Antioxidant-Rich Fraction of *Urtica dioica* Mediated Rescue of Striatal Mito-Oxidative Damage in MPTP-Induced Behavioral, Cellular, and Neurochemical Alterations in Rats. *Mol. Neurobiol.* **2017**, *54*, 5632–5645. [CrossRef]
44. Di Sotto, A.; Mazzanti, G.; Savickiene, N.; Staršelskytė, R.; Baksenskaite, V.; Di Giacomo, S.; Vitalone, A. Antimutagenic and antioxidant activity of a protein fraction from aerial parts of *Urtica dioica*. *Pharm. Biol.* **2015**, *53*, 935–938. [CrossRef]
45. Fattahi, S.; Ardekani, A.M.; Zabihi, E.; Abedian, Z.; Mostafazadeh, A.; Pourbagher, R.; Akhavan-Niaki, H. Antioxidant and apoptotic effects of an aqueous extract of *Urtica dioica* on the MCF-7 human breast cancer cell line. *Asian Pac. J. Cancer Prev.* **2013**, *14*, 5317–5323. [CrossRef]
46. Kukric, Z.; Topalić-Trivunović, L.; Kukavica, B.; Matos, S.; Pavicic, S.; Boroja, M.; Savić, A. Characterization of antioxidant and antimicrobial activities of nettle leaves (*Urtica dioica* L.). *Acta Period. Technol.* **2012**, *43*, 257–272. [CrossRef]
47. Ghaima, K.K.; Hashim, N.M.; Ali, S.A. Antibacterial and antioxidant activities of ethyl acetate extract of nettle (*Urtica dioica*) and dandelion (*Taraxacum officinale*). *J. Pharm. Sci.* **2013**, *3*, 96–99.
48. Kregiel, D.; Pawlikowska, E.; Antolak, H. *Urtica* spp.: Ordinary Plants with Extraordinary Properties. *Molecules* **2018**, *23*, 1664. [CrossRef] [PubMed]
49. Batool, R.; Salahuddin, H.; Mahmood, T.; Ismail, M. Study of anticancer and antibacterial activities of Foeniculum vulgare, Justicia adhatoda and *Urtica dioica* as natural curatives. *Cell. Mol. Biol.* **2017**, *63*, 109–114. [CrossRef] [PubMed]
50. Ranjbari, A.; Azarbayjani, M.A.; Yusof, A.; Halim Mokhtar, A.; Akbarzadeh, S.; Ibrahim, M.Y.; Tarverdizadeh, B.; Farzadinia, P.; Hajiaghaee, R.; Dehghan, F. In vivo and in vitro evaluation of the effects of *Urtica dioica* and swimming activity on diabetic factors and pancreatic beta cells. *BMC Complement. Altern. Med.* **2016**, *16*, 101. [CrossRef]
51. Domola, M.S.; Vu, V.; Robson-Doucette, C.A.; Sweeney, G.; Wheeler, M.B. Insulin mimetics in *Urtica dioica*: Structural and computational analyses of *Urtica dioica* extracts. *Phytother. Res.* **2010**, *24* (Suppl. 2), S175–S182. [CrossRef]
52. Franišković, M.; Gonzalez-Pérez, R.; Orčić, D.; Sánchez de Medina, F.; Martínez-Augustin, O.; Svirčev, E.; Simin, N.; Mimica-Dukić, N. Chemical Composition and Immuno-Modulatory Effects of *Urtica dioica* L. (Stinging Nettle) Extracts. *Phytother. Res.* **2017**, *31*, 1183–1191. [CrossRef]
53. Siegel, R.L.; Miller, K.D.; Jemal, A. Cancer statistics, 2016. *CA Cancer J. Clin.* **2016**, *66*, 7–30. [CrossRef]
54. Torre, L.A.; Siegel, R.L.; Ward, E.M.; Jemal, A. Global Cancer Incidence and Mortality Rates and Trends–An Update. *Cancer Epidemiol. Biomark. Prev.* **2016**, *25*, 16–27. [CrossRef]
55. Ferlay, J.; Soerjomataram, I.; Dikshit, R.; Eser, S.; Mathers, C.; Rebelo, M.; Parkin, D.M.; Forman, D.; Bray, F. Cancer incidence and mortality worldwide: Sources, methods and major patterns in GLOBOCAN 2012. *Int. J. Cancer* **2015**, *136*, E359–E386. [CrossRef]
56. Gelmann, E.P.; Sawyers, C.L.; Rauscher, F.J.I. *Molecular Oncology: Causes of Cancer and Targets for Treatment*; Cambridge University Press: Cambridge, UK, 2013.

57. Steward, W.P.; Brown, K. Cancer chemoprevention: A rapidly evolving field. *Br. J. Cancer* **2013**, *109*, 1–7. [CrossRef]
58. Kucuk, O. New opportunities in chemoprevention research. *Cancer Investig.* **2002**, *20*, 237–245. [CrossRef]
59. Kocasli, S.; Demircan, Z. Herbal product use by the cancer patients in both the pre and post surgery periods and during chemotherapy. *Afr. J. Tradit. Complement. Altern. Med.* **2017**, *14*, 325–333. [CrossRef]
60. Liao, G.S.; Apaya, M.K.; Shyur, L.F. Herbal medicine and acupuncture for breast cancer palliative care and adjuvant therapy. *Evid. Based Complement. Altern. Med.* **2013**, *2013*, 437948. [CrossRef]
61. Nahata, A.; Saxena, A.; Suri, N.; Saxena, A.K.; Dixit, V.K. Sphaeranthus indicus induces apoptosis through mitochondrial-dependent pathway in HL-60 cells and exerts cytotoxic potential on several human cancer cell lines. *Integr. Cancer Ther.* **2013**, *12*, 236–247. [CrossRef]
62. Olaku, O.; White, J.D. Herbal therapy use by cancer patients: A literature review on case reports. *Eur. J. Cancer* **2011**, *47*, 508–514. [CrossRef]
63. Balunas, M.J.; Kinghorn, A.D. Drug discovery from medicinal plants. *Life Sci.* **2005**, *78*, 431–441. [CrossRef]
64. D'Incalci, M.; Steward, W.P.; Gescher, A.J. Use of cancer chemopreventive phytochemicals as antineoplastic agents. *Lancet Oncol.* **2005**, *6*, 899–904. [CrossRef]
65. Paterson, I.; Anderson, E.A. The renaissance of natural products as drug candidates. *Science* **2005**, *310*. [CrossRef]
66. Smit, A.J. Medicinal and pharmaceutical uses of seaweed natural products: A review. *J. Appl. Phycol.* **2004**, *16*. [CrossRef]
67. Jung Park, E.; Pezzuto, J.M. Botanicals in cancer chemoprevention. *Cancer Metastasis Rev.* **2002**, *21*. [CrossRef]
68. Mann, J. Natural products in cancer chemotherapy: Past, present and future. *Nat. Rev. Cancer* **2002**, *2*. [CrossRef]
69. Kandaswami, C.; Kanadaswami, C.; Lee, L.T.; Lee, P.P.; Hwang, J.J.; Ke, F.C.; Huang, Y.T.; Lee, M.T. The antitumor activities of flavonoids. *In Vivo* **2005**, *19*, 895–909.
70. Kawaii, S.; Tomono, Y.; Katase, E.; Ogawa, K.; Yano, M. Antiproliferative activity of flavonoids on several cancer cell lines. *Biosci. Biotechnol. Biochem.* **1999**, *63*. [CrossRef]
71. Cragg, G.M.; Newman, D.J. Plants as a source of anti-cancer agents. *J. Ethnopharmacol.* **2005**, *100*, 72–79. [CrossRef]
72. Jin, Z.; El-Deiry, W.S. Overview of cell death signaling pathways. *Cancer Biol. Ther.* **2005**, *4*, 139–163. [CrossRef]
73. Ciniglia, C.; Mastrobuoni, F.; Scortichini, M.; Petriccione, M. Oxidative damage and cell-programmed death induced in Zea mays L. by allelochemical stress. *Ecotoxicology* **2015**, *24*, 926–937. [CrossRef]
74. Florea, A.M.; Busselberg, D. Cisplatin as an anti-tumor drug: Cellular mechanisms of activity, drug resistance and induced side effects. *Cancers* **2011**, *3*, 1351. [CrossRef]
75. Siddik, Z.H. Cisplatin: Mode of cytotoxic action and molecular basis of resistance. *Oncogene* **2003**, *22*. [CrossRef]
76. Ozkol, H.; Musa, D.; Tuluce, Y.; Koyuncu, I.; Asadi-Samani, M.; Rafieian-Kopaei, M.; Lorigooini, Z.; Shirzad, H. Ameliorative influence of *Urtica dioica* L against cisplatin-induced toxicity in mice bearing Ehrlich ascites carcinoma. *Drug Chem. Toxicol.* **2012**, *35*, 251–257. [CrossRef]
77. Nollet, L.M.L.; Gutierrez-Uribe, J.A. *Phenolic Compounds in Food: Characterization and Analysis*; CRC Press: Boca Raton, FL, USA, 2018.
78. Faramarzi, S.; Pacifico, S.; Yadollahi, A.; Lettieri, A.; Nocera, P.; Piccolella, S. Red-fleshed Apples: Old Autochthonous Fruits as a Novel Source of Anthocyanin Antioxidants. *Plant Foods Hum. Nutr.* **2015**, *70*, 324–330. [CrossRef]
79. Keskin-Šašic, I.; Tahirovic, A.; Top˘cagic, A.; Klepo, L.; Salihovic, M.; Ibragic, S.; Toromanovi´, J.; Ajanovic, A.; Velispahic, E. Total phenolic content and antioxidant capacity of fruit juices. *Bull. Chem. Technol. Bosnia Herzeg.* **2012**, *39*, 25–28.
80. Wang, C.; Yu, J.; Wang, H.; Zhang, J.; Wu, N. Lipid peroxidation and altered anti-oxidant status in breast adenocarcinoma patients. *Drug Res.* **2014**, *64*, 690–692. [CrossRef]
81. Ray, G.; Batra, S.; Shukla, N.K.; Deo, S.; Raina, V.; Ashok, S.; Husain, S.A. Lipid peroxidation, free radical production and antioxidant status in breast cancer. *Breast Cancer Res. Treat.* **2000**, *59*, 163–170. [CrossRef]
82. Khanzode, S.S.; Muddeshwar, M.G.; Khanzode, S.D.; Dakhale, G.N. Antioxidant Enzymes and Lipid Peroxidation in Different Stages of Breast Cancer. *Free Radic. Res.* **2004**, *38*, 81–85. [CrossRef]

83. Benz, E.J.; Nathan, D.G.; Amaravadi, R.K.; Danial, N.N. Targeting the cell death-survival equation. *Clin. Cancer Res.* **2007**, *13*, 7250–7253. [CrossRef]

84. Taraphdar, A.K.; Roy, M.; Bhattacharya, R.K. Natural products as inducers of apoptosis: Implication for cancer therapy and prevention. *Curr. Sci.* **2001**, *80*, 1387–1396.

85. Wagner, H.; Willer, F.; Samtleben, R.; Boos, G. Search for the antiprostatic principle of stinging nettle (*Urtica dioica*) roots. *Phytomedicine* **1994**, *1*, 213–224. [CrossRef]

86. Ghasemi, S.; Moradzadeh, M.; Mousavi, S.H.; Sadeghnia, H.R. Cytotoxic effects of *Urtica dioica* radix on human colon (HT29) and gastric (MKN45) cancer cells mediated through oxidative and apoptotic mechanisms. *Cell. Mol. Biol.* **2016**, *62*, 90–96.

87. Mohammadi, A.; Mansoori, B.; Aghapour, M.; Baradaran, P.C.; Shajari, N.; Davudian, S.; Salehi, S.; Baradaran, B. The Herbal Medicine Utrica Dioica Inhibits Proliferation of Colorectal Cancer Cell Line by Inducing Apoptosis and Arrest at the G2/M Phase. *J. Gastrointest. Cancer* **2016**, *47*, 187–195. [CrossRef]

88. Lichius, J.J.; Muth, C. The inhibiting effects of *Urtica dioica* root extracts on experimentally induced prostatic hyperplasia in the mouse. *Planta Med.* **1997**, *63*, 307–310. [CrossRef]

89. Konrad, L.; Müller, H.H.; Lenz, C.; Laubinger, H.; Aumüller, G.; Lichius, J.J. Antiproliferative effect on human prostate cancer cells by a stinging nettle root (*Urtica dioica*) extract. *Planta Med.* **2000**, *66*, 44–47. [CrossRef]

90. Durak, I.; Biri, H.; Devrim, E.; Sözen, S.; Avci, A. Aqueous extract of *Urtica dioica* makes significant inhibition on adenosine deaminase activity in prostate tissue from patients with prostate cancer. *Cancer Biol. Ther.* **2004**, *3*, 855–857. [CrossRef]

91. Nahata, A.; Dixit, V.K. Ameliorative effects of stinging nettle (*Urtica dioica*) on testosterone-induced prostatic hyperplasia in rats. *Andrologia* **2012**, *44* (Suppl. 1), 396–409. [CrossRef]

92. Mohammadi, A.; Mansoori, B.; Aghapour, M.; Baradaran, B. *Urtica dioica* dichloromethane extract induce apoptosis from intrinsic pathway on human prostate cancer cells (PC3). *Cell. Mol. Biol.* **2016**, *62*, 78–83.

93. Wang, G.; Reed, E.; Li, Q.Q. Molecular basis of cellular response to cisplatin chemotherapy in non-small cell lung cancer (Review). *Oncol. Rep.* **2004**, *12*. [CrossRef]

94. Abdelrahim, M.; Newman, K.; Vanderlaag, K.; Samudio, I.; Safe, S. 3, 3′-Diindolylmethane (DIM) and its derivatives induce apoptosis in pancreatic cancer cells through endoplasmic reticulum stress-dependent upregulation of DR5. *Carcinogenesis* **2006**, *27*. [CrossRef]

95. Oyadomari, S.; Mori, M. Roles of CHOP/GADD153 in endoplasmic reticulum stress. *Cell Death Differ.* **2004**, *11*. [CrossRef]

96. Yamaguchi, H.; Wang, H.G. CHOP is involved in endoplasmic reticulum stress-induced apoptosis by enhancing DR5 expression in human carcinoma cells. *J. Biol. Chem.* **2004**, *279*. [CrossRef]

97. Kelley, S.K.; Ashkenazi, A. Targeting death receptors in cancer with Apo2L/TRAIL. *Curr. Opin. Pharmacol.* **2004**, *4*. [CrossRef]

98. Randa, A.; Kamel, S. Bioactive oxylipins from the endophyte Khuskia oryzae isolated from the medicinal plant Bidens alba. *Eur. J. Biomed. Pharm. Sci.* **2015**, *2*, 630–639.

99. Romano, G. Design and synthesis of pro-apoptotic compounds inspired by diatom oxylipins. *Mar. Drugs* **2013**, *11*, 4527. [CrossRef]

100. Gerwick, W.H. Biologically active oxylipins from seaweeds. *Hydrobiologia* **1993**, *260–261*. [CrossRef]

101. Hwang, J.H.; Voortman, J.; Giovannetti, E.; Steinberg, S.M.; Leon, L.G.; Kim, Y.T.; Funel, N.; Park, J.K.; Kim, M.A.; Kang, G.H.; et al. Identification of microRNA-21 as a biomarker for chemoresistance and clinical outcome following adjuvant therapy in resectable pancreatic cancer. *PLoS ONE* **2010**, *5*, e10630. [CrossRef]

102. Verpoorte, R.; Choi, Y.H.; Kim, H.K. NMR-based metabolomics at work in phytochemistry. *Phytochem. Rev.* **2007**, *6*. [CrossRef]

103. Gabbs, M.; Leng, S.; Devassy, J.G.; Monirujjaman, M.; Aukema, H.M. Advances in Our Understanding of Oxylipins Derived from Dietary PUFAs. *Adv. Nutr.* **2015**, *6*, 513–540. [CrossRef]

104. Baldin, V.; Lukas, J.; Marcote, M.J.; Pagano, M.; Draetta, G. Cyclin D1 is a nuclear protein required for cell cycle progression in G1. *Genes Dev* **1993**, *7*, 812–821. [CrossRef]

105. Altiok, N.; Mezzadra, H.; Patel, P.; Koyuturk, M.; Altiok, S. A plant oxylipin, 12-oxo-phytodienoic acid, inhibits proliferation of human breast cancer cells by targeting cyclin D1. *Breast Cancer Res. Treat.* **2008**, *109*. [CrossRef]

106. Does, M.P.; Ng, D.K.; Dekker, H.L.; Peumans, W.J.; Houterman, P.M.; Van Damme, E.J.; Cornelissen, B.J. Characterization of *Urtica dioica* agglutinin isolectins and the encoding gene family. *Plant Mol. Biol.* **1999**, *39*, 335–347. [CrossRef]

107. Wagner, H.; Geiger, W.N.; Boos, G.; Samtleben, R. Studies on the binding of *Urtica dioica* agglutinin (UDA) and other lectins in an in vitro epidermal growth factor receptor test. *Phytomedicine* **1995**, *1*, 287–290. [CrossRef]

108. Steers, W.D. 5alpha-reductase activity in the prostate. *Urology* **2001**, *58*, 17–24. [CrossRef]

109. Valitova, J.N.; Sulkarnayeva, A.G.; Minibayeva, F.V. Plant Sterols: Diversity, Biosynthesis, and Physiological Functions. *Biochemistry* **2016**, *81*, 819–834. [CrossRef]

110. Berges, R.R.; Windeler, J.; Trampisch, H.J.; Senge, T. Randomised, placebo-controlled, double-blind clinical trial of beta-sitosterol in patients with benign prostatic hyperplasia. Beta-sitosterol Study Group. *Lancet* **1995**, *345*, 1529–1532. [CrossRef]

111. Wilt, T.J.; MacDonald, R.; Ishani, A. beta-sitosterol for the treatment of benign prostatic hyperplasia: A systematic review. *BJU Int.* **1999**, *83*, 976–983. [CrossRef]

112. Bourgaud, F.; Hehn, A.; Larbat, R.; Doerper, S.; Gontier, E.; Kellner, S.; Matern, U. Biosynthesis of coumarins in plants: A major pathway still to be unravelled for cytochrome P450 enzymes. *Phytochem. Rev.* **2006**, *5*, 293–308. [CrossRef]

113. Ding, Z.; Dai, Y.; Hao, H.; Pan, R.; Yao, X.; Wang, Z. Anti-inflammatory effects of scopoletin and underlying mechanisms. *Pharm. Biol.* **2008**, *46*, 854–860. [CrossRef]

114. Liu, X.L.; Zhang, L.; Fu, X.L.; Chen, K.; Qian, B.C. Effect of scopoletin on PC3 cell proliferation and apoptosis. *Acta Pharmacol. Sin.* **2001**, *22*, 929–933.

115. Kohler, B.A.; Sherman, R.L.; Howlader, N.; Jemal, A.; Ryerson, A.B.; Henry, K.A.; Boscoe, F.P.; Cronin, K.A.; Lake, A.; Noone, A.-M.; et al. Annual Report to the Nation on the Status of Cancer, 1975–2011, Featuring Incidence of Breast Cancer Subtypes by Race/Ethnicity, Poverty, and State. *JNCI J. Natl. Cancer Inst.* **2015**, *107*. [CrossRef]

116. DeSantis, C.; Ma, J.; Bryan, L.; Jemal, A. Breast cancer statistics, 2013. *CA Cancer J. Clin.* **2014**, *64*, 52–62. [CrossRef]

117. Sharma, G.N.; Dave, R.; Sanadya, J.; Sharma, P.; Sharma, K.K. Various types and management of breast cancer: An overview. *J. Adv. Pharm. Technol. Res.* **2010**, *1*, 109–126.

118. Nounou, M.I.; ElAmrawy, F.; Ahmed, N.; Abdelraouf, K.; Goda, S.; Syed-Sha-Qhattal, H. Breast Cancer: Conventional Diagnosis and Treatment Modalities and Recent Patents and Technologies. *Breast Cancer* **2015**, *9*, 17–34. [CrossRef]

119. Mitra, S.; Dash, R. Natural Products for the Management and Prevention of Breast Cancer. *Evid. Based Complement. Altern. Med.* **2018**, *2018*, 8324696. [CrossRef]

120. Lin, B.; Kolluri, S.K.; Lin, F.; Liu, W.; Han, Y.H.; Cao, X.; Dawson, M.I.; Reed, J.C.; Zhang, X.K. Conversion of Bcl-2 from protector to killer by interaction with nuclear orphan receptor Nur77/TR3. *Cell* **2004**, *116*, 527–540. [CrossRef]

121. Fattahi, S.; Ghadami, E.; Asouri, M.; Motevalizadeh Ardekanid, A.; Akhavan-Niaki, H. *Urtica dioica* inhibits cell growth and induces apoptosis by targeting Ornithine decarboxylase and Adenosine deaminase as key regulatory enzymes in adenosine and polyamines homeostasis in human breast cancer cell lines. *Cell. Mol. Biol.* **2018**, *64*, 97–102. [CrossRef]

122. Tullo, A.; Mastropasqua, G.; Bourdon, J.C.; Centonze, P.; Gostissa, M.; Costanzo, A.; Levrero, M.; Del Sal, G.; Saccone, C.; Sbisà, E. Adenosine deaminase, a key enzyme in DNA precursors control, is a new p73 target. *Oncogene* **2003**, *22*, 8738–8748. [CrossRef]

123. Zhu, Q.; Jin, L.; Casero, R.A.; Davidson, N.E.; Huang, Y. Role of ornithine decarboxylase in regulation of estrogen receptor alpha expression and growth in human breast cancer cells. *Breast Cancer Res. Treat.* **2012**, *136*, 57–66. [CrossRef]

124. Apasov, S.G.; Blackburn, M.R.; Kellems, R.E.; Smith, P.T.; Sitkovsky, M.V. Adenosine deaminase deficiency increases thymic apoptosis and causes defective T cell receptor signaling. *J. Clin. Investig.* **2001**, *108*, 131–141. [CrossRef]

125. Nowotarski, S.L.; Woster, P.M.; Casero, R.A. Polyamines and cancer: Implications for chemotherapy and chemoprevention. *Expert Rev. Mol. Med.* **2013**, *15*, e3. [CrossRef]

126. Gerner, E.W.; Meyskens, F.L. Polyamines and cancer: Old molecules, new understanding. *Nat. Rev. Cancer* **2004**, *4*, 781–792. [CrossRef]

127. Jalili, C.; Salahshoor, M.R.; Yousefi, D.; Khazaei, M.; Shabanizadeh Darehdori, A.; Mokhtari, T. Morphometric and Hormonal Study of the Effect of Utrica diocia Extract on Mammary Glands in Rats. *Int. J. Morphol.* **2015**, 983–987. [CrossRef]

128. Abu-Dahab, R.; Afifi, F. Antiproliferative activity of selected medicinal plants of Jordan against a breast adenocarcinoma cell line (MCF7). *Sci. Pharm.* **2007**, *75*. [CrossRef]

129. Mohammadi, A.; Mansoori, B.; Goldar, S.; Shanehbandi, D.; Khaze, V.; Mohammadnejad, L.; Baghbani, E.; Baradaran, B. Effects of *Urtica dioica* dichloromethane extract on cell apoptosis and related gene expression in human breast cancer cell line (MDA-MB-468). *Cell. Mol. Biol.* **2016**, *62*, 62–67.

130. Mohammadi, A.; Mansoori, B.; Aghapour, M.; Shirjang, S.; Nami, S.; Baradaran, B. The *Urtica dioica* extract enhances sensitivity of paclitaxel drug to MDA-MB-468 breast cancer cells. *Biomed. Pharmacother.* **2016**, *83*, 835–842. [CrossRef]

131. Mansoori, B.; Mohammadi, A.; Hashemzadeh, S.; Shirjang, S.; Baradaran, A.; Asadi, M.; Doustvandi, M.A.; Baradaran, B. *Urtica dioica* extract suppresses miR-21 and metastasis-related genes in breast cancer. *Biomed. Pharmacother.* **2017**, *93*, 95–102. [CrossRef]

132. Mohammadi, A.; Mansoori, B.; Baradaran, P.C.; Khaze, V.; Aghapour, M.; Farhadi, M.; Baradaran, B. *Urtica dioica* Extract Inhibits Proliferation and Induces Apoptosis and Related Gene Expression of Breast Cancer Cells In Vitro and In Vivo. *Clin. Breast Cancer* **2017**, *17*, 463–470. [CrossRef]

133. Telo, S.; Halifeoglu, I.; Ozercan, I.H. Effects of Stinging Nettle (*Urtica dioica* L.,) on Antioxidant Enzyme Activities in Rat Model of Mammary Gland Cancer. *Iran. J. Pharm. Res.* **2017**, *16*, 164–170.

134. Mekhail, T.M.; Markman, M. Paclitaxel in cancer therapy. *Expert Opin. Pharmacother.* **2002**, *3*, 755–766.

135. Yarrow, J.C.; Perlman, Z.E.; Westwood, N.J.; Mitchison, T.J. A high-throughput cell migration assay using scratch wound healing, a comparison of image-based readout methods. *BMC Biotechnol.* **2004**, *4*, 21. [CrossRef]

136. Savary, K.; Caglayan, D.; Caja, L.; Tzavlaki, K.; Bin Nayeem, S.; Bergström, T.; Jiang, Y.; Uhrbom, L.; Forsberg-Nilsson, K.; Westermark, B.; et al. Snail depletes the tumorigenic potential of glioblastoma. *Oncogene* **2013**, *32*, 5409. [CrossRef]

137. Waldmann, J.; Feldmann, G.; Slater, E.P.; Langer, P.; Buchholz, M.; Ramaswamy, A.; Saeger, W.; Rothmund, M.; Fendrich, V. Expression of the zinc-finger transcription factor Snail in adrenocortical carcinoma is associated with decreased survival. *Br. J. Cancer* **2008**, *99*, 1900–1907. [CrossRef]

138. Batlle, E.; Sancho, E.; Francí, C.; Domínguez, D.; Monfar, M.; Baulida, J.; García De Herreros, A. The transcription factor snail is a repressor of E-cadherin gene expression in epithelial tumour cells. *Nat. Cell Biol.* **2000**, *2*, 84–89. [CrossRef]

139. Tran, H.D.; Luitel, K.; Kim, M.; Zhang, K.; Longmore, G.D.; Tran, D.D. Transient SNAIL1 expression is necessary for metastatic competence in breast cancer. *Cancer Res.* **2014**, *74*, 6330–6340. [CrossRef]

140. Lin, H.-C.; Cheng, Y.-W.; Hsu, N.-Y. The association of miR-21, HER-2/neu, and PTEN expression and clinical outcome of breast cancer. *Cancer Res.* **2014**, *74* (Suppl. 19), 1470.

141. Liu, Z.L.; Wang, H.; Liu, J.; Wang, Z.X. MicroRNA-21 (miR-21) expression promotes growth, metastasis, and chemo- or radioresistance in non-small cell lung cancer cells by targeting PTEN. *Mol. Cell. Biochem.* **2013**, *372*, 35–45. [CrossRef]

142. Cheng, H.; Shi, S.; Cai, X.; Long, J.; Xu, J.; Liu, C.; Yu, X. microRNA signature for human pancreatic cancer invasion and metastasis. *Exp. Ther. Med.* **2012**, *4*, 181–187. [CrossRef]

143. Medina, P.P.; Nolde, M.; Slack, F.J. OncomiR addiction in an in vivo model of microRNA-21-induced pre-B-cell lymphoma. *Nature* **2010**, *467*, 86–90. [CrossRef]

144. Huang, T.H.; Wu, F.; Loeb, G.B.; Hsu, R.; Heidersbach, A.; Brincat, A.; Horiuchi, D.; Lebbink, R.J.; Mo, Y.Y.; Goga, A.; et al. Up-regulation of miR-21 by HER2/neu signaling promotes cell invasion. *J. Biol. Chem.* **2009**, *284*, 18515–18524. [CrossRef]

145. Nicoloso, M.S.; Spizzo, R.; Shimizu, M.; Rossi, S.; Calin, G.A. MicroRNAs—The micro steering wheel of tumour metastases. *Nat. Rev. Cancer* **2009**, *9*, 293–302. [CrossRef]

146. He, L.; Hannon, G.J. MicroRNAs: Small RNAs with a big role in gene regulation. *Nat. Rev. Genet.* **2004**, *5*, 522–531. [CrossRef]

147. Merdad, A.; Karim, S.; Schulten, H.J.; Dallol, A.; Buhmeida, A.; Al-Thubaity, F.; Gari, M.A.; Chaudhary, A.G.; Abuzenadah, A.M.; Al-Qahtani, M.H. Expression of matrix metalloproteinases (MMPs) in primary human